AMERICAN CONTRIBUTIONS
TO MATHEMATICAL STATISTICS
IN THE NINETEENTH CENTURY

This is a volume in the Arno Press collection

THREE CENTURIES OF SCIENCE IN AMERICA

Advisory Editor
I. Bernard Cohen

Editorial Board
Anderson Hunter Dupree
Donald H. Fleming
Brooke Hindle

See last pages of this volume for a complete list of titles

AMERICAN CONTRIBUTIONS TO MATHEMATICAL STATISTICS IN THE NINETEENTH CENTURY

Stephen M. Stigler, editor

Vol. II

ARNO PRESS
A New York Times Company
New York • 1980

Publisher's Note: This book has been reproduced from the best available copy.

Editorial Supervision: Steve Bedney

Reprint Edition 1980 by Arno Press Inc.

Copyright © 1980 by Arno Press Inc.

"Measures of Association for Cross-Classification," by L.A. Goodman and W.H. Krushkal, has been reprinted from the *Journal of the American Statistical Association*, vol. 54, 1959. "Simon Newcomb, Percy Daniell, and the History of Robust Estimation," by Stephen Stigler, has been reprinted from the *Journal*, vol. 68, 1973. Both articles have been reprinted by permission of the managing director of the *Journal*.

THREE CENTURIES OF SCIENCE IN AMERICA
ISBN for complete set: 0-405-12525-9
See last pages of this volume for titles.

Manufactured in the United States of America

Library of Congress Cataloging in Publication Data

Main entry under title:

American contributions to mathematical statistics in the nineteenth century.

  (Three centuries of science in America)
  A collection of previously published articles.
  Bibliography: p.
  1. Mathematical statistics--United States--History
--19th century--Sources. I. Stigler, Stephen M.
II. Series.
QA276.15.A43   1980   519.5   79-8002
ISBN 0-405-12590-9

CONTENTS

Volume II

Newcomb, Simon
Notes on the Theory of Probabilities (Reprinted from *Mathematical Monthly*, vols. 1-3), Cambridge, England, 1859-1861

Newcomb, Simon
Note on the Frequency of Use of the Different Digits in Natural Numbers (Reprinted from *American Journal of Mathematics*, vol. 4), 1881

Newcomb, Simon
A Generalized Theory of the Combination of Observations so as to Obtain the Best Result (Reprinted from *American Journal of Mathematics*, vol. 8), Baltimore, Md., 1886

Newcomb, Simon
The Reminiscences of an Astronomer. New York, 1903

Newcomb, Simon
A Statistical Inquiry into the Probability of Causes of the Production of Sex in Human Offspring. Washington, D.C., 1904

Pearson, Karl
On Skew Probability Curves (Reprinted from *Nature*, vol. 52), London and New York, August 1, 1895

Peirce, Benjamin
Criterion for the Rejection of Doubtful Observations (Reprinted from *The Astronomical Journal*, vol. 2, no. 21), Cambridge, England, July 24, 1852

Peirce, Benjamin and Chas. Schott
On Peirce's Criterion (Reprinted from *Proceedings of the American Academy of Arts and Sciences*, vol. 13), 1878

Peirce, C.S.
On the Theory of Errors of Observations (Reprinted from Appendix 21, *Report of the Superintendent of the United States Coast Survey*), Washington, D.C., 1873

Peirce, C.S.
Note on the Theory of the Economy of Research (Reprinted from Appendix 14, *Report of the Superintendent of the United States Coast Survey*), Washington, D.C., 1879

Peirce, C.S.
The Numerical Measure of the Success of Predictions (Reprinted from *Science*, vol. 4), Cambridge, Mass., November 14, 1884

Peirce, C.S. and J. Jastrow
On Small Differences of Sensation (Reprinted from *Memoirs of the National Academy of Sciences for 1884*, vol. 3), 1885

Stigler, Stephen M.
Simon Newcomb, Percy Daniell, and the History of Robust Estimation, 1885-1920 (Reprinted from *Journal of American Statistics*, vol. 68, no. 344), December, 1973

Wigglesworth, Edward
A Table Showing the Probability of the Duration, the Decrement, and the Expectation of Life, in the States of Massachusetts and New Hampshire (Reprinted from *Memoirs of the American Academy of Arts and Sciences*, vol. 2), Boston, 1793

Wilson, Edwin B., and Margaret M. Hilferty
Note on C.S. Peirce's Experimental Discussion of the Law of Errors (Reprinted from *Proceedings of the National Academy of Science*, vol. 15), 1929

Winlock, Joseph
On Professor Airy's Objections to Peirce's Criterion (Reprinted from *The Astronomical Journal*, vol. 4, no. 19), Cambridge, England, March 19, 1856

Wolfenden, H.H.
On the Development of Formulae for Graduation by Linear Compounding, With Special Reference to the Work of Erastus L. De Forest (Reprinted from *Transactions of the Actuarial Society of America*, vol. 26), 1925

Woodward, R.S.
Some Practical Features of a Field Time Determination with a Meridian Transit [abstract] (Reprinted from *Bulletin of the Philosophical Society of Washington*, vol 8), Washington, D.C., 1885

Wright, T.W.
On the Computation of Probable Error (Reprinted from *The Analyst*, vol. 9, no. 3), May, 1882

NOTES
ON
THE THEORY OF PROBABILITIES

Simon Newcomb

NOTES ON THE THEORY OF PROBABILITIES.

By Simon Newcomb. Naut. Alm. Office, Cambridge, Mass.

1. The theory of probabilities, or doctrine of chances, as it has sometimes been called, being one of those subjects on which little has yet been written among us, it is proposed to develop its elementary principles through the pages of this Journal, and to apply them to a number of such problems as are suggested either by the observation of the material world, or by the affairs of society and of daily life.

2. Against the logical accuracy of the results of this calculus many objections have been raised. In the latter part of the last century, the Rev. Mr. Mitchell computed, that if the stars visible to the naked eye, had been scattered fortuitously over the heavens, there were five hundred thousand chances to one against any six of them falling together in so small a space as that occupied by the Pleiades. The correctness of this conclusion has lately been disputed by an eminent English scientist, who thinks that in the case supposed, such groupings of the stars are precisely what ought to be expected. This problem will be examined in the course of these notes as a particular case of one more general, and we shall find by a mode of reasoning entirely different from that of Mr. Mitchell, that he is substantially correct.

3. The object of the theory of probabilities is to determine the amount of reason which we have to expect any event when we have not the data to determine with certainty whether it will occur or not; and when the data of which we are possessed are such as to admit of the application of mathematical analysis. This amount of reason is not any quality of the event itself, but only a degree of rational expectation in the mind of the individual.

It is considered a well-established philosophical principle that the same causes will always produce the same effects; so that an intellect having a perfect knowledge of the former, and possessed of sufficient reasoning power, could always predict the latter. But in the external world, we can never arrive at any knowledge of the philosophical efficient cause, but can only observe that certain events have always been followed by certain other events. The preceding event is, then, generally denominated the cause, and the event which follows it the effect. Using the term cause in this sense, it generally happens, that so much of the cause as we are cognizant of may be followed by a great number of effects; and then we can only determine of the latter that they are more or less probable.

4. Hereafter, the operation of so much of the cause as we are cognizant of will be called a *trial*; and the effects of which we wish to find the probabilities will be designated as *events*. Generally the number of separate possible results of the trial will be greater than the number of events which we consider. The former will then be designated as *cases*; such of the cases as necessarily involve or produce the event in question are called *favorable*, while those which necessarily negate or prevent it are termed *unfavorable*. If the separate cases are all equally probable, the probability of the event is represented by the ratio of the number of the favorable cases to the whole possible number of cases.

5. To illustrate the above principles and definitions: suppose that a cube has the sides numbered from one to six; that sides one and two are white, and the remaining four black. Suppose, also, that the cube is thrown from a box at random, and the probability that a white side will fall upward is required. Then —

The given cause or *trial* is the throwing of the cube.

The *event* is the falling upward of a white side.

The number of possible *cases* is six; namely, the falling up of sides 1, 2, 3, &c., to 6 respectively.

The *favorable* cases are two, the falling of side 1, and of side 2.

Wherefore the *probability* of the event is $\frac{2}{6} = \frac{1}{3}$.

6. In general, if there are m cases equally probable, of which one must, and only one can, occur; and if p of these cases are favorable to the production of an event, the probability of that event is $\frac{p}{m}$. We see that if the probability is $\frac{1}{2}$, the event is just as likely to occur as to fail; and that if the probability is 1, the event becomes certain. All degrees of probability must be included between the limits 0 and 1; 0 being the symbol of impossibility and 1 of certainty.

7. In enumerating the different possible cases, we must take care so to subdivide them that they shall all be equally probable. Through not taking this precaution, an eminent mathematician of the last century fell into an error with regard to a very simple problem, of the following nature: If a coin is thrown twice, what is the probability that a head will fall upward? D'ALEMBERT reasoned as follows: If head is thrown the first time, then the event occurs, and the throw need not be repeated; only three cases are therefore to be considered —

> Heads the first time,
> Tails the first time and heads the second,
> Tails both times.

Two of these cases being favorable, the probability of the event is $\frac{2}{3}$. This result, however, is incorrect, because, as is easily seen, the probability of the first case alone is $\frac{1}{2}$, and therefore twice as great as that of either of the others. It ought, therefore, to be considered as subdivided into two separate cases, making the probability of the event $\frac{3}{4}$, which is the correct answer.

8. The following theorem will frequently save us the trouble of enumerating all the possible cases.

The probability that both of two independent events will happen, is equal to the product of their individual probabilities.

For the first event, let there be m cases, of which p are favorable, and for the second, n cases, of which q are favorable. If we give both events a trial, each of the m cases may be combined with any one of the n cases, making in all $m \times n$ possible combinations, each of which will be equally probable. The combinations favorable to *both* events will be those only in which one of the p cases, favorable to the first, is combined with one of the q cases, favorable to the second, the number of which will be $p \times q$. The probability of the compound event is therefore $\frac{pq}{mn}$, which is equal to the product of the separate probabilities.

As an example, suppose that with the cube spoken of in § 5 is put an octahedron, of which three sides only are white. If the two solids are thrown down at random, each of the six sides of the cube may be combined with any one of the eight sides of the octahedron, making the possible combinations $(1,1)$, $(1,2)$, $(1,3)$, &c. to $(1,8)$; $(2,1)$, $(2,2)$, $(2,3)$, &c. to $(2,8)$; .... $(6,1)$, $(6,2)$, &c. to $(6,8)$; in all, 48 combinations. If we determine the result of each of these combinations, we shall find that the upper side of

The cube will be white and the octahedron white in 6 cases.
" " " " white " " " black " 10 "
" " " " black " " " white " 12 "
" " " " black " " " black " 20 "

Dividing each of these results by 48, the whole number of cases, we shall find the probabilities of the several compound events to be equal to the product of the probabilities of the simple events which compose them; that is, $\frac{6}{48} = \frac{2}{6} \times \frac{3}{8}$; $\frac{10}{48} = \frac{2}{6} \times \frac{5}{8}$; $\frac{12}{48} = \frac{4}{6} \times \frac{3}{8}$; $\frac{20}{48} = \frac{4}{6} \times \frac{5}{8}$.

NOTES ON THE THEORY OF PROBABILITIES.

By Simon Newcomb, Nautical Almanac Office, Cambridge, Mass.

9. The theorem given in § 8 may be extended to any number of independent events. Let there be three events, E_1, E_2, E_3, of which

the respective probabilities are p_1, p_2, p_3. Call p' the probability of the compound event ($E_1 E_2$). Then by the theorem the probability that this compound event and the event E_3 will *both* occur will be $p' \times p_3 = p_1 \times p_2 \times p_3$, since $p' = p_1 \times p_2$. By repeating the same process it may be shown that in general if there are n events, of which the probabilities are p_1, p_2, $p_3 \ldots p_n$, the probability of the concurrence of all these events is the continued product $p_1 \times p_2 \times \ldots p_n$. If the p's are mostly small fractions, and n large, this product will be very small.

10. If the events are not entirely independent, that is, if the probability of the second is affected by the occurrence or non-occurrence of the first; then, the probability of the compound event is equal to the probability of the first multiplied into the probability of the second *on the supposition that the first occurs*.

Let c be the whole number of possible cases, — let p of these cases be favorable to the first event, and of these p cases let q be favorable to the second. The probability that both events will occur is then $\frac{q}{c}$. But $\frac{p}{c}$ is the probability of the first event, and if this event occurs, the probability of the second becomes $\frac{q}{p}$, the product of which into $\frac{p}{c}$ gives $\frac{q}{c}$, which proves the proposition.

It may be remarked that this theorem applies also to independent events, and includes that of § 8, the probability of the second event being then the same, whether the first event be supposed to occur or fail.

Example. What is the probability that out of a bag which contains two white and two black balls, a white ball will be drawn at the first drawing and a black ball at the second? In this case, if the first event occur, the probability of the second will be $\frac{2}{3}$, since two out of the three balls remaining will be black. The probability of the first event being $\frac{1}{2}$, the probability of the compound event will be $\frac{1}{2} \times \frac{2}{3} = \frac{1}{3}$.

11. *Conflicting events.* When a system of events are so connected that one *must* and only one *can* occur, they are called a *system of conflicting events.*

The sum of the probabilities of a system of conflicting events is equal to unity; since, every separate case being favorable to one of them, and to one only, the sum of the numerators of the different fractions expressing the probabilities (§ 6) is equal to the common denominator.

12. *Value of expectation.* Let there be n men, q of whom to be determined by lot, are certain of receiving the sum s each. What is each man's chance worth? If each man were to sell out his chance, the purchaser should give in all the sum qs which he would then be sure of receiving, and since each man's chance is equal, he ought to give each man $\frac{q}{n}s$. But $\frac{q}{n}$ is the probability for each man that he will receive the sum s, and the value of his chance is therefore equal to the sum that he may receive multiplied by the probability that he will receive it.

PROBLEMS IN PROBABILITIES.

A has the reputation of telling the truth as often as three times in four; B as often as four times in five; and C as often as five times in seven. When A and B agree in affirming what C denies, what is the probability that A and B tell the truth? — Communicated by Professor Root.

A person goes on throwing a common die until he throws an ace; at whatever throw this occurs (the nth) he is to receive the nth of a dollar. What is the value of his expectation? — Communicated by Asher B. Evans.

NOTES ON THE THEORY OF PROBABILITIES.

By Simon Newcomb, Nautical Almanac Office, Cambridge.

13. Some theorems relating to combinations will be given here for convenience of reference.

If we have a set of n things, the number of different ways in which a collection of s things can be taken from it, is called the number of combinations of s in n, and is represented by the notation $\overset{s}{\underset{n}{C}}$, or $C(s, n-s)$. Since, for every collection of s things which can be taken, there will remain a collection of $n-s$ things not taken, we may consider the number of combinations of s in n as being the number of ways in which a collection of s things can be divided into two sets, one containing s things, the other $n-s$. It is shown in treatises on Algebra, that

(1) $\quad \overset{s}{\underset{n}{C}} = \dfrac{n(n-1)(n-2)\ldots(n-s+1)}{1.2.3\ldots s} = \dfrac{n!}{(n-s)!\,s!}$.

From the above, we may easily deduce the following equations:

(2) $\quad \overset{s+1}{\underset{s}{C}} = \dfrac{n-s}{s+1} \overset{s}{\underset{n}{C}}; \quad \overset{s-1}{\underset{n}{C}} = \dfrac{s}{n-s+1} \overset{s}{\underset{n}{C}}; \quad \overset{s}{\underset{n}{C}} = \overset{n-s}{\underset{n}{C}}; \quad \overset{0}{\underset{n}{C}} = \overset{n}{\underset{n}{C}} = 1.$

Using the notation of combinations, the binominal theorem may be expressed in the form

(3) $\quad (a+x)^n = \overset{0}{\underset{n}{C}} a^n + \overset{1}{\underset{n}{C}} a^{n-1} x + \overset{2}{\underset{n}{C}} a^{n-2} x^2 + \ldots = \overset{\infty}{\underset{0}{\Sigma_s}} \overset{s}{\underset{n}{C}} a^{n-s} x^s.$

Suppose that the set is divided into two classes, white and black, for example; then a collection of s things may be composed of any s white things,

or any $s-1$ white combined with any 1 black,
" $s-2$ "   "   2   "
"   &c.   "   "   &c.   "
"   1   "   " $s-1$ "
"   0   "   " s "

from which we deduce the general theorem

(4) $\quad \overset{s}{\underset{n}{C}} = \overset{s}{\underset{n-l}{C}} + 1\overset{s-1}{\underset{n-l}{C}} + \overset{2}{\underset{l}{C}}\overset{s-2}{\underset{n-l}{C}} + \text{&c.} \ldots + \overset{s-1}{\underset{l}{C}}\overset{1}{\underset{n-l}{C}} + \overset{s}{\underset{l}{C}} = \Sigma_r \overset{r}{\underset{l}{C}}\overset{s-r}{\underset{n-l}{C}},$

where l may be any number (at pleasure) less than n.

If we represent the number of ways in which a number of things equal to $(s + s' + s'' + \ldots \text{&c.})$ can be divided into sets, of which one shall contain s things, another s', &c., by $C(s, s', s'' \ldots)$, we have

$$C(s, s', s'' \ldots) = \frac{(s + s' + s'' + \ldots)!}{s!\, s'!\, s''! \ldots}.$$

14. **Illustrative Problems.**

I. A bag contains b black balls, and b' white ones. A number s of balls being drawn, what is the probability that they will all be white?

$\overset{s}{\underset{b+b'}{C}}$ will be the whole number of combinations of s balls in the bag, any one of which may be drawn. Of these combinations, only $\overset{s}{\underset{b'}{C}}$ will consist entirely of white balls. The probability required is therefore $\overset{s}{\underset{b'}{C}} \div \overset{s}{\underset{b+b'}{C}}$. In the same way the probability of drawing s black balls is found to be $\overset{s}{\underset{b}{C}} \div \overset{s}{\underset{b+b'}{C}}.$

EXAMPLE. If there are two white balls and eight black ones, of which two are drawn, there will be forty-five combinations of 2 in 10, any one of which may be drawn. But only one of these combinations will consist entirely of white balls, and the probability of drawing that combination will be $\frac{1}{45}$, and the odds against it 44 to 1. We might arrive at the same result by the principle of § 10. If we suppose the balls to be drawn in succession, the probability that the first ball drawn will prove white is $\frac{1}{5}$, since of 10 balls 2 are white. On the supposition that this event occurs, the probability that the second ball will also be white is $\frac{1}{9}$, since of 9 balls one will be white. The probability of the compound event is therefore $\frac{1}{3} \times \frac{1}{5} = \frac{1}{15}$.

II. To find the probability, that, at whist, a player other than the dealer holds any given number of trumps.

The trump turned up belongs to the dealer, there will remain 51 cards, 12 trumps, and 39 non-trumps to be divided at random, the dealer taking 12, and each of the players 13. In order that a player may have no trump, his 13 cards must come entirely from the 39 non-trumps, and the number of ways in which this result can happen is $\overset{13}{\underset{39}{C}}$. A hand containing one trump only may be composed of any one of the 12 trumps combined with any combination of 12 in the 39 non-trumps, and such a hand can happen in $12 \times \overset{12}{\underset{39}{C}}$ different ways. In general, a hand containing s trumps may consist of any s of the 12 trumps combined with any $13 - s$ of the 39 non-trumps, and there will be $\overset{s}{\underset{12}{C}} \times \overset{13-s}{\underset{39}{C}}$ such hands. The probability of holding s trumps will therefore be $\overset{s}{\underset{12}{C}} \times \overset{13-s}{\underset{39}{C}} \div \overset{13}{\underset{51}{C}}$.

If, in the above expression, we give to s the successive values 0, 1, 2, &c., we shall find

a probability of .02 that the player has no trump,
" " .10 " " 1 "
" " .23 " " 2 trumps,
" " .30 " " 3 "
" " .22 " " 4 "
" " .10 " " 5 "
" " .03 " " 6 "
a very small probability " 7 or more.

By a process of reasoning similar to the above we shall find the probability that the dealer has s trumps to be

$$\underset{12}{\overset{s-1}{C}} \times \underset{39}{\overset{13-s}{C}} \div \underset{51}{\overset{12}{C}}.$$

III. A set of dice being thrown from a box, to find the probability that any given system of numbers will be thrown.

Suppose for the sake of clearness, that the several dice are distinguished as the first, second, &c. Then any one of the six sides of the first die may be combined with any one of the six sides of the second, making 6^2 on two dice. Any one of these combinations may be combined with any side of the third, making 6^3 combinations on three dice; and by continuing the same process it is seen, that, in general, the number of combinations on n dice is 6^n. The given system of numbers may be thrown in as many ways as there are permutations of the numbers composing it; and if we represent the number of permutations by P, the probability required will be $\frac{P}{6^n}$. This will be easily understood by comparing the following examples:

1. Suppose that there are three dice of different colors, — say white, yellow, red, and let it be required to find the probability that the numbers 1, 2, 3, will be thrown. This system can be thrown on the three dice in the six following ways:

White,	1	1	2	2	3	3
Yellow,	2	3	1	3	1	2
Red,	3	2	3	1	2	1

The probability of each way being $\frac{1}{216}$, the probability required will be $\frac{1}{36}$.

2. If two aces and a deuce were required to be thrown, it could be done only in the following ways:

White,	1	1	2
Yellow,	1	2	1
Red,	2	1	1

The probability of its being thrown is therefore $\frac{1}{72}$.

3. The probability of throwing three aces is $\frac{1}{216}$, since they can be thrown in only one way.

The proposition, that an ace, deuce, and three are six times as likely to be thrown as three aces, might at first sight seem paradoxical. A comparison of the above will, however, make it quite clear.

NOTES ON THE THEORY OF PROBABILITIES.

By Simon Newcomb, Cambridge, Mass.

15. The problem which we are about to consider is one of the most fruitful in the theory of probabilities, as out of it grow the theory of errors, the theory of chance distribution, the law of averages, and the estimation of the probability that an observed concurrence of events is the result of a law of nature.

To find the probability that an event of which the probability on a single trial is p *will happen* s *times on* n *trials.*

The probability that it will fail on every trial is $(1-p)^n$, $1-p$ being the probability that it will fail on any single trial.

The probability that it will happen on the first trial and fail on the $n-1$ following ones is $p(1-p)^{n-1}$. But as the single event is as likely to occur on the 2d, 3d, ... nth trial as on the first, the probability that it will occur just once is $np(1-p)^{n-1}$.

The probability that the event will occur on the first two trials and fail on the $n-2$ subsequent ones is $p^2(1-p)^{n-2}$. But the two events can equally occur on the $(1,3), (1,4) \ldots (1,n)$, or the $(2,3), (2,4)$, &c. trials; in fact there will be C_n^2 pairs of trials on which the two events can occur, so that $C_n^2 p^2(1-p)^{n-2}$ is the probability that it will occur twice.

By a process of reasoning exactly like the last, we find the probability that it will occur s times to be

$$(1) \qquad P_s = C_n^s p^s (1-p)^{n-s},$$

which is the $(s+1)$st term in the development of the binomial $[(1-p)+p]^n$. The sum of the probabilities of all the possible results of the n trials is therefore 1, as it ought to be.

As an example to elucidate the above, suppose that a cent is so formed that a head is twice as likely to be thrown as a tail, so that the probability of the former on each throw is $\frac{2}{3}$. If the coin is thrown four times, the results of the four throws may be as follows. After each separate result is written the fraction expressing the probability of that result.

No heads, $tttt$ $\frac{1}{3^4} = \frac{1}{81}$,
1 head, $httt, thtt, ttht, ttth$, each $\frac{2}{81}$; $\times 4 = \frac{8}{81}$,
2 heads, $hhtt, htht, htth, thht, thth, tthh$, $6 \times \frac{4}{81} = \frac{24}{81}$,
3 heads, $hhht, hhth, hthh, thhh, = \frac{32}{81}$,
4 heads, $hhhh = \frac{16}{81}$.

If we supposed heads as likely to be thrown as tails, we should find these probabilities to be $\frac{1}{16}, \frac{4}{16}, \frac{6}{16}, \frac{4}{16}, \frac{1}{16}$, respectively. The result would evidently be the same if we supposed that four coins were thrown from a box together.

16. To resume the general discussion, let us see what value of s is the most probable. This value we will determine by the condition that its probability must be greater than that of the next smaller number, and also greater than that of the next greater number, or

$$\overset{\cdot}{\underset{n}{C}} p^s (1-p)^{n-s} > \overset{\cdot}{\underset{n}{\overset{s-1}{C}}} p^{s-1} (1-p)^{n-s+1};$$

$$\overset{\cdot}{\underset{n}{C}} p^s (1-p)^{n-s} > \overset{\cdot}{\underset{n}{\overset{s+1}{C}}} p^{s+1} (1-p)^{n-s-1}.$$

Since $\qquad \overset{\cdot}{\underset{n}{C}} = \frac{n-s+1}{s} \overset{\cdot}{\underset{n}{\overset{s-1}{C}}}; \qquad \overset{\cdot}{\underset{n}{\overset{s+1}{C}}} = \frac{n-s}{s+1} \overset{\cdot}{\underset{n}{C}};$

we have from the division of the first inequality by $\overset{\cdot}{\underset{n}{\overset{s-1}{C}}} p^{s-1}(1-p)^{n-s}$,

$$1-p < \frac{n-s+1}{s} p, \quad \text{which gives } s < p(n+1);$$

and from the division of the second inequality by $\overset{\cdot}{\underset{n}{C}} p^s (1-p)^{n-s-1}$ we have

$$1-p > \frac{n-s}{s+1}p, \qquad \text{which gives } s > p(n+1)-1;$$

s is therefore the greatest whole number in $p(n+1)$. If s and n are very large numbers, we have very nearly

$$(2) \qquad \frac{s}{n} = p.$$

It follows, therefore, that in a great number of trials events are more likely to occur a number of times proportional to their respective probabilities than any other number. Thus if a cent is thrown one hundred times, heads are more likely to be thrown fifty than any other single number of times. But it must not be supposed that they are therefore more likely to be thrown fifty times than not, for it is *almost* as likely to be thrown 49 times or 51 times as fifty times. The chances that it would be thrown *exactly* fifty times would be quite small, because there are so many other numbers that might be thrown.

17. Another deduction from the expression (1) is the following: however small the probability of an event on a single trial, by increasing the number of trials we can render the probability that the event will occur at least once as great as we please. For the probability that it will fail on every one of the n trials being $(1-p)^n$; however small p may be, we may make n so great that $(1-p)^n$ shall be as small as we please.

18. Suppose now that n is infinitely great, and p infinitely small, and that $np = a$, a being a finite quantity. We may then put $n = n-1 = n-2$, &c. We shall then have, while s is finite,

$$\overset{s}{\underset{n}{C}} p^s = \frac{n^s p^s}{s!} = \frac{a^s}{s!},$$

$$(1-p)^{n-s} = (1-p)^n = (1-p)^{\frac{a}{p}} = e^{-a},$$

e being the Napierian base. Substituting these values in (1) we obtain for the probability that the event will occur s times

(3) $$P_s = \frac{a^s e^{-a}}{s!}.$$

The probability that the event will fail, is therefore e^{-a}; that it will occur once only, $a\, e^{-a}$; twice only, $\frac{a^2}{2} e^{-a}$, &c.

The sum of this series of probabilities continued to infinity is

$e^{-a}(1 + a + \frac{a^2}{2!} + \frac{a^3}{3!} + \&\text{c.},) = e^{-a} \cdot e^{a} = 1$, as it ought to be.

19. We may apply this equation to the determination of the probability that, if the stars were scattered at random over the heavens, any small space selected at random would contain s stars. Let N be the whole number of stars, h the number of units of space in the heavens, then $\frac{N}{h} dh$ may be taken to represent the infinitely small probability that the infinitely small space dh contains a star. Moreover, if l represents the extent of space selected at random which we consider, we may consider the examination of each dh as a trial, and the number n of trials will then be $\frac{l}{dh}$. The value of a will then become $\frac{N}{h} l$, and by substitution in (3) we have

(4) $$P = \frac{N^s l^s e^{-\frac{Nl}{h}}}{h^s s!}$$

for the probability that the space l contains s stars. Suppose, as a numerical example, that l is a square degree, and $N = 1500$, which is about the number of stars of the fifth and higher magnitudes; $s = 6$. We then have $\frac{l}{h} = \frac{1}{41253}$; by the substitution of these values in (4) we shall have the probability that any square degree selected at random in the heavens contains six stars. Multiplying this probability by 41253, the number of square degrees in the whole heavens, we obtain the probability that, *if the heavens were divided at random into square degrees, some one of those square degrees would contain six stars.* This probability we find to be

$$\frac{1500^6}{41253^5 \cdot 6!} \cdot e^{-.03636} = .000000128.$$

This, however, is evidently rather smaller than the probability that six stars should be found so near together that a square degree could be fitted on so as to include them.

Mr. MITCHELL committed an error in his solution, the effect of which, if I mistake not, is to make his probability too great. His general method is, however, better applicable to this particular problem than that given above, but as there is a margin of vagueness and uncertainty about the problem in question, so that the answer does not admit of being expressed in exact numbers without an excessively complicated process of reasoning, I have preferred to deduce an approximate solution from the general formulæ, to be used in so many more problems.

20. Let us now consider Prof. FORBES's objections to the above results of the calculus of probabilities.

He scattered paint from a brush upon a wall, and found double and triple spots and groups innumerable. This is about as decisive as an attempt to disprove the Pythagorean proposition by measuring the squares described on a triangle without knowing whether it had or had not a right angle, and finding that one square was not equal to the sum of the others. As a mathematician would answer this objection by saying that his result only proved that he had either made a mistake in his measurements, or had not measured a right triangle; so Prof. FORBES's result only proves that either he was mistaken as to the marked character of the groupings, or that the proximity of the components of each group was the effect of their positions being determined *by the action of the same cause*, which is all that the theory of probabilities claims for the Pleiades. The latter supposition is by no means improbable, because a group of spots might be formed by the breaking up of a drop after it had left the brush.

21. Prof. FORBES remarked that an exactly uniform distribution of stars would not be expected as the result of a random distribution. In this he is correct; and as it is an interesting problem, we shall here determine what law a random distribution may be expected to follow. It may appear paradoxical to assert that the results of chance can be expected to follow any law; but such is really the case, and the formula (3) determines the law. As an example, suppose that the heavens are divided into 1500 equal portions, and that 1500 stars are distributed at random, or, to speak with more philosophical accuracy, that the causes which determine the position of each separate star are entirely independent of those which determine the position of any other. Then by reasoning as in § 19 we find $a = 1$; and by formula (3) the probability that a unit of space selected at will contains no star, will be $\frac{1}{e} = \frac{1}{2.718\ldots}$*; one star, $\frac{1}{e}$; two stars, $\frac{1}{2e}$; three stars, $\frac{1}{2.3e}$; &c. If we then select the whole 1500 units we ought to expect the number which would be found to contain the several numbers of stars to be somewhere near 1500 multiplied by the respective probabilities, or

about $\frac{1500}{e} = 552$ portions containing no star,

" $\frac{1500}{e} = 552$ " " 1 star,

" $\frac{1500}{2e} = 276$ " " 2 stars,

" $\frac{1500}{2.3e} = 92$ " " 3 stars,

" $\frac{1500}{4!e} = 23$ " " 4 stars,

" $\frac{1500}{5!e} = 4(+)$ " " 5 stars,

" $\frac{1500}{6!e} = 1$ " " 6 stars,

---

* The acute reader will perceive that the solutions in §§ 19 and 21 are those of a problem slightly (though not materially) different from that actually propounded.

and it would be quite improbable (about 1 chance to 8) that any space would be found to contain more than six stars.

If any one wishes an experimental illustration of the principle let him take a pint of rice, color a hundred grains of it black, mix the black grains thoroughly with the remainder, and stir the mixture till he finds six or eight of the black grains to form a group by themselves. Before this result arrives he will in all probability be willing to admit that should he ever see such a group in such a mixture, he would not believe that it was formed by indiscriminate mixing.

NOTES ON PROBABILITIES.

By SIMON NEWCOMB, Nautical Almanac Office, Cambridge, Mass.

22. THE *a priori* probability of an event is its probability deduced solely from circumstances which preceded or might have preceded it. In the theory of probabilities these circumstances are regarded as causes.

The *a posteriori* probability of an event or hypothesis is its probability deduced from events which have followed it, which events are regarded as its possible effects.

Theorem. — The *a priori* probability of an event is equal to the sum of the products obtained by multiplying the probability of each distinct cause by the probability that the event would occur on the hypothesis that the cause operates.

Let m be the whole number of separate cases. Let n of these cases be favorable to the first cause, n' to the second, &c. Their probabilities will then be $\frac{n}{m}, \frac{n'}{m}, \frac{n''}{m}$, &c. Also, let h of the n cases,

h' of the n' cases, &c., be favorable to the production of the event. Its probability, on the several hypotheses of the occurrence of each separate cause, will then be $\frac{h}{n}$, $\frac{h'}{n'}$, $\frac{h''}{n''}$, &c. Multiplying the corresponding fractions in the two series, and taking the sum of the products we obtain $\frac{h + h' + h'' + \&c.}{m}$. But this is the probability of the event, because $h + h' + h'' +$ &c. is the whole number of cases favorable to its production.

23. *Problem.* — Let E, E', E'', &c., represent a *portion* of the events which may occur on a certain trial, and let their respective probabilities be p, p', p'', &c. Suppose now that after I have estimated these probabilities, I am informed by a person who knows the result of the trial that one of the events E, E', E'', &c., has actually occurred. What ought to be their respective probabilities in my mind *after I have received this information?*

Let m be the whole number of possible cases; let n of these cases be favorable to E, n' to E', n'' to E'', &c. Then we have

$$p = \frac{n}{m}; \qquad p' = \frac{n'}{m'}, \text{ &c.}$$

But when it is ascertained that one of the events E has actually occurred, the number of possible cases is *reduced* from m to $n + n' + n'' +$ &c., those favorable to the production of some one of the events E. The probabilities of these events will therefore become

$$\frac{n}{n + n' + \&c.}, \qquad \frac{n'}{n + n' + \&c.};$$

that is, each of the former probabilities will be multiplied by the factor $\frac{m}{n + n' + n'' + \&c.}$; or, *they will each be increased in the same ratio* to such a degree as to make their sum equal to unity.

Example. — On a table are fifty small boxes, of which three each contain a gold coin, seven a silver coin, thirteen a copper coin, and twenty-seven nothing. The probability that one selected at random contains a gold coin is $\frac{3}{50}$, silver $\frac{7}{50}$, &c. But if by shaking the box it is found that it contains a coin, it will be known that one of the 23 boxes containing coins must have been selected. The probability that the box contains gold will then be increased to $\frac{3}{23}$, that it contains silver to $\frac{7}{23}$, that it contains copper to $\frac{13}{23}$.

24. *Problem.* — To find the probability of an event or hypothesis from both *a priori* and *a posteriori* circumstances.

Suppose that a possible event E, if it occur at all, must be preceded by *one* and *one only* of the causes or circumstances C, C', C'', &c. Let p, p', p'', &c., be the probabilities of C, C', C'', &c., before it is known whether the event E has occurred. Let q be the probability of E on the hypothesis that C occurs, q' its probability on the hypothesis that C' occurs, &c. The *a priori* probabilities of the compound events $CE, C'E'$, &c., will then be $pq, p'q'$, &c. If, now, it is ascertained that the event E has actually occurred, it becomes certain that *one* of these compound events has occurred; and, by § 23, their probabilities will be increased to

(1) $$\frac{pq}{pq + p'q' + \text{&c.}}, \quad \frac{p'q'}{pq + p'q' + \text{&c.}}, \text{ &c.}$$

But, the event E being now certain, the probabilities of the causes C, C', &c., are the same as the probabilities of the compound events $CE, C'E'$, &c.; so that the above fractions represent the probabilities of the causes or hypotheses, deduced from all the circumstances in our knowledge, which bear on the question.

Example I. — A, handling a die, entertains a suspicion of $\frac{1}{3}$ that the die is so loaded that a six is as likely to be thrown as not. He throws it twice, and each time throws a six. What is the probability that his suspicion is correct? The solution is exhibited as follows: —

C	p	q	pq
die not loaded,	$\frac{2}{3}$	$\frac{1}{6^2} = \frac{1}{36}$	$\frac{1}{54}$,
die loaded,	$\frac{1}{3}$	$\frac{1}{2^2} = \frac{1}{4}$	$\frac{1}{12}$.

The probability that the die is loaded is therefore

$$\frac{\frac{1}{12}}{\frac{1}{54}+\frac{1}{12}} = \frac{9}{11}.$$

Had he thrown six a third time this probability would have increased to $\frac{27}{29}$.

Example II. — An urn contains 99 white balls, and 1 black one. An individual who falsifies once in ten times draws a ball at random, and asserts that it is black. What is the probability that he tells the truth?

The *a priori* probability that the black ball is drawn, and that he tells the truth respecting it, is $\frac{1}{100} \times \frac{9}{10}$. The *a priori* probability that a white ball is drawn, and that he falsifies, is $\frac{99}{100} \times \frac{1}{10}$. The probability that he tells the truth is therefore

$$\frac{\frac{9}{1000}}{\frac{9}{1000}+\frac{99}{1000}} = \frac{1}{12}.$$

Note.*— It is not necessary, as has been supposed by Mr. John Stuart Mill, that C and E should be connected by the *philosophical relation* of cause and effect. No such relationship has been supposed in our reasoning. The fraction pq does not, as Mr. Mill supposes, express the probability of E on the hypothesis that C is true, but the probability of the compound event CE. The groundlessness of the former opinion may be shown by solving the following problem. Find the *a priori* probability that it will rain to-morrow, considered as an effect of the planet Jupiter being inhabited or uninhabited.

Let p be the probability that Jupiter is inhabited, r the probability that it will rain to-morrow if Jupiter is inhabited, r also the probability of the same supposing Jupiter to be uninhabited.

By § 22 the probability required will be $pr + (1-p)r = r$.

This is the common-sense result, so that although we learn nothing new by bringing in irrelevant circumstances as causes, yet neither are we led into error by so doing.

Suppose now that to-morrow arrives, and it does rain; substituting in (1) we have for the probability that Jupiter is inhabited

$$\frac{pr}{pr+(1-p)r} = p.$$

If it does not rain, we have, by a corresponding supposition,

$$\frac{p(1-r)}{p(1-r)+(1-p)(1-r)} = p.$$

---

* Since this note was in type, we have found that Mr. Mill has corrected this error.

NOTES ON PROBABILITIES.

By SIMON NEWCOMB, Cambridge, Mass.

THE problem with which the article on probabilities, in the Mathematical Monthly for May, 1860, was closed, belongs to a class which has been much discussed. We shall here repeat it, and then show a variation in it which leads to a result, which, at first sight, seems paradoxical.

An urn contains 100 balls, 1 of which is black, and 99 white; an individual, who falsifies one time in ten, draws a ball, and asserts that it is black. What is the probability that he tells the truth? We have found that it is $\frac{1}{12}$.

But suppose that the balls are numbered from 1 to 100, and that he asserts that the ball No. 79 is drawn. What is the probability that he tells the truth?

The solution is exhibited in tabular form, as follows. The first column contains the separate hypotheses which it is possible to make, column p gives the *a priori* probabilities of these hypotheses, and column q the probability of the observed event on the several hypotheses, as explained in the number of the Mathematical Monthly above referred to.

c	p	q	pq
No. 79 drawn	$\frac{1}{100}$	$\frac{9}{10}$	$\frac{9}{1000}$
Some other No. drawn,	$\frac{99}{100}$	$\frac{1}{990}$	$\frac{1}{1000}$

Whence the probability that No. 79 was drawn, or that the truth is told, is, by § 24, $\dfrac{\frac{9}{1000}}{\frac{9}{1000} + \frac{1}{1000}} = \frac{9}{10}$.

The value of q, on the second hypothesis, is obtained as follows:— Suppose that some other number than 79 is drawn. Then, if he is going to falsify, he has the 99 numbers which were not drawn, to choose from. Since there is supposed to be no reason why he should say 79 rather than any number, the probability that he will select this number from the 99 not drawn is $\frac{1}{99}$. But the probability that he will falsify in the first place is $\frac{1}{10}$. Whence the probability of saying 79 falsely is $\frac{1}{990}$.

Now the apparent paradox is, that the probability that he has told the truth is so vastly different when he says that a single black ball has been drawn from 100, from what it is when he says that the single number 79 has been so drawn, being $\frac{1}{12}$ in the first case, and $\frac{9}{10}$ in the second. Several forms of explanation may be given, and the reader can select that which seems to him most clear.

1. The difference of the results proceeds from the circumstance that he is supposed to have but two colors to choose from if he falsifies, viz. white and black, while he has 100 numbers. Since a white ball is drawn 99 times out of a hundred, and since he falsifies about one time in ten, he will have said black *about* 11 times in

every hundred, ten times falsely, and once truly. Whence I conclude that when he says black he does not tell the truth, for if I believed him to tell the truth every time, I should have to believe that an excessively improbable number of black balls had been drawn; instead of which I knew beforehand that he would make some ten or twelve false statements respecting the drawing of the black ball. But if 100 numbered balls have been drawn, his ten falsifications are supposed to have been divided amongst the 100 numbers I know not how, whence the chance that any one number was called falsely is only $\frac{1}{10}$.

If, however, I found that some one number, such as 79, had been called 10 or 12 times, I should conclude that the witness was disposed to falsify by saying 79, and thus I would not believe him, and the second case would be reduced to the first. Again, if there were a hundred colors from which he might choose, I would believe him when he said black, because I should conclude that if he were going to falsify, he would probably select some of the other 99 colors, and thus the first case of the problem would become identical with the second.

The following is an interesting problem, and might be useful. A bag contains an unknown number of tickets, numbered from one upwards. A person draws a ticket, notes the number, and replaces it; then repeats the operation till he has drawn s times. What judgment can he form as to the number of tickets in the bag?

Let p_n represent the *a priori* probability that the bag contains just n tickets, and no more. Let i be the largest number drawn in the s drawings. The number of tickets, then, cannot be less than i. We need not know any of the drawn numbers except the largest, because after we know this, every combination of smaller numbers will be equally probable on every admissible hypothesis, and will therefore be of no assistance in judging of these hypotheses. We may then be supposed to know that s tickets have been drawn, and that the largest number observed is i. Now let us make our hypotheses. Suppose that the bag contains just i tickets. The probability that No. i would *not* be drawn on any one drawing is $\frac{i-1}{i}$. Hence the probability that it would not be drawn at all is $\frac{(i-1)^s}{i^s}$; hence the probability that it would be drawn on one or more trials is $\frac{i^s-(i-1)^s}{i^s}$. Again, if we suppose the bag to contain $i+h$ tickets, the probability that i would be the largest number drawn would be equal to the product of the probability that none of the h numbers above i would be drawn into the probability that i would be drawn if none of the h were drawn. The first of these probabilities is $\frac{i^s}{(i+h)^s}$; whence the probability that i would be the largest number is $\frac{i^s-(i-1)^s}{(i+h)^s}$. We can now arrange the quantities as before. We put for brevity $i^s-(i-1)^s=\triangle$.

c	p	q	pq
i balls	p_i	$\frac{\triangle}{i^s}$	$\frac{\triangle p_i}{i^s}$
$(i+1)$ balls	p_{i+1}	$\frac{\triangle}{(i+1)^s}$	$\frac{\triangle p_{i+1}}{(i+1)^s}$
$(i+2)$ balls	p_{i+2}	$\frac{\triangle}{(i+2)^s}$	$\frac{\triangle p_{i+2}}{(i+2)^s}$
&c.	&c.	&c.	&c.

Put P_n for the final probability that the bag contains just n tickets, and we have, by the general formula,

$$(1) \quad P_{i+h} = \frac{\triangle p_{i+h}}{(i+h)^s \left\{ \frac{\triangle p_i}{i^s} + \frac{\triangle p_{i+1}}{(i+1)^s} + \&c. \right\}} = \frac{p_{i+h}}{(i+h)^s \cdot \Sigma}.$$

putting, for brevity,

$$\Sigma = \frac{p_i}{i^s} + \frac{p_{i+1}}{(i+1)^s} + \frac{p_{i+2}}{(i+2)^s} + \cdots$$

By taking the sum of a sufficient number of the values of P_{i+h} from $h=0$ upwards to make $\frac{1}{2}$, we should find that number which it was an even chance whether the balls would fall short of, or exceed. In getting the value of Σ, the series extends to all the numbers which it is possible that the bag should contain.

As a particular case, suppose that all numbers up to a certain limit are, before the drawing, equally probable. The values of p_n will then be all equal; and by putting

$$S = \frac{1}{i^s} + \frac{1}{(i+1)^s} + \frac{1}{(i+2)^s} + \&c.,$$

for all possible values, we shall have

(2) $\quad P_i = \frac{1}{Si^s}, \quad P_{i+1} = \frac{1}{S(i+1)^s}, \ldots P_{i+h} = \frac{1}{S(i+h)^s}.$

We see from these expressions that the highest number drawn is the most probable of *any single number*, so that if we were required to guess the exact number contained in the bag, we ought to guess this one. But if i and h are not very small, several numbers above i will each be almost as probable as i itself, and the guess would therefore probably be wrong. But it would be more satisfactory to find the probability that the number is contained between given limits. Thus, the probability that it does not exceed $i+k$ is, by summing the separate probabilities,

(3) $\quad \frac{1}{S}\left\{\frac{1}{i^s} + \frac{1}{(i+1)^s} + \frac{1}{(i+2)^s} + \cdots + \frac{1}{(i+k)^s}\right\}.$

As a numerical example, suppose that all numbers are equally before drawing, and that upon drawing four times, the highest number drawn is 10. Then we have $i = 10$, $s = 4$, $h = \infty$.

Also $\quad S = \sum_{1}^{\infty} \frac{1}{x^4} - \sum_{1}^{9} \frac{1}{x^4} = \frac{\pi^4}{90} - \sum_{1}^{9} \frac{1}{x^4} = 0.000386650,$

$\frac{1}{i^s} = 0.0001 = \frac{1}{10^4}, \quad \frac{1}{(i+1)^s} = 0.0000683, \quad \frac{1}{(i+2)^s} = 0.0000482.$

The sum of the first two numbers is rather more than $\frac{1}{2}S$, and the sum of the first three rather greater. It is therefore rather less than an even chance that the bag contains 11 tickets, or less, but rather more than an even chance that it does not contain more than 12. If we make $k=10$ in (3), the resulting value of the probability is $\frac{3481}{3856}$. We might therefore wager more than 8 against 1 that the bag did not contain more than 20 balls.

As another numerical example, suppose that I read in a paper the addresses of ten persons, all at the same post-office, but having different boxes, of which the numbers are given. The highest number was 200. What is the probable number of boxes which the office contains, supposing that one box is as likely to be let to any person as another?

To solve this, we observe that the sum of the series

$$1 + \frac{1}{2^{10}} + \frac{1}{3^{10}} + \ldots + \frac{1}{(x-1)^{10}}$$

is $\quad \dfrac{\pi^{10}}{93555} - \dfrac{1}{9\,x^9} - \dfrac{1}{2\,x^{10}} \cdot - \dfrac{10}{12\,x^{11}} + \dfrac{132}{72\,x^{13}} - \&c.$

The sum from x to infinity is, therefore,

$$\frac{1}{x^{10}} + \frac{1}{(x+1)^{10}} + \frac{1}{(x+2)^{10}} + \ldots = \frac{1}{9\,x^9} + \frac{1}{2\,x^{10}} + \frac{10}{12\,x^{11}} - \frac{11}{6\,x^{13}} + \&c.$$

and by taking the difference of these values for different values of x, we can find the sum of the series between finite limits. Representing by $\Sigma_{i,\,i'}$ the sum of the series between the limits i and i' inclusive, we have

$$\Sigma_{200,\,\infty} = 22194, \qquad\qquad \Sigma_{200,\,215} = 11110,$$

in units of the twenty-sixth place of decimals. There is, therefore, about an even chance that the post-office does not contain more than 215 boxes.

When i is so large that we may neglect all terms after the first in the expressions for the sum of the series, we shall have

$$S = \Sigma_{i\infty} = \frac{1}{(S-1)\,i^{s-1}} \qquad \Sigma_{i,\,i'} = \frac{1}{(S-1)\,i^{s-1}} - \frac{1}{(S-1)\,(i'-1)^{s-1}};$$

which, from equation (3), gives for the probability that the number required is contained between the limits i and i' inclusive,

$$P_{i,\,i'} = 1 - \frac{i^{s-1}}{(i'+1)^{s-1}}.$$

If we wish to find the limits within which there is a given probability that the number is contained, we obtain from the last equation,

$$i' = \frac{i}{(1-P_{i,\,i})^{s-1}} - 1.$$

NOTES ON THE THEORY OF PROBABILITY.—INSURANCE.

By Simon Newcomb, Nautical Almanac Office, Cambridge, Mass.

In the first volume of the Mathematical Monthly it was remarked, that the value of a probability of obtaining any sum of money was equal to the product of the sum into the fraction expressing the probability. But this was only on the hypothesis that the desirableness of the money was the same, whether it was all in the hands of one man, or divided among several. This hypothesis is by no means universally true; hence the real value of the expectation of receiving money cannot always be estimated by this rule.

The value or desirableness of a small sum, say one dollar, is much greater to a man who is poor than to one who is wealthy; and as the same man grows richer, the value of one dollar to him will grow less and less. The value of money is to be estimated by the desirableness of the commodities which are purchased with it. A man will first spend his money for articles which are quite indispensable. As he is able to spend more, he spends it in articles which are less and less desirable. If, now, we had for any man some continuous function which should express a relationship between the desirableness and the money value of commodities in general, we should be able to express the value of any sum in terms of its nominal amount. It is obviously impossible in any case to obtain such a function by any rigorous process. But an hypothesis, near enough for practical purposes, has been framed. This hypothesis is, that the value of one dollar to any man is inversely proportional to his wealth; or, perhaps better still, inversely proportional to his annual income, or to the amount of money which he can afford to spend in a year.

If, independently of his income, he possesses productive property, it is perhaps fair to presume that the value of a dollar to him will be the same as if his annual income were increased by one tenth of the amount of the property. On all these hypotheses, if we represent by v the moral value, or desirableness, of a unit of money, say one dollar, and by s the absolute amount of money, reckoned by dollars, and by i the annual income, we shall have

$$dv = \frac{i\,ds}{i + \frac{1}{10}s}. \tag{1}$$

The unit of absolute value is here taken to be the value of one dollar when the individual has nothing but his annual income. s being then zero, we have $dv = ds$. As s increases, the value of $\frac{dv}{ds}$ will diminish in the same proportion that $i + \frac{1}{10}s$ increases, as is required by hypothesis. Integrating (1), we have

$$v = \log c(s + 10\,i)^{10 i},$$

c being an arbitrary constant. Taking this so that v shall represent the moral value of s, we have

$$v = \log\left(1 + \frac{s}{10\,i}\right)^{10 i}, \tag{2}$$

which, when s is small, gives, very nearly, $v = s$, as it should. But when s grows very large, v does not increase so rapidly as s. But if we make s negative, we find that the negative value of v will be much greater than that of s. Both of these results are perfectly in accordance with the judgment of common sense. If a man loses a considerable amount, his increased poverty increases the moral value of a given amount of money, and thus his moral loss exceeds the loss in dollars. If he gains a large sum, his increased wealth diminishes the moral value of money, and thus his moral gain is less than that in dollars.

From these facts arise the benefit of insurance and the evil of gambling. Everything which causes arbitrary transfers of money from one person to another is necessarily an injury, even when the more remote evil effects are left out of consideration. From what has been seen, it will be perceived that the actual loss of the loser is *greater* than the value of the money he parts with, while the winning of the gainer is *less*. Thus, on the whole, there is a moral loss.

Again, everything which tends to prevent sudden and disastrous fluctuations of wealth is beneficial. Especially is this the case where an unforeseen loss can be equitably divided among a large number, instead of falling on a single person. This is the effect of insurance. Thus the difference between those contingent payments which are beneficial and those which are injurious admits of perfect mathematical definition. The class to which they belong depends on whether they are designed to prevent or to cause fluctuations in individual wealth.

Life Insurance.

The advantage of life insurance is founded on the fact, that to the heirs of a person who has no resource but an annual income, money will, after his death, be more valuable than during his life. He therefore lays up a certain sum annually. But it is entirely uncertain to what his savings may amount at his death, and in order to avoid this uncertainty, he compounds with the insurance company by paying it the amount of his annual saving which he wishes to devote to his family after his decease, on condition that they shall, on his decease, pay his family the probable accumulation of his earnings, be the actual accumulation great or small. To show clearly the manner in which this is effected, let us consider successively several possible forms of mutual life-insurance companies. We shall first proceed on the supposition that the actual experience coincides exactly with the probable law of mortality.

1. Out of 1,000 persons aged 25, about 8 die annually. That number of persons at that age may then form a company, and, desiring that each shall receive $1,000 on his death, agree to pay annually the product of $1,000, in the probability that he will die during the year. Each will therefore commence with the annual payment of eight dollars, which will just suffice to pay the policies of the eight deaths which will occur during each year. As the living ones grow older, a larger proportion of them will die; the premium must therefore be increased. At the age of forty it will amount to ten dollars, and at the age of eighty to about a hundred. In this form of company there would be no accumulation of funds, since the amount of premiums each year would only suffice to pay the losses. The policies at the ends of the several years would always be valueless; hence a member could withdraw at any time without suffering any loss, except what arose from his fortune in continuing to live. He would, in fact, have been insuring his life from year to year in the same manner that he would insure a house.

This form of company would have the disadvantage of obliging those who attained a considerable age either to give up their policies or pay enormous premiums. To avoid this, the earlier premiums are increased, and the premium made constant during life; its amount being so regulated that the accumulations of the earlier years with their interest shall just suffice to make up the deficiencies of the later ones. In the case which we have supposed, the constant annual payment, reckoning interest at four per cent, would be about fourteen dollars. The six thousand dollars extra per annum which would thus at first accumulate, would be put at interest for the benefit of those who should live to old age. As the members of the company died off, the annual receipts would be continually lessened, while at the same time, owing to the greater number of deaths, the disbursements would increase. Thus, in thirty years the payments would exceed the income, the accumulated fund would be drawn upon, and finally, when the last man died, it would be exhausted in paying his policy.

The amount of the accumulated fund at any time shows the *value of the policies*. It is, indeed, the difference between the present value of all the future annual payments and the present value of the expectation of the value of the policy at the time of decease. To explain this mathematically, let $\varphi.t$ represent the probability that an individual will be alive at the end of t years. $\varphi.t$ must be equal to unity when t is zero, and vanish for $t = \infty$. It cannot be exactly expressed by any simple function, but is tabulated in tables of mortality. The present value of a premium, a, to be paid in t years, will be $\frac{a}{(1+r)^t}$, r being the rate of interest. But if the payment is to be made only on condition that an individual is living at that time, the present value of the expectation will be equal to the above multiplied by the probability that the man will then be alive, or by $\varphi.t$. It is therefore equal to $\frac{a \cdot \varphi.t}{(1+r)^t}$. The present value of the expectation of a series of annual premiums, the first one being paid down, is

$$a\left\{1 + \frac{\varphi.1}{1+r} + \frac{\varphi.2}{(1+r)^2} + \frac{\varphi.3}{(1+r)^3} + \&c.\right\} \qquad (3)$$

The probability that the individual will die between the times t and $t+dt$ is $d.\varphi t$ or $\varphi'.t\,dt$. The present value of the expectation of a sum, S, to be paid if he dies at that time, is $S\frac{\varphi't\,dt}{(1+r)^t}$. The sum being payable whenever the man dies, the complete present value of the expectation is the integral of this expression, or

$$S\int_0^\infty \frac{\varphi't}{(1+r)^t}\,dt. \qquad (4)$$

If the man is but entering the company, his annual premium is determined by the condition that the two expressions above shall be equal, which gives

$$\frac{a}{S} = \int_0^\infty \frac{\varphi' t}{(1+r)^t} \, dt \div \sum_0^\infty {}_t \frac{\varphi \cdot t}{(1+r)^t}. \tag{5}$$

The values of these integrals are determined numerically from tables of mortality, by the method of mechanical integration.

The form of the function φ is such, that the numerical value of $\varphi' t$ becomes greater the older the man grows, whence a larger annual premium is requisite. Since, in the company which we have supposed, the expectation of final profit is nothing, the present value of the expectation of the policy payable on death must exceed the present value of the expectation of future premiums by that portion of the accumulation which belongs to the policy. This accumulation, therefore, represents the value of the policy.

This form of company requires a modification, from two causes. In the first place, it is probable that the actual rate of mortality will differ slightly from that of the tables, a contingency which must be provided against; and again, the actual rate of interest may differ from that used in calculating the premiums. In the next place, expenses must be provided for. Hence the theoretical premium is increased so as to provide for expenses and for possible depreciations in the rate of interest, and for possible extraordinary numbers of deaths. This increase is termed the loading, and usually amounts to about one fourth the amount of the theoretical premium.

Owing to the fact, that the actual rate of interest is generally much higher than that on which the premiums are calculated (the latter being usually but four per cent), it is almost certain that there will be a considerable accumulation in the treasury over and above the value of the policies. This surplus ought, at stated intervals, to be paid back to, or at least placed to the credit of, those policy-holders from whose premiums it arose. If this is not done, or if the entire amount of this extra accumulation is not thus disposed of, it will be unfair to the earlier members, whose money will go to the benefit of future members; if more than the surplus is divided, it will be unfair to future policy-holders, who will ultimately be obliged to make up the deficiency. The amount which ought to be distributed is always susceptible of pretty exact mathematical calculations.

NOTE ON THE FREQUENCY OF USE OF THE DIFFERENT DIGITS IN NATURAL NUMBERS

Simon Newcomb

Note on the Frequency of Use of the Different Digits in Natural Numbers.

By Simon Newcomb.

That the ten digits do not occur with equal frequency must be evident to any one making much use of logarithmic tables, and noticing how much faster the first pages wear out than the last ones. The first significant figure is oftener 1 than any other digit, and the frequency diminishes up to 9. The question naturally arises whether the reverse would be true of logarithms. That is, in a table of anti-logarithms, would the last part be more used than the first, or would every part be used equally? The law of frequency in the one case may be deduced from that in the other. The question we have to consider is, what is the probability that if a natural number be taken at random its first significant digit will be n, its second n', etc.

As natural numbers occur in nature, they are to be considered as the ratios of quantities. Therefore, instead of selecting a number at random, we must select two numbers, and inquire what is the probability that the first significant digit of their ratio is the digit n. To solve the problem we may form an indefinite number of such ratios, taken independently; and then must make the same inquiry respecting their quotients, and continue the process so as to find the limit towards which the probability approaches.

Let us suppose the numbers with which we commence to be arranged in periods according to the number of their digits, or, which is the same thing, according to the characteristics of their logarithms on the scale of which the basis is i, (i being 10 in the common system). Then, if two numbers are i^{c+s} and $i^{c'+s}$, c and c' being integers, the significant figures of the ratio will be independent of c and c', since changing these integers will only change the decimal point. We may, therefore, take both numerator and denominator of the ratio out of the same period.

Moreover, since both numerator and denominator are formed by the same process, we may suppose the law of distribution of the numbers from which they are selected to be the same. Our problem is thus reduced to the following:

We have a series of numbers between 1 and i, represented by fractional powers of i, say i^s, the distribution of the exponents s, and therefore of the numbers, being according to any arbitrary law. Since these exponents are formed by casting off all the integers from a series of numbers, we may suppose them arranged around a circle according to some law. Then, if we select 2^n exponents at random and call them s', s'', s''', etc., the final ratio, obtained in the manner we have described, will be

$$i^{s' - s'' + s''' - s'''' + \text{etc.}}$$

The question is, what is the probability that the positive fractional portion of $s' - s'' + s''' - s'''' + $ etc., will be contained between the limits s and $s + ds$. It is evident that, whatever be the original law of arrangement, the fractions will approach to an equal distribution around the circle as n is increased, or the required probability will be equal to ds. But, the fractional part of $s' - s'' + s''' - $ etc. is the mantissa of the logarithm of the limiting ratio. We thus reach the conclusion:

The law of probability of the occurrence of numbers is such that all mantissæ of their logarithms are equally probable.

In other words, every part of a table of anti-logarithms is entered with equal frequency. We thus find the required probabilities of occurrence in the case of the first two significant digits of a natural number to be:

Dig.	First Digit.	Second Digit.
0		0.1197
1	0.3010	0.1139
2	0.1761	0.1088
3	0.1249	0.1043
4	0.0969	0.1003
5	0.0792	0.0967
6	0.0669	0.0934
7	0.0580	0.0904
8	0.0512	0.0876
9	0.0458	0.0850

In the case of the third figure the probability will be nearly the same for each digit, and for the fourth and following ones the difference will be inappreciable.

It is curious to remark that this law would enable us to decide whether a large collection of independent numerical results were composed of natural numbers or logarithms.

A GENERALIZED THEORY OF THE COMBINATION OF OBSERVATIONS SO AS TO OBTAIN THE BEST RESULT

Simon Newcomb

A Generalized Theory of the Combination of Observations so as to Obtain the Best Result.

By Simon Newcomb.

§1. *Introductory Remarks.*

The accepted practice of combining observations rests upon the hypothesis that the frequency of errors follows a certain well-known law which may be expressed as follows: Let Δ be the amount by which the result of an observation may differ from the value obtained by taking the mean result of an infinity of similar observations. Δ will then be the *error* of the observation. The infinitesimal probability that an error will be contained between the limits Δ and $\Delta + d\Delta$ is supposed to be given by the equation

$$dp = \frac{h}{\sqrt{\pi}} e^{-h^2 \Delta^2} d\Delta,$$

in which h is the "modulus of precision" depending upon the accuracy of the observations, and e is the Naperian base.

The modulus h is commonly replaced by a probable error r, which term signifies such a magnitude that the number of errors less than r in absolute value is equal to the number which exceed r. The value of r is given in terms of the modulus h by the equation

$$r = \pm \frac{0.4769}{h}.$$

When the errors really follow the law in question, they diminish with extreme rapidity as Δ increases. For example, only one per cent. of the errors should fall without the limits $\pm 4r$.

As a matter of fact, however, the cases are quite exceptional in which the errors are found to really follow the law. The general rule is that much more than one per cent. of the errors exceed four times the probable error. In other words, it is nearly always found that some of the outstanding errors seem abnormally large. The method of dealing with these abnormal errors has always been one of the most difficult questions in the treatment of observations. The common practice has been to consider the observations affected by them as abnormal, and to reject them in obtaining the final result. But we here meet with the difficulty that no positive criterion for determining whether an observation should or should not be treated as abnormal is possible. Several attempts have indeed been made to formulate such a criterion, the best known of which is that of Peirce.*

Peirce's criterion has always seemed to me subject to two serious objections. One is that it takes no account of any probable error of the observations under consideration which may be known beforehand, but proceeds as if the value of the probable error could be deduced from the comparison of the observations *inter se*. An immediate general consequence of this is that if all the errors of a system are multiplied by the same factor, the same observations are rejected as before, how small or great soever the factor may be.†

The second objection is that it takes no account of the fact that the *a priori* probability that an observer should make an abnormal observation varies with the observer, and places all observers on a level by regarding that probability as determined by a general mathematical principle applicable to all cases.

It is, however, well known that some observers make very few abnormal observations, while others are extremely liable to them. It is evident that if we are dealing with an observation whose error is so large that we doubt whether it should or should not be considered abnormal, our judgment must depend very largely upon any knowledge we may have of the carefulness of the observer.

The fact is, however, that any system of rejecting supposed abnormal observations is subject to the objection of leading to a result which is a discontinuous function of the separate errors of observation, and hence to results

---

* Gould's Astronomical Journal, Vol. II, p. 161.

† Certain results of Peirce's criterion in special cases, when applied to sets of three or four observations, do not seem to have been hitherto noticed. The following are cases in point:

Of a set of three observations none are ever rejected by it, no matter how much one may deviate from the mean of the other two.

In a set of four observations, if three agree exactly, the fourth will always be rejected if it differs ever so little from the others. More generally, if no one of the three results which agree best among themselves differs from the mean of the three by more than ε, then a fourth, which differs from that mean by more than 4ε, will be rejected. For example, if the results of four observations with a meridian circle were $0''.3$; $0''.4$; $0''.5$; $0''.8$, the last would be rejected.

which are sometimes indeterminate. Suppose, for example, that we are dealing with the mean of three observations, two of which are closely accordant, while the third differs from the mean of the other two by the quantity x. Let us represent the mean of the two accordant ones by the symbol m'; then, if we include the discordant observation, the general expression for the mean result in terms of x will be

$$m = m' + \tfrac{1}{3}x.$$

In ordinary astronomical practice we retain this value of m so long as x does not exceed the limit which we consider that of a normal error. But, as soon as this limit is reached, we drop x entirely and take m' for the value of m. In other words, if we consider x to increase from zero, the adopted value of m will increase one-third as fast until the assigned limit is reached, and will then suddenly spring back from $m' + \tfrac{1}{3}x$ to m'. If the critical point at which x is to be rejected could be satisfactorily defined, this course would be less objectionable. But, as a matter of fact, it is to be determined by the judgment of the investigator, with the result that between certain wide limits the investigator must himself be doubtful whether he should take m' or $m' + \tfrac{1}{3}x$ as his result. Of course different investigators would reach different conclusions in special cases, and thus the most probable result is frequently indeterminate.

There are classes of important observations in which the proportion of large errors is so great that no separation into normal and abnormal observations is possible. This is the case in observations of transits of Venus and Mercury over the sun. A noteworthy instance has been given by the writer in his discussion of transits of Mercury.* By a comparison of 684 observations it was found that the errors of one-half of them were contained between the limits $\pm 6''.8$. If the errors followed the commonly assumed law, then only 5 of them should have exceeded ± 27 seconds. As a matter of fact, however, it was found that 49 exceeded these limits. Yet these 49 observations cannot be considered as wholly worthless, because their results are not scattered entirely at random, and are mostly included between comparatively narrow limits. They differ from the other observations only in having a larger probable error.

The case may be made clearer by reflecting that the law in question presupposes that the observations under consideration are all of the same general quality as regards liability to error; in other words, that they are all liable to

---

* Astronomical Papers of the American Ephemeris, Vol. I, pp. 379-383.

the same errors, and differ only in the accidental circumstances which give rise to the errors. If, however, this is not true; if, for example, we are furnished with a system of observations of which one portion have a small probable error, another a larger probable error, a third a yet larger one, and so on, then the errors of the whole system will not follow the law in question, but we shall find that large errors are disproportionately frequent. Now, this must be the case in nearly all astronomical and physical work.

From this another conclusion follows. In such a mixed system of observations the most probable result will be, not the arithmetical mean, but a mean obtained by giving less weight to the more discordant observations. This will be evident on reflecting that in such a case the more discordant results will probably belong to the observations having a larger probable error and therefore the less weight.

§ 2. *Modified Curves of Probability.*

The preceding considerations lead us to the further conclusion that the commonly received theory which presupposes that there must always be some one "most probable value" of a quantity determined by observations, lacks generality. The fact is that, in special cases, owing to a possibility of abnormal observations, the curve of probability may have a great variety of forms. As one example, let it be supposed that two mean declinations of a star, determined with a good meridian circle the micrometer-head of which is numbered at intervals of 5″, differ from each other by a quantity approximating to 5″. We then may make three hypotheses: that the observations are both normal, or that one or the other of them is in error by 5″ through a mistake in recording.

According to the probability of the first hypothesis, and of either of the other two, we may have different curves. Assuming the instrument and observer to be so accurate that a difference of 5″ between two normal observations is nearly out of the question, we shall have a curve of the form A. As the probability of the first hypothesis increases, the curve may assume the form B. If the observer is one never known to make mistakes in reading, the curve will approximate to its usual form.

Fig. A.   Fig. B.

Now, it is evident that in such a case as that indicated by curve A, we can assign no one most probable value of the observed quantity. The only complete statement we could make would be embodied in a table showing for each separate possible value of the required quantity the probability that the quantity had a value not differing from it by more than a small arbitrary amount. Assuming the intervals of the table to be taken to equal double this amount, the sum of all these probabilities would be unity.

Looking further into the matter, we see that this is a general method of expressing the conclusion to be derived in all cases from an observation or a series of observations. No matter how definite the primary value given by an observation may be, the actual conclusion to be drawn is always a series of separate probabilities that the quantity observed has some one of an infinite series of values. If the law of probability is that commonly assumed, then the probability of each assignable value is completely determined when the most probable value and the probable error are given. But such is not the case when the law of probability deviates from this form.

§ 3. *Evil and Worth of Erroneous Results.*

The question now arises whether, when we consider the most general case, in which there may be several maxima of probability, and when, therefore, no one most probable value can be assigned, it is possible to formulate any general principle by which a single value shall be preferred for acceptance above all others. Taking as an example such a case as A just given, it is clear that no such principle is possible without some antecedent hypothesis determining a law of choice between errors of different magnitudes. If, to fix the ideas, we suppose that in case A the results of the two separate observations were $0''.0$ and $5''.0$, then the three hypotheses will give us $0''.0$; $2''.5$; $5''.0$, as three values between which we are to choose. If we choose either the first or third, we shall have a probability of slightly less than one-half of being very near the truth, and an equal probability of being $5''$ in error, together with a very small probability of being about $2''.5$ in error. If, on the other hand, we take $2''.5$ as our result, we shall be almost sure of being between $2''$ and $3''$ in error, and no more. Our choice, then, must depend on whether a certainty of being $2''.5$ in error, or an even chance of each of the errors $0''$ and $5''$, is preferable. This again turns upon the question whether an error of $5''$ involves more or less than twice the evil of an error of $2''.5$.

The ordinary requirements of practical life are in favor of the view that the evil increases in a higher ratio than the simple magnitude of the error. As examples, if it is a case of an error in the position of a ship arising from an error of the Nautical Almanac, we readily see that the probability of the error leading to a shipwreck increases in a higher ratio than that of the simple error itself. Again, in the case in which, by the labor of continually increasing observations of a single quantity, we lessen the probable error, we know that it requires fourfold labor to reduce the probable error to one-half. It would seem, therefore, that the best hypothesis that we can adopt is that the evil of an error is proportional to the square of its magnitude.

A determination has more or less value according as it is less or more liable to errors. The simplest definition of the value of an observation that we can adopt is that it is inversely as the sum total of the evils to which it is subjected, each evil being multiplied by its probability. This also is in strict accord with the ordinary law of probability of a number of observations, since it involves the hypothesis that the value of a result is proportional to the number of observations on which it depends. As, however, the word "value," if used to express this conception, would be ambiguous in consequence of being applied to the simple amount of a quantity, we shall use the term *worth* to express the economic value of an assigned value as just described. We therefore have the definitions:

The *evil* of any value assigned to a quantity is equal to the sum of all the products obtained by multiplying the square of each possible error of that assigned value by the probability of its occurrence.*

The *worth* of any such value is inversely proportioned to its evil.

The value to which we are to give the preference is that whose worth is a maximum, or, which amounts to the same thing, whose evil is a minimum. This value, and the magnitude of the evil with which it is affected, will be two elements corresponding to "most probable value" and "probable error" in the usual theory.

§ 4. *Algebraic Expressions for the Evil.*

The general expression for these elements is readily obtained. Let us represent all possible values which the required quantity can have by the series

$$x_1, x_2, x_3, \ldots x_n,$$

and let

$$p_1, p_2, p_3, \ldots p_n$$

---

* This idea of an evil attached to each error, and proportional to the square of its magnitude, is due to Gauss (Theoria Combinationis Observationum, etc., Pars prior, §6), who applies the term *jactura* to the conception here called evil.

be the several probabilities of these values; we necessarily have in this case

$$\sum_{1}^{n} p_i = 1.$$

Putting x for any value of the quantity in question, the evil of this value will be, by definition,

$$e = p_1 (x-x_1)^2 + p_2 (x-x_2)^2 + \ldots + p_n (x-x_n)^2$$
$$= x^2 - 2Ax + B,$$

where
$$A = \sum_{1}^{n} p_i x_i,$$
$$B = \sum_{1}^{n} p_i x_i^2.$$

This quantity is a minimum when
$$x = A,$$

which is, therefore, the value we are to prefer. The evil of this value is
$$e_0 = B - A^2.$$

The respective values of A^2 and B may be written, since $\Sigma p = 1$,

$$A^2 = \sum_{1}^{n} {}^2_{i,j} p_i p_j x_i x_j,$$
$$B = \sum_{1}^{n} {}^2_{i,j} p_i p_j x_i^2.$$

The value of the minimum evil then becomes

$$e_0 = \frac{1}{2} \sum_{1}^{n} {}^2_{i,j} p_i p_j (x_i - x_j)^2.$$

It therefore appears that the inverse of this expression is the worth of the best value of the required quantity, which best value is given by the equation

$$x_0 = \sum_{1}^{n} {}_i p_i x_i. \qquad (1)$$

In what precedes, the form of our equations is based upon the supposition that $x_1, x_2 \ldots x_n$ are a finite number of discrete values which x may have. In the usual case, however, the unknown quantity may have all values between

certain wide limits, and the probability that it is contained between the limits x and $x + dx$ is given by an equation of the form
$$dp = \phi(x)\,dx,$$
dp being an infinitesimal probability. Since this is a pure number, it follows from this equation that, whatever physical quantity may be represented by x, $\phi(x)$ must be of the dimension -1 in this quantity.

Reducing the formula (1) to the present case, we find that the preferable value of x is given by the equation
$$x_0 = \int_{-\infty}^{+\infty} x\phi(x)\,dx. \tag{2}$$
The evil of this value is
$$e_0 = \frac{1}{2}\int_{-\infty}^{+\infty}\int_{-\infty}^{+\infty} \phi(y)\cdot\phi(z)(y-z)^2\,dy\,dz \tag{3}$$
$$= \int_{-\infty}^{+\infty} y^2 \phi y\,dy - \int_{-\infty}^{+\infty}\int_{-\infty}^{+\infty} yz\phi(y)\phi(z)\,dy\,dz.$$

From the definition of evil, it is a quantity of the dimensions $+2$ in those of the physical quantity in question. Its square root is therefore a definite quantity of the magnitude to be determined, which we may regard as an error.

An example of the results of the present theory will be found by applying it to the case of the usually assumed law of error, namely:
$$\phi(x) = \frac{h}{\sqrt{\pi}} e^{-h^2 x^2}$$
We then have $x_0 = 0$ as the preferable value of x, in accordance with the usual theory. In the expression for the evil of this value we have
$$A = x_0 = 0,$$
$$B = \frac{h}{\sqrt{\pi}}\int_{-\infty}^{+\infty} y^2 e^{-h^2 y^2}\,dy = \frac{1}{2h^2}. \tag{4}$$
This value of B is identical with the square of what is commonly called the *mean error*, which again is equal to $1.5 \times$ probable error nearly. Hence, in this special case, the evil is identical with the square of the mean error.

If, instead of taking zero as the value of x, we wish to express the evil of any other assumed value, we have the expression
$$e = x^2 + B = \varepsilon^2 + x^2,$$
ε being the mean error. If, instead of ε, we use r, the probable error, the expression will be
$$e = x^2 + 2.198 r^2.$$

We readily find that if, instead of using the most probable value of a quantity, we adopt a value differing from it by its probable error, the evil will be increased by a little less than half its whole amount.

It appears, therefore, that, whatever the law of error, we may always find two quantities corresponding to "most probable value," and "probable error" of that value in the usual theory. One of these quantities will naturally be the best value of the required magnitude, or A itself. The other may be either the evil of the value A, or the change in the value of the magnitude required to increase its evil in a definite ratio, these last quantities being functions of the same quantity, and therefore of each other. If we present the result in the form
$$x = A \pm \sqrt{B - A^2},$$
the last term will be the "mean error" in the usual theory, and the change in x which would double its evil in the generalized theory. If we wish to express the quantity corresponding to the "probable error," we write
$$x = A \pm 0.6745\sqrt{B - A^2}.$$

§ 5. *Modified Law of Probability.*

The whole problem now before us is reduced to finding a curve of probability in the case of a number of observations of the same quantity. This problem naturally involves that of the law of error of the separate observations, and leads us to inquire what modification should be made in the usually assumed law in order that it may be applicable to all cases whatever.

The defect of the commonly assumed law, as represented by the equation
$$\phi(x) = \frac{h}{\sqrt{\pi}} e^{-h^2 x^2},$$
is that, in practice, large errors are more frequent than this equation would indicate them to be. This defect might be remedied by substituting some other function of x than $h^2 x^2$ as the exponent of e. The requirements of this function would be:

1. That it should be an even function of x, or of the form $f(x^2)$.
2. That it should become infinite when x did.
3. That it should increase less rapidly than $h^2 x^2$, or, more exactly, that the second derivative $\frac{\partial^2 f}{\partial x^2}$, instead of being a constant, should diminish with an increase

of x. Such a function can be formed by writing, instead of h^2, an expression of the form
$$\frac{h^2(1+h^2x^2)}{1+h'^2x^2},$$
so that we should have, for the exponent $f(x^2)$, an expression of the form
$$-\frac{h^2(x^2+h^2x^4)}{1+h'^2x^2}.$$

The management of such an exponent might, however, prove inconvenient, and I shall adopt a law of error founded on the very probable hypothesis that we are dealing with a mixture of observations having various measures of precision. Let us put
$$h_1, h_2, \ldots h_n,$$
the possible separate values of the different measures of precision;
$$p_1, p_2, \ldots p_n,$$
the corresponding probabilities that an observation selected at random has any one of these several measures of precision.

Then, for an observation selected at random, the law of error will be

$$\phi(x) = \frac{1}{\sqrt{\pi}} \{ p_1 h_1 e^{-h_1^2 x^2} + p_2 h_2 e^{-h_2^2 x^2} + \ldots + p_n h_n e^{-h_n^2 x^2} \}. \tag{5}$$

§6. Deduction of Best Result.

If we have m observations of the same quantity giving $x_1, x_2 \ldots x_m$ as the observed values, then, assuming the law of error expressed in (5), the probability that the quantity is contained between the limits η and $\eta + d\eta$ is given by the equation
$$dp = a\psi(\eta) d\eta,$$
in which
$$\psi(\eta) = \phi(x_1-\eta)\phi(x_2-\eta)\ldots\phi(x_m-\eta), \tag{6}$$
while a is a constant so chosen as to make
$$a \int_{-\infty}^{+\infty} \psi(\eta) d\eta = 1.$$

The formula (2) then gives for the best value of x the expression

$$x = \int \eta \, dp = a \int_{-\infty}^{+\infty} \eta \psi(\eta) d\eta = \frac{\int_{-\infty}^{+\infty} \eta \psi(\eta) d\eta}{\int_{-\infty}^{+\infty} \psi(\eta) d\eta}. \tag{7}$$

If $\psi(\eta)$ be multiplied by any constant factor, it will disappear from this expression; we may therefore disregard all such factors in forming $\psi(\eta)$. We may therefore take for $\psi(\eta)$ the product of the m quantities

$$h_1' e^{-h_1^2(x_1-\eta)^2} + h_2' e^{-h_2^2(x_1-\eta)^2} + \ldots + h_n' e^{-h_n^2(x_1-\eta)^2}$$
$$\times\ h_1' e^{-h_1^2(x_2-\eta)^2} + h_2' e^{-h_2^2(x_2-\eta)^2} + \ldots + h_n' e^{-h_n^2(x_2-\eta)^2}$$
$$\times\ \ldots\ldots\ldots\ldots\ldots\ldots\ldots\ldots\ldots\ldots\ldots$$
$$\times\ h_1' e^{-h_1^2(x_m-\eta)^2} + h_2' e^{-h_2^2(x_m-\eta)^2} + \ldots + h_n' e^{-h_n^2(x_m-\eta)^2},$$

in which we have written, for brevity,
$$h' \equiv hp.$$

This product will be formed of n^m terms, each of the form
$$Pe^{-k^2\eta^2 + 2b\eta - c},$$
where
$$\begin{aligned}P &\equiv h_i' h_j' h_l' \ldots h_r', \\ k^2 &\equiv h_i^2 + h_j^2 + h_l^2 + \ldots + h_r^2, \\ b &= h_i^2 x_1 + h_j^2 x_2 + h_l^2 x_3 + \ldots + h_r^2 x_m, \\ c &= h_i^2 x_1^2 + h_j^2 x_2^2 + h_l^2 x_3^2 + \ldots + h_r^2 x_m^2, \end{aligned} \qquad (8)$$

$i, j, l \ldots r$ being any m of the n indices $1, 2, 3 \ldots n$, with repetition. We therefore write
$$\psi(\eta) = \Sigma P e^{-k^2\eta^2 + 2b\eta - c}. \qquad (9)$$

We then have, by integration,
$$\int_{-\infty}^{+\infty} \psi(\eta)\, d\eta = \sum \frac{P\sqrt{\pi}}{k} e^{\frac{b^2}{k^2} - c}, \qquad (10)$$
$$\int_{-\infty}^{+\infty} \eta\psi(\eta)\, d\eta = \sum \frac{Pb\sqrt{\pi}}{k^3} e^{\frac{b^2}{k^2} - c}.$$

If we distinguish the n^m values of each of the quantities P, k, b and c by the suffixes $1, 2, 3 \ldots l$, so that $l = n^m$, and if we put, for perspicuity,

$$\begin{aligned}\eta_t &= \frac{b_t}{k_t^2}, \\ w_t &= \frac{P_t}{k_t} e^{\frac{b_t^2}{k_t^2} - c_t} \quad (t = 1, 2, 3 \ldots l),\end{aligned} \qquad (11)$$

the equation (7) will give
$$x = \frac{w_1 \eta_1 + w_2 \eta_2 + \ldots + w_l \eta_l}{w_1 + w_2 + \ldots + w_l}. \qquad (12)$$

This result admits of a statement which will make the principles of the method quite clear, independently of the analytic processes. The n^m quantities η_1, η_2, etc., are so many *means by weights* of the observed quantities $x_1, x_2, x_3 \ldots x_m$, each mean being obtained by making an hypothesis respecting the distribution of the measures of precision $h_1, h_2 \ldots h_n$ among the m separate observations. Since each observation may, independently of all the others, have any one of the n measures of precision, there will be n^m such hypotheses, each leading to a different mean, η. The final value of x is again a mean by weights of the results of the different hypotheses, the weight of each result being proportional to the probability of the hypothesis on which it depends, which is represented by w. This probability is a product of two factors, of which one, $\dfrac{P}{k}$, is proportional to the probability of the combination, while the other, $e^{\frac{b^2}{k^2}-c}$, is the probability of the combination of outstanding errors to which the hypothesis leads.

§ 7. Application to an Example.

Before showing how the preceding method may be simplified in practice, it may be of interest to give a simple numerical example of its rigorous application. Let it be granted that we have three observations of a class for which there is a probability of $\frac{2}{3}$ that an observation is good, and of $\frac{1}{3}$ that it is poor. Let the measure of precision of a good observation be 4, and of a poor one 1. Let the results of the three observations be

$$\text{I}, 0; \text{II}, 0; \text{III}, 1.$$

Since we have, *a priori*, no reason to distinguish between these results, the usual method of treatment would lead either to $\frac{1}{3}$ as the best result, or to the rejection of the third observation, and hence to the result 0.

From the point of view of the present paper, the agreement of the first two observations and the discordance of the third give color to the hypothesis that the first two observations are good and the third poor. On this hypothesis the best result would be $\frac{1}{33}$, the weights of the results being 16, 16 and 1. But, since every other hypothesis we can make would lead to a larger result, the best result must be greater than this.

The rigorous treatment of the problem gives

$$\phi(x) = \frac{1}{3\sqrt{\pi}} \left\{ 8 e^{-16x^2} + e^{-x^2} \right\}.$$

Hence, when $x_1 = x_2 = 0$ and $x_3 = 1$,

$$\begin{aligned}
\psi(\eta) = \; & 512 e^{-48\eta^2 + 32\eta - 16} + 64 e^{-33\eta^2 + 2\eta - 1} \\
& + 128 e^{-33\eta^2 + 32\eta - 16} + 16 e^{-18\eta^2 + 2\eta - 1} \\
& + 8 e^{-18\eta^2 + 32\eta - 16} + e^{-3\eta^2 + 2\eta - 1}.
\end{aligned}$$

Observations so as to Obtain the Best Result.

These six terms correspond respectively to the six essentially different hypotheses which can be made respecting the distribution of the measures of precision among the different observations, it being observed that among the eight hypotheses there are two pairs such that the members of each pair lead to identical results. The results are tabulated as follows:

Hypothesis.	Results. $\dfrac{b}{k^2}$	Probability. $\dfrac{P}{k} e^{\frac{b^2}{k^2}-c}$	Product.
I, II and III all good,	$\dfrac{1}{3} = 0.3333$	0.003	0.001
I $\begin{cases} \text{good} \\ \text{bad} \end{cases}$; II $\begin{cases} \text{bad} \\ \text{good} \end{cases}$; III good,	$\dfrac{16}{33} = 0.4848$	0.006	0.003
I and II bad; III good,	$\dfrac{8}{9} = 0.8889$	0.318	0.283
I and II good; III bad,	$\dfrac{1}{33} = 0.0303$	4.224	0.128
I $\begin{cases} \text{good} \\ \text{bad} \end{cases}$; II $\begin{cases} \text{bad} \\ \text{good} \end{cases}$; III bad,	$\dfrac{1}{18} = 0.0555$	1.467	0.082
I, II and III all bad,	$\dfrac{1}{3} = 0.3333$	0.296	0.099
		$\Sigma = 6.314$	0.596

Hence, for the value of maximum worth, we have
$$x = 0.0944.$$

§8. *Modification of the Method when the Observations are Numerous.*

In order to apply the preceding method, it is necessary to know the respective probabilities that the measure of precision of any one observation has the several values $h_1, h_2 \ldots h_n$. These probabilities are determined from the actual distribution of the residuals with respect to magnitude as found by the study of large masses of observations. If it were found that in any class of observations the magnitudes of the residuals followed the commonly assumed law, we should have but one value of h. If, as will commonly be the case, we find a larger number of large residuals than would be given by the common theory, we assume one, two, three or more additional values of h, and determine how many observations we must assign to each class in order that the distribution may be represented by an equation of the form (5).

To carry out the rigorous process of finding the best mean value of x, we should form, by the equations (8) and (11), n^m different values of P, k, b, c, η and w, and thence, from (12), the required value of x.

To effect this, let us attach, or suppose to be attached, to each of the quantities P, k, etc., a system of m indices, i, j, k, etc., each index taking the values $1, 2 \ldots n$. The system of indices
$$i, j, k \ldots q$$
attached to a quantity will then indicate the special value of that quantity which results from assigning

to x_1 the precision h_i,
to x_2 " " h_j,
to x_3 " " h_k,
. . . .
. . . .
to x_m " " h_q.

Moreover, we shall, for brevity, represent the combination of indices
$$(i, j, k \ldots q)$$
by the single symbol t.

Any one value of w in (11) may then be written in the form
$$w_t = \frac{P_t}{k_t} e^{-\frac{\Delta_t}{k_t^2}},$$
where we put, for brevity,
$$\Delta = k^2 c - b^2.$$

If we here substitute for k^2, b and c their values from (8), we find that this expression reduces to
$$\Delta = \sum_{i,j}^{2} h_i^2 h_j^2 (x_i - x_j)^2,$$
$$(i = 1, 2, 3 \ldots m-1; \; j = i+1, i+2 \ldots m),$$
where h_i for the moment indicates the special value of h which, in any combination, is assigned to x_i. We may equally represent Δ in the form
$$\frac{1}{2} \sum_{i=1}^{i=m} \sum_{j=1}^{j=m} h_i^2 h_j^2 (x_i - x_j)^2.$$

The first form will consist of $\frac{m(m-1)}{2}$ terms; the second of m^2 terms, if we count those which vanish through $i = j$.

Since Δ depends only on the differences between the x's, it will remain unchanged when we subtract any constant from all of them. Let us then subtract η from all of them, putting
$$\xi_i = x_i - \eta;$$

we shall then have, for each combination of indices,
$$\frac{\Delta}{k^2} = h_1^2\xi_1^2 + h_2^2\xi_2^2 + h_3^2\xi_3^2 + \ldots + h_m^2\xi_m^2.$$

In strictness, each of the values of ξ_i should be affected by the m indices $i, j, k \ldots r$, because the values of η are so affected. But, when m is large, the different values of η arising from different combinations of indices will differ but slightly in the large majority of combinations. We may therefore, to form the m ξ's, take one general mean value of η for all the combinations of the indices.

We have now to substitute the above value of Δ in the exponential expression for w. The result may be expressed in the following form. If we put

$$\begin{aligned} \mathrm{w}_i^{(1)} &= h_i' e^{-h_i^2 \xi_1^2}, \\ \mathrm{w}_i^{(2)} &= h_i' e^{-h_i^2 \xi_2^2} \quad (i = 1, 2, 3 \ldots n), \\ \mathrm{w}_i^{(m)} &= h_i' e^{-h_i^2 \xi_m^2} \end{aligned} \qquad (13)$$

(in which, it will be noted, the suffix i has its original signification), we shall have
$$w_{(i,j,k\ldots r)} = \mathrm{w}_i^{(1)} \mathrm{w}_j^{(2)} \mathrm{w}_k^{(3)} \ldots \mathrm{w}_r^{(m)} \div k_{(i,j,k\ldots r)},$$
$$\eta_{(i,j,k\ldots r)} = h_i^2 x_1 + h_j^2 x_2 + \ldots + h_r^2 x_m \div k^2_{(i,j,k\ldots r)}.$$

We have now to introduce another approximate hypothesis, namely, that the various values of k are so nearly equal that they may be regarded as having a common value. We note that k^2 is the sum of the squares of the m h's, and that, in the large majority of cases, the sum will approximate to a certain mean value, found by distributing the m values among the n classes proportionally to their several probabilities.

If we now substitute the preceding values of w and η in (12) we find, on the above hypothesis, that the value of x may be reduced to the form

$$x = \frac{W_1 x_1 + W_2 x_2 + W_3 x_3 + \ldots + W_m x_m}{W_1 + W_2 + W_3 + \ldots + W_m},$$

where
$$\begin{aligned} W_1 &= h_1^2 \mathrm{w}_1^{(1)} + h_2^2 \mathrm{w}_2^{(1)} + h_3^2 \mathrm{w}_3^{(1)} + \ldots + h_n^2 \mathrm{w}_n^{(1)} \div \Sigma \mathrm{w}_i^{(1)}, \\ W_2 &= h_1^2 \mathrm{w}_1^{(2)} + h_2^2 \mathrm{w}_2^{(2)} + h_3^2 \mathrm{w}_3^{(2)} + \ldots + h_n^2 \mathrm{w}_n^{(2)} \div \Sigma \mathrm{w}_i^{(2)}, \\ &\qquad \ldots \ldots \ldots \ldots \ldots \ldots \ldots \ldots \ldots \\ W_m &= h_1^2 \mathrm{w}_1^{(m)} + h_2^2 \mathrm{w}_2^{(m)} + h_3^2 \mathrm{w}_3^{(m)} + \ldots + h_n^2 \mathrm{w}_n^{(m)} \div \Sigma \mathrm{w}_i^{(m)}. \end{aligned} \qquad (14)$$

These values of W are the modified weights of the several results $x_1, x_2 \ldots x_m$ arising from the probable variability of h from one observation to another. Were there no such variability; were there but one value of h, then these weights would all be equal. But, in the case for which the preceding theory is constructed, each result x_i may have any one of the weights h_1^2, h_2^2, etc.; and the equation (14) determines a certain mean among these weights which we are to assign to x_i. The coefficients W are functions of ξ, and admit of being tabulated as such in any special case.

In what precedes we have presupposed no difference of weights among the results $x_1 \ldots x_m$ to be known *a priori*. But, since each observation may have any one of the weights $h_1 \ldots h_n$, a certain mean weight w of each is determined *a posteriori*, as a function of its deviation from the general mean, by the equations (14). This mean weight can be tabulated as a function of ξ, and thus taken out from a table with a single argument.

If, however, we have some knowledge of an observation which leads us to assign it one precision rather than another, we may utilize this knowledge so as to modify the values of $w_j^{(i)}$. If h_1, h_2, etc., are taken in the descending order of magnitude, then h_1^2 will be the weight of each observation of the best class. The theoretically best mode of dealing with such cases will, however, depend upon the circumstances of that case. Simplicity is so important an advantage that it will probably be found well to adopt the rule of replacing W by its product by the weight fixed from *a priori* considerations.

§ 9. *Application to Transits of Mercury over the Sun's Disc.*

I now propose to apply the preceding theory to the case of observed contacts of the inferior planets, Mercury and Venus, with the limb of the Sun. A peculiarity of the observations of these phenomena is the great number of them which investigators have had to reject on account of discordance of individual observations from the general mean. By suitable rejection very different final results may be obtained, and it is impossible to draw any line between those observations which should be rejected and those which should be retained.

In my discussion of observations of transits of Mercury,* I have shown that the residuals of 684 observations of the interior contact of the limbs of the Sun and Mercury are distributed as follows, the value of each residual being considered only to the nearest round 5 seconds:

---

* Astronomical Papers of the American Ephemeris, Vol. I.

Below — 27 seconds were . . . . 20 residuals.
of — 27ˢ, — 26ˢ, — 25ˢ, — 24ˢ, — 23ˢ were 11 "
 " — 22, — 21, — 20, — 19, — 18   "   15 "
 " — 17, — 16, — 15, — 14, — 13   "   44 "
 " — 12, — 11, — 10, — 9, — 8   "   77 "
 " — 7, — 6, — 5, — 4, — 3   " 132 "
 " — 2, — 1, — 0, + 1, + 2   " 147 "
 " + 3, + 4, + 5, + 6, + 7   "   89 "
 " + 8, + 9, + 10, + 11, + 12   "   52 "
 " + 13, + 14, + 15, + 16, + 17   "   33 "
 " + 18, + 19, + 20, + 21, + 22   "   23 "
 " + 23, + 24, + 25, + 26, + 27   "   12 "
Exceeding + 27 . . . . "   29 "

Collecting those of equal absolute value and classifying them according to the middle value of each group, we have the following comparison of actual and probable numbers, the latter being obtained on the usual theory, assuming a probable error of $\pm\ 6.67^{s}$, or a value of $\frac{1}{h} = 14.0^{s}$.

Mean Values of Residual.	Actual Number.	Probable Number.	$A - P$.
0ˢ	147	137	+ 10
5	221	240	— 19
10	129	166	— 37
15	77	88	— 11
20	38	36	+ 2
25	23	12	+ 11
> 27	49	5	+ 44.

There is, therefore, a large excess of both small and large residuals, which would have been yet more pronounced had the mean error been determined from the sum of the squares of all the residuals.

I find, by several trials, that the residuals which do not exceed 27^{s} can be well represented by the following distribution of precisions and probable errors:

110 observations of $1:h = 6^{s}$ or probable error $= 2.9^{s}$,
100   "   " "   10 "   "   "   4.8,
400   "   " "   18 "   "   "   8.6,
 50   "   " "   36 "   "   "   17.2.

The comparison of actual and probable numbers of residuals will then be as follows:

s.	Actual.	Probable.	A − P.
0	147	143	+ 4
5	221	220	+ 1
10	129	128	+ 1
15	77	76	+ 1
20	38	44	− 6
25	23	23	0
> 27	49	26	+ 23.

It must be understood that these four values of $1:h$ and of the consequent probable errors are not four entirely determinate quantities. Really we should consider that the precision has all values between the extreme limits; but it is not at all necessary to consider it as what it really is, a continuously varying quantity. All we have to do is to form an expression which shall represent the relation between the number and magnitude of the residuals; and this we do most conveniently by assuming three, four or more separate values of h, and then finding how many observations we have to assign to each class in order to represent the observed relation. From the above table we may infer that about one-third the observations of transits of Mercury belong to classes which might be called good or very good, the probable error ranging from $2\frac{1}{2}^s$ to 6^s; that more than half belong to an average class, of which the probable error may range from 6 to 12 seconds, and that about one-twelfth are made under such unfavorable circumstances that their probable error averages 17^s. Even with this large probable error, we see that there is an excess of 23 residuals exceeding 27 seconds, so that we should have increased the number of this imperfect class. But I suspect that many of these arose from errors of a minute in the record, or from other pure blunders.

I am also inclined to think that the comparative excess of very small residuals, indicating that one-fifth of all the observations had a probable error as small as 3^s, may be partly due to the fact that many of the residuals are deviations from the mean of a small number of observations, and that no comparison of the separate observations with the final theory founded on the whole mass of observations was made. On the whole, we may suppose that of the actual observations,

0.30 have a precision $h_1 = 1:10^s$,
0.60 " " " $h_2 = 1:18$,
0.10 " " " $h_3 = 1:36$.

Observations so as to Obtain the Best Result.

This will give for the law of probability (*Eq.* 5):

$$\phi(x) = \frac{1}{\sqrt{\pi}}\left\{.030 e^{-\left(\frac{x}{10}\right)^2} + .0333\ldots e^{-\left(\frac{x}{18}\right)^2} + .00277\ldots e^{-\left(\frac{x}{36}\right)^2}\right\}.$$

We have, for the values of w_1, w_2 and w_3, as given by (13),

$$w_1 = .03000 e^{-\left(\frac{\xi}{10}\right)^2}; \quad h_1^2 = 1 : 100,$$
$$w_2 = .03333 e^{-\left(\frac{\xi}{18}\right)^2}; \quad h_2^2 = 1 : 324,$$
$$w_3 = .00277 e^{-\left(\frac{\xi}{36}\right)^2}; \quad h_3^2 = 1 : 1296.$$

In the following table the quantities required are tabulated as a function of the residual ξ of an observation. The w's are multiplied by 1,000, and the W's by 10,000, so as to express them in convenient units:

ξ s.	w_1	w_2	w_3	W
0	30.0	33.3	2.8	6.1
2	28.8	32.9	2.8	6.1
4	25.6	31.7	2.7	5.9
6	20.9	29.8	2.7	5.7
8	15.8	27.4	2.6	5.3
10	11.0	24.5	2.6	4.9
12	7.1	21.4	2.5	4.5
14	4.2	18.2	2.4	4.0
16	2.3	15.1	2.3	3.6
18	1.2	12.3	2.2	3.3
20	0.55	9.70	2.04	3.0
22	0.23	7.49	1.91	2.8
24	0.09	5.63	1.78	2.6
26	0.03	4.14	1.65	2.5
28	0.01	2.96	1.52	2.3
30	0.00	2.07	1.39	2.2

If we could be sure that any one observation belonged to the best class, its weight on the above scale would be 10; were we sure it belonged to the intermediate class, its weight would be 3, and if it belonged to the worst class, it would be 0.77. The value of W for $\xi = 0$, namely, 6.1, falls below 10 in consequence of the probability that an observation of residual zero may belong to one of the poorer classes, and the value of W for an observation of residual

30 is above 0.77 on account of the possibility that such an observation may belong to the intermediate class.

§10. *Approximation Expression for the Evil.*

It remains to find an expression for the evil of the best result, as obtained by the preceding method. As already defined and shown, if the probability that the value of the observed quantity is contained between the limits x and $x + dx$ be
$$\theta x dx,$$
then the evil of any assumed value x_0 of the required quantity is given by the equation
$$E = \int_{-\infty}^{+\infty} (x - x_0)^2 \theta x dx.$$

In the case of m observed values of $x, x_1, x_2, \ldots x_m$, following the general law of error, we have
$$\theta x = \alpha \phi(x_1 - x) \phi(x_2 - x) \ldots \phi(x_m - x),$$
the coefficient α being a constant, determined by the condition
$$\int_{-\infty}^{+\infty} \theta x dx = 1.$$

If we take for x_0 the best value of x, namely, the value which satisfies the condition
$$x_0 = \int_{-\infty}^{+\infty} x \theta x dx,$$
we have for its evil
$$E = \int_{-\infty}^{+\infty} x^2 \theta x dx - x_0^2.$$

It will be noticed that the function θx differs from $\psi(x)$ in (6) only in containing the factor α; that is, we have
$$\theta x = \alpha \psi(x) = \alpha \Sigma P e^{-k^2 x^2 + 2bx - c}.$$

In consequence of α we may omit from ψx any constant factor, as $\sqrt{\pi}$. Then from

(9) $$\psi(\eta) = \Sigma P e^{-k^2 \eta^2 + 2b\eta - c}$$

we have
$$\int_{-\infty}^{+\infty} x^2 \psi(x) dx = \Sigma P \frac{k^2 + 2b^2}{2k^5} e^{\frac{b^2}{k^2} - c}.$$

α is determined by the condition
$$\alpha \int_{-\infty}^{+\infty} \psi(x) dx = 1 = \alpha \Sigma w.$$

Comparing with the equations (9) and (11), we find

$$\int_{-\infty}^{+\infty} x^2 \theta x dx = \frac{\Sigma w \eta^2}{\Sigma w} + \frac{\frac{1}{2}\Sigma \frac{w}{k^2}}{\Sigma w},$$

the sign Σ extending to the n^m distributions of the measures of precision. We thus have, for the minimum evil,

$$E = \frac{\Sigma w \eta^2}{\Sigma w} - \frac{(\Sigma w \eta)^2}{(\Sigma w)^2} + \frac{\frac{1}{2}\Sigma \frac{w}{k^2}}{\Sigma w} = \frac{\Sigma_{ij}^2 w_i w_j (\eta_i - \eta_j)^2}{(\Sigma w)^2} + \frac{\Sigma \frac{w}{2k^2}}{\Sigma w}$$
$$(i = 1, 2 \ldots n^m - 1; j = i+1, \ldots n^m) \qquad (15)$$

The second term of the right-hand member of this equation is a certain mean among the various values of $\frac{1}{2k^2}$, and coincides with the square of the "mean error" of the usual theory, which, as already shown, coincides with the "evil," as that term is here defined. If there is but one distribution of the h's among the x's, then there will be but one value of η, and the first term of the evil will vanish, so that we shall have no evil left except the "mean error." But when, as here supposed, the weights of the observations are themselves uncertain, then the last equation expresses the logical conclusion that, in order to obtain the total evil, we must add to the result of the mean uncertainty of the observations a quantity depending upon the uncertainty in the weights we should individually or collectively assign to them.

The first term of (15) comprises $\frac{\mu(\mu-1)}{2}$ terms $(\mu = n^m)$; its actual computation is therefore out of the question. We may, however, remark that it admits of being expressed in the form

$$\Sigma a_{i,j}(x_i - x_j)^2,$$

$a_{i,j}$ being numerical coefficients. This form contains only $\frac{m(m-1)}{2}$ terms.

To show this, we remark that each value of η may be written in the form

$$\eta = p_1 x_1 + p_2 x_2 + \ldots + p_m x_m,$$

where
$$p_1 + p_2 + p_3 + \ldots + p_m = 1.$$

The difference of any two values of η multiplied by any factor, such as $\frac{\sqrt{w_i w_j}}{(\Sigma w)}$,

will therefore be of the form
$$\mu_1 x_1 + \mu_2 x_2 + \ldots + \mu_m x_m,$$
where
$$\mu_1 + \mu_2 + \mu_3 + \ldots + \mu_m = 0, \qquad (16)$$
and where each of the coefficients μ_1, μ_2, etc., takes $\dfrac{n^m(n^m-1)}{2}$ values, corresponding to the number of differences of the η's. The sum of the squares of these differences will be
$$x_1^2 \Sigma \mu_1^2 + x_2^2 \Sigma \mu_2^2 + \ldots + x_m^2 \Sigma \mu_m^2$$
$$+ 2 x_1 x_2 \Sigma \mu_1 \mu_2 + 2 x_1 x_3 \Sigma \mu_1 \mu_3 + 2 x_2 x_3 \Sigma \mu_2 \mu_3 +, \text{ etc.}$$
On account of the condition (16) this expression may be transformed into
$$\Sigma^2 A_{ij}(x_i - x_j)^2,$$
where
$$A_{i,j} = -\Sigma \mu_i \mu_j.$$
Here the sign Σ of summation extends to all the $\dfrac{n^m(n^m-1)}{2}$ products of μ having the same constant suffixes i and j.

Although the value of this expression admits of rigorous algebraic determination, it is doubtful whether there would be any advantage in computing it in any special case. I shall therefore seek only for a rough approximation to its probable value. Returning to the equation (15), we first note that the expression
$$\frac{\Sigma^2 w_i w_j (\eta_i - \eta_j)^2}{(\Sigma w_i)^2} \quad (i = 1, 2, \ldots n^m-1; j = i+1, \ldots n^m)$$
is one-half a weighted mean value of $(\eta_i - \eta_j)^2$, the weight of each being the product $w_i w_j$, and the zero terms $(\eta_i - \eta_i)^2$ being allowed to enter with half weight in taking the mean. Instead of this weighted mean we may take the general mean formed by giving all the differences equal weight.

Now, when the number of observations treated is large, we may consider the amount by which any one value of η differs from the mean of all the values, or x, to be the result of an accumulation of accidental errors; namely, if we put
$$\frac{h_i^2}{k^2} \equiv q_i \qquad (17)$$
(where h_i means the precision assigned to x_i), we shall have
$$\eta = q_1 x_1 + q_2 x_2 + \ldots + q_m x_m,$$
while we have for x an expression of the form
$$x = q_1' x_1 + q_2' x_2 + \ldots + q_m' x_m,$$

q'_i being the weighted mean of all the n^m values of q_i. If we now put, as before, ξ_i for the deviation of x_i from the general mean x, we have, on account of
$$\Sigma q_i = \Sigma q'_i = 1,$$
$$\eta - x = (q_1 - q'_1)\xi_1 + (q_2 - q'_2)\xi_2 + \ldots + (q_m - q'_m)\xi_m.$$

Now, since k^2 is, in all cases, the sum of *some* m values of h^2, the mean value of q given by (17) is $\frac{1}{m}$. The actual special values can never reach zero as their lower limit, and will seldom exceed $\frac{2}{m}$ as their upper limit. The range of value will, however, depend upon the range of values taken by the precisions h. Unless in extreme cases, the mean deviation of q from $\frac{1}{m}$ cannot exceed $\frac{1}{2m}$; that is, the mean value of $q - q'$ will, in general, be less than $\frac{1}{2m}$. Assigning this mean value, we may regard $\eta - x$ as made up of the probable accumulation of terms
$$\frac{1}{2m}(\pm \xi_1 \pm \xi_2 \pm \xi_3 \pm \ldots \pm \xi_m).$$

If we put Δ^2 for the mean value of ξ^2, then, by the theory of errors, we shall have, for the mean value of $(\eta - x)^2$,
$$\text{mean } (\eta - x)^2 = \frac{\Delta^2}{4m}.$$

Hence
$$\text{mean } (\eta_i - \eta_j)^2 = \frac{\Delta^2}{2m} = \frac{\Sigma \xi^2}{2m^2}.$$

To compare this with the last term of (15), let us suppose the most probable distribution of precisions with respect to their magnitude to be

m_1 precisions of value h_1;
m_2 " " h_2;
. . . .
. . . .
. . . .
m_n " " h_n.

We shall then have as a close approximation to the mean value of k^2,
$$k^2 = m_1 h_1^2 + m_2 h_2^2 + \ldots m_n h_n^2,$$
one-half the reciprocal of which may be taken for the last term of (15). Thus we may take for the amount of the evil, in all ordinary cases
$$E = \frac{1}{2} \frac{1}{m_1 h_1^2 + \ldots + m_n h_n^2} + \frac{\Delta^2}{2m}.$$

As already shown, this evil will be the square of the "mean error" to be expected of the usual theory; so that we may take
$$\varepsilon = \pm \sqrt{E}$$
as the mean error to be expected.

I remark, in conclusion, that this theory and method may be extended to the case of several unknown quantities without any other difficulty than that of a resolution of the equations of condition with the *a posteriori* weights. We should first solve the equations if necessary, using equal weights for all, or such system of weights as might be deemed most probable. From the residuals thus obtained we should deduce the law of error, and in practice we should, in order to determine such law, combine with the residuals in question all others that astronomy could furnish pertaining to the same class of observations. Then we should re-solve the equations using the modified weights, which re-solution would give us the definitive result.

THE REMINISCENCES OF AN ASTRONOMER

Simon Newcomb

THE REMINISCENCES
OF
AN ASTRONOMER

BY

SIMON NEWCOMB

BOSTON AND NEW YORK
HOUGHTON, MIFFLIN AND COMPANY
The Riverside Press, Cambridge
1903

COPYRIGHT, 1903, BY SIMON NEWCOMB
ALL RIGHTS RESERVED

Published October, 1903

PREFACE

THE earlier chapters of this collection are so much in the nature of an autobiography that the author has long shrunk from the idea of allowing them to see the light during his lifetime. His repugnance has been overcome by very warm expressions on the subject uttered by valued friends to whom they were shown, and by a desire that some at least who knew him in youth should be able to read what he has written.

The author trusts that neither critic nor reader will object because he has, in some cases, strayed outside the limits of his purely personal experience, in order to give a more complete view of a situation, or to bring out matters that might be of historic interest. If some of the chapters are scrappy, it is because he has tried to collect those experiences which have afforded him most food for thought, have been most influential in shaping his views, or are recalled with most pleasure.

CONTENTS

I

THE WORLD OF COLD AND DARKNESS

Ancestry. — Squire Thomas Prince. — Parentage. — Early Education. — Books read 1

II

DR. FOSHAY

A Long Journey on Foot. — A Wonderful Doctor. — The Botanic System of Medicine. — Phrenology. — A Launch into the World. — A Disillusion. — Life in Maryland. — Acquaintance with Professor Henry. — Removal to Cambridge 23

III

THE WORLD OF SWEETNESS AND LIGHT

The American Astronomical Ephemeris. — The Men who made it. — Harvard in the Middle of the Century. — A Librarian of the Time. — Professor Peirce. — Dr. Gould, the "Astronomical Journal," and the Dudley Observatory. — W. P. G. Bartlett. — John D. Runkle and the "Mathematical Monthly." — A Mathematical Politician. — A Trip to Manitoba and a Voyage up the Saskatchewan. — A Wonderful Star 62

IV

LIFE AND WORK AT AN OBSERVATORY

A Professor, United States Navy. — The Naval Observatory in 1861. — Captain Gilliss and his Plans. — Admiral Davis. — A New Instrument and a New Departure. — Astronomical Activity. — The Question of Observatory Administration. — Visit from the Emperor of Brazil. — Admiral John Rodgers. — Efforts to improve the Work of the Observatory 97

CONTENTS

V

GREAT TELESCOPES AND THEIR WORK

Curious Origin of the Great Washington Telescope. — Congress is induced to act. — A Case of Astronomical Fallibility. — The Discovery of the Satellites of Mars. — The Great Telescope of the Pulkova Observatory. — Alvan Clark and his Sons. — A Sad Astronomical Accident 128

VI

THE TRANSITS OF VENUS

Old Transits of Venus. — An Astronomical Expedition in the 18th Century. — Father Hell and his Observations. — A Suspected Forger vindicated. — The American Commission on the Transit of Venus. — The Photographic Method to be applied. — Garfield and the Appropriation Committee. — Weather Uncertainties. — Voyage to the Cape of Good Hope. — The Transit of 1882. — Our Failure to publish our Observations 151

VII

THE LICK OBSERVATORY

James Lick and his Ideas. — Mr. D. O. Mills. — Plans for the Lick Observatory. — Edward E. Barnard. — Professor Holden. — Wonderful Success of the Observatory 182

VIII

THE AUTHOR'S SCIENTIFIC WORK

The Orbits of the Asteroids. — The Problems of Mathematical Astronomy. — The Motion of the Moon and its Perplexing Inequalities. — A Visit to the Paris Observatory to search for Forgotten Observations. — Wonderful Success in finding Them. — The Paris Commune. — The History of the Moon's Motion carried back a Century. — The Harvard Observatory. — The "Nautical Almanac" Office and its Work. — Mr. George W. Hill and his Work. — A Wonderful Algebraist. — The Meridian Conference of 1884, and the Question of Universal Time. — Tables of the Planets completed. — The Astronomical Constants. — Work unfinished 195

IX

Scientific Washington

Professor Henry and the Smithsonian Institution. — Alumni Associations. — The Scientific Club. — General Sherman. — Mr. Hugh McCulloch. — A Forgotten Scientist. — The National Academy of Sciences. — The Geological Survey of the Territories. — The Government Forestry System. — Professor O. C. Marsh. — Scientific Humbugs. — Life on the Plains 234

X

Scientific England

My First Trip to Europe. — Mr. Thomas Hughes. — Mr. John Stuart Mill. — Mr. Gladstone and the Royal Society Dinner. — Other Eminent Englishmen. — Professors Cayley and Adams. — Professor Airy and the Greenwich Observatory. — A Visit to Edinburgh 271

XI

Men and Things in Europe

A Voyage to Gibraltar with Professor Tyndall. — The Great Fortress. — "Whispering Boanerges." — A Winter Voyage in the Mediterranean. — Malta and Messina. — Advantage of not understanding a Language. — German Astronomers. — The Pulkova Observatory. — A Meeting which might have been Embarrassing. — From Germany to Paris at the Close of the War. — Experiences at Paris during the Commune. — The Greatest Astronomer of France. — The Paris Observatory 302

XII

The Old and the New Washington

Washington during the Civil War. — Secretary Stanton. — The Raid of General Early. — A Presidential Levee in 1864. — The Fall of Richmond. — The Assassination of President Lincoln. — Negro Traits and Education. — Senator Sumner. — An Ambitious Academy. — President Garfield and his Assassination. — Cooling the White House during his Illness. — The Shepherd Régime in Washington 334

XIII

MISCELLANEA

The Great Star-Catalogue Case. — Professor Peters and the Almagest of Ptolemy. — Scientific Cranks. — The Degrees of the French Universities. — A Virginia Country School. — Political Economy and Education. — Exact Science in America before the Johns Hopkins University. — Professor Ely and Economics. — Spiritualism and Psychic Research. — The Georgia Magnetic Girl 372

THE REMINISCENCES OF AN
ASTRONOMER

THE REMINISCENCES OF AN ASTRONOMER

I

THE WORLD OF COLD AND DARKNESS

I DATE my birth into the world of sweetness and light on one frosty morning in January, 1857, when I took my seat between two well-known mathematicians, before a blazing fire in the office of the "Nautical Almanac" at Cambridge, Mass. I had come on from Washington, armed with letters from Professor Henry and Mr. Hilgard, to seek a trial as an astronomical computer. The men beside me were Professor Joseph Winlock, the superintendent, and Mr. John D. Runkle, the senior assistant in the office. I talked of my unsuccessful attempt to master the "Mécanique Céleste" of Laplace without other preparation than that afforded by the most meagre text-books of elementary mathematics of that period. Runkle spoke of the translator as "the Captain." So familiar a designation of the great Bowditch — LL. D. and member of the Royal Societies of London, Edinburgh, and Dublin — quite shocked me.

I was then in my twenty-second year, but it was the first time I had ever seen any one who was familiar with the " Mécanique Céleste." I looked with awe upon the assistants who filed in and out as upon men who had all the mysteries of gravitation and the celestial motions at their fingers' ends. I should not have been surprised to learn that even the Hibernian who fed the fire had imbibed so much of the spirit of the place as to admire the genius of Laplace and Lagrange. My own rank was scarcely up to that of a tyro; but I was a few weeks later employed on trial as computer at a salary of thirty dollars a month.

How could an incident so simple and an employment so humble be in itself an epoch in one's life — an entrance into a new world? To answer this question some account of my early life is necessary. The interest now taken in questions of heredity and in the study of the growing mind of the child may excuse a word about my ancestry and early training.

Though born in Nova Scotia, I am of almost pure New England descent. The first Simon Newcomb, from whom I am of the sixth generation, was born in Massachusetts or Maine about 1666, and died at Lebanon, Conn., in 1745. His descendants had a fancy for naming their eldest sons after him, and but for the chance of my father being a younger son, I should have been the sixth Simon in unbroken lineal descent.[1]

Among my paternal ancestors none, so far as I

[1] The actual sixth was my late excellent and esteemed cousin, Judge Simon Bolivar Newcomb, of New Mexico.

know, with the exception of Elder Brewster, were what we should now call educated men. Nor did any other of them acquire great wealth, hold a high official position, or do anything to make his name live in history. On my mother's side are found New England clergymen and an English nonconformist preacher, named Prince, who is said to have studied at Oxford towards the end of the seventeenth century, but did not take a degree. I do not know of any college graduate in the list.

Until I was four years old I lived in the house of my paternal grandfather, about two miles from the pretty little village of Wallace, at the mouth of the river of that name. He was, I believe, a stonecutter by trade and owner of a quarry which has since become important; but tradition credits him with unusual learning and with having at some time taught school.

My maternal grandfather was "Squire" Thomas Prince, a native of Maine, who had moved to Moncton, N. B., early in life, and lived there the rest of his days. He was an upright magistrate, a Puritan in principle, and a pillar of the Baptist Church, highly respected throughout the province. He came from a long-lived family, and one so prolific that it is said most of the Princes of New England are descended from it. I have heard a story of him which may illustrate the freedom of the time in matters of legal proceedings before a magistrate's court. At that time a party in a suit could not be a witness. In the terse language of the common people, "no man could swear money into his own pocket." The plaintiff in the case advised the magistrate in advance that he had no legal proof of the debt, but that defendant freely acknowledged it in private conversation.

"Well," said the magistrate, "bring him in here and get him to talk about it while I am absent."

The time came.

"If you hadn't sued me I would have paid you," said the defendant.

On the moment the magistrate stepped from behind a door with the remark, —

"I think you will pay him now, whether or no."

My father was the most rational and the most dispassionate of men. The conduct of his life was guided by a philosophy based on Combe's "Constitution of Man," and I used to feel that the law of the land was a potent instrument in shaping his paternal affections. His method of seeking a wife was so far unique that it may not be devoid of interest, even at this date. From careful study he had learned that the age at which a man should marry was twenty-five. A healthy and well-endowed offspring should be one of the main objects in view in entering the marriage state, and this required a mentally gifted wife. She must be of different temperament from his own and an economical housekeeper. So when he found the age of twenty-five approaching, he began to look about. There was no one in Wallace who satisfied the requirements. He therefore set out afoot to discover his ideal. In those days and regions the professional tramp and mendicant were unknown, and every farmhouse dispensed its hospitality with an Arcadian simplicity little known in our times. Wherever he stopped overnight he made a critical investigation of the housekeeping, perhaps rising before the family for this purpose. He searched in vain until his road carried him out of the province. One young woman spoiled any possible chance she might have had by a lack of economy in the making of bread. She was asked what she did with an unnecessarily large remnant of dough which she left sticking to the sides of the pan. She replied that she fed it to the horses. Her case received no further consideration.

The search had extended nearly a hundred miles when, early one evening, he reached what was then the small village of Moncton. He was attracted by the strains of music from a church, went into it, and found a religious meeting in progress. His eye was at once arrested by the face and head of a young woman playing on a melodeon, who was leading the singing. He sat in such a position that he could carefully scan her face and movements. As he continued this study the conviction grew upon him that here was the object of his search. That such should have occurred before there was any opportunity to inspect the dough-pan may lead the reader to conclusions of his own. He inquired her name — Emily Prince. He cultivated her acquaintance, paid his addresses, and was accepted. He was fond of astronomy, and during the months of his engagement one of his favorite occupations was to take her out of an evening and show her the constellations. It is even said that, among the daydreams in which they indulged, one was that their firstborn might be an astronomer. Probably this was only a passing fancy, as I heard nothing of it during my childhood. The marriage was in all respects a happy one, so far as congeniality of nature and mutual regard could go. Although the wife died at the early age of thirty-seven, the husband never ceased to cherish her memory, and, so far as I am aware, never again thought of marrying.

My mother was the most profoundly and sincerely religious woman with whom I was ever intimately acquainted, and my father always entertained and expressed the highest admiration for her mental gifts, to which he attributed whatever talents his children might have possessed. The unfitness of her environment to her constitution is the saddest memory of my childhood. More I do not trust myself to say to the public, nor will the reader expect more of me.

My father followed, during most of his life, the precarious occupation of a country school teacher. It was then, as it still is in many thinly settled parts of the country, an almost nomadic profession, a teacher seldom remaining more than one or two years in the same place. Thus it happened that, during the first fifteen years of my life, movings were frequent. My father tried his fortune in a number of places, both in Nova Scotia and Prince Edward Island. Our lot was made harder by the fact that his ideas of education did not coincide with those prevalent in the communities where he taught. He was a disciple and admirer of William Cobbett, and though he did not run so far counter to the ideas of his patrons as to teach Cobbett's grammar at school, he always recommended it to me as the one by which alone I could learn to write good English. The learning of anything, especially of arithmetic and grammar, by the glib repetition of rules was a system that he held in contempt. With the public, ability to recite the rules of such subjects as those went farther than any actual demonstration of the power to cipher correctly or write grammatically.

So far as the economic condition of society and the general mode of living and thinking were concerned, I might claim to have lived in the time of the American Revolution. A railway was something read or heard about with wonder; a steamer had never ploughed the waters of Wallace Bay. Nearly everything necessary for the daily life of the people had to be made on the spot, and even at home. The work of the men and boys was " from sun to sun," — I might almost say from daylight to darkness, — as they tilled the ground, mended the fences, or cut lumber, wood, and stone for export to more favored climes. The spinning wheel and the loom were almost a necessary part of the furniture of any well-ordered house; the exceptions were among people rich enough to buy their own clothes, or so poor and miserable that they had to wear the cast-off rags of their more fortunate neighbors. The women and girls sheared the sheep, carded the wool, spun the yarn, wove the homespun cloth, and made the clothes. In the haying season they amused themselves by joining in the raking of hay, in which they had to be particularly active if rain was threatened; but any man would have lost caste who allowed wife or daughter to engage in heavy work outside the house.

The contrast between the social conditions and those which surround even the poorest classes at the present day have had a profound influence upon my views of economic subjects. The conception which the masses of the present time have of how their ancestors lived in the early years of the century are so vague and shadowy as not to influence their conduct at the present time.

What we now call school training, the pursuit of fixed studies at stated hours under the constant guidance of a teacher, I could scarcely be said to have enjoyed. For the most part, when I attended my father's school at all, I came and went with entire freedom, and this for causes which, as we shall see, he had reasons for deeming good.

It would seem that I was rather precocious. I was taught the alphabet by my aunts before I was four years old, and I was reading the Bible in class and beginning geography when I was six.

One curious feature of my reading I do not remember to have seen noticed in the case of children. The printed words, for the most part, brought no well-defined images to my mind; none at least that were retained in their connection. I remember one instance of this. We were at Bedeque, Prince Edward Island. During the absence of my father, the school was kept for a time by Mr. Bacon. The class in reading had that chapter in the New Testament in which the treason of Judas is described. It was then examined on the subject. To the question what Judas did, no one could return an answer until it came my turn. I had a vague impression of some one hanging himself, and so I said quite at random that he hanged himself. It was with a qualm of conscience that I went to the head of the class.

Arithmetic was commenced at the age of five, my father drawing me to school day by day on a little sled during the winter. Just what progress I made at that time I do not recall. Long years afterward, my father, at my request, wrote me a letter describing my early education, extracts from which I shall ask permission to reproduce, instead of attempting to treat the matter myself. The letter, covering twelve closely written foolscap pages, was probably dashed off at a sitting without supposing any eye but my own would ever see it: —

June 8th, '58.

I will now proceed to write, according to your request, about your early life.

While in your fifth year, your mother spoke several times of the propriety of teaching you the first rudiments of book-learning; but I insisted that you should not be taught the first letter until you became five.[1] I think, though, that at about four, or four and a half I taught you to count, as far, perhaps, as 100.

When a little over four and a half, one evening, as I came home from school, you ran to me, and asked, "Father, is not 4 and 4 and 4 and 4, 16?" "Yes, how did you find it out?" You showed me the counterpane which was napped. The spot of four rows each way was the one you had counted up. After this, for a week or two, you spent a considerable number of hours every day, making calculations in addition and multiplication. The rows of naps being crossed and complexed in various ways, your greatest delight was to clear them out, find how many small ones were equal to one large one, and such like. After a space of two or three weeks we became afraid you would calculate yourself "out of your head," and laid away the counterpane.

Winter came, and passed along, and your birthday came; on that day, having a light hand-sled prepared, I fixed you on it, and away we went a mile and a half to school.

According to my belief in educational matters "that the slate should be put into the child's hands as soon as the book is," you of course had your slate, and commenced making figures and letters the first day.

In all cases, after you had read and spelled a lesson, and made some figures, and worked a sum, suppose one

[1] He had evidently forgotten the home instruction from my aunts, received more than a year previous to the date he mentions.

THE WORLD OF COLD AND DARKNESS 11

hour's study, I sent you out, telling you to run about and play a "good spell." To the best of my judgment you studied, during the five months that this school lasted, nearly four hours a day, two being at figures.

. . . .

During the year that I taught at Bedeque, you studied about five hours a day in school; and I used to exercise you about an hour a day besides, either morning or evening. This would make six hours per day, nearly or quite two and a half hours of that time at numbers either at your slate or mentally. When my school ended here, you were six and a half years of age, and pretty well through the arithmetic. You had studied, I think, all the rules preceding including the cube root. . . .

I had frequently heard, during my boyhood, of a supposed mental breakdown about this period, and had asked my father for a description of it in the letter from which I am quoting. On this subject the letter continues : —

You had lost all relish for reading, study, play, or talk. Sat most of the day flat on the floor or hearth. When sent of an errand, you would half the time forget what you went for. I have seen you come back from Cale Schurman's crying,[1] and after asking you several times you would make out to answer, you had not been all the way over because you forgot what you went for. You would frequently jump up from the corner, and ask some peculiar question. I remember three you asked me.

[1] The grandfather of President Schurman of Cornell University. I retain a dreamy impression of two half-grown or nearly grown boys, perhaps between fourteen and eighteen years of age, one of whom became, I believe, the father of the president.

1st. Father, does form mean shape? Yes. Has everything some shape? Yes. Can it be possible for anything to be made that would not have any shape? I answered no; and then showed you several things, explaining that they all had some shape or form. You now brightened up like a lawyer who had led on a witness with easy questions to a certain point, and who had cautiously reserved a thunderbolt question, to floor the witness at a proper time; proceeded with, "Well, then, how could the world be without form when God made it?"

. . . .

3d. Does Cale Schurman's big ram know that he has such big crooked horns on him? Does he know it himself, I mean? Does he know himself that he has such horns on him?

You were taken down suddenly I think about two or three days from the first symptoms until you were fairly in the corner. Your rise was also rapid, I think about a week (or perhaps two weeks) from your first at recovery, until you seemed to show nothing unusual. From the time you were taken down until you commenced recovery was about a month.

We returned to Prince Edward Island, and after a few weeks I began to examine you in figures, and found you had forgotten nearly all you had ever learned.

. . . .

While at New London I got an old work on Astronomy; you were wonderfully taken with it, and read it with avidity. While here you read considerable in "Goldsmith's History of England." We lived two years in New London; I think you attended school nearly one year there. I usually asked you questions on the road going to school, in the morning, upon the history you had read, or something you had studied the day previous. While there, you made a dozen or two of the folks raise a terrible laugh. I one evening lectured on astronomy at home; the house was pretty well filled, I suppose about twenty were present. You were not quite ten years old and small at that. Almost as soon as I was done you said: "Father, I think you were wrong in one thing." Such a roar of laughter almost shook the house.

You were an uncommon child for *truth*. I never knew you to deviate from it in one single instance, either in infancy or youth.

From your infancy you showed great physical courage in going along the woods or in places in the dark among cattle, and I am surprised at what you say about your fears of a stove-pipe and trees.

Perhaps I should have said "mental" instead of physical courage, for in one respect you were uncommonly deficient in that sort of courage necessary to perform bodily labor. Until nine or ten years of age you made a most pitiful attempt at any sort of bodily or rather "handy" work.

. . . .

An extraordinary peculiarity in you was never to leap past a word you could not make out. I certainly never gave you any particular instructions about this, or the fact itself would not at the time have appeared so strange to me. I will name one case. After a return to Wallace (you were eleven) I, one day, on going from home for an hour or so, gave you a borrowed newspaper, telling you there was a fine piece; to read it, and tell me its contents when I returned. On my return you were near the house chopping wood. "Well, Simon, did you read the piece?" "No, sir." "Why not?" "I came to a word I did not know." This word was just about four lines from the commencement.

At thirteen you read Phrenology. I now often impressed upon you the necessity of bodily labor; that you might attain a strong and healthy physical system, so as to be able to stand long hours of study when you came to manhood, for it was evident to me that you would not labor with the hands for a business. On this account, as much as on account of poverty, I hired you out for a large portion of the three years that we lived at Clements.

At fifteen you studied Euclid, and were enraptured with it. It is a little singular that all this time you never showed any self-esteem; or spoke of getting into employment at some future day, among the learned. The pleasure of intellectual exercise in demonstrating or analyzing a geometrical problem, or solving an algebraic equation, seemed to be your only object. No Junior, Seignour or Sophomore class, with annual honors, was ever, I suppose, presented to your mind.

Your almost intuitive knowledge of geography, navigation, and nautical matters in general caused me to think most ardently of writing to the Admiral at Halifax, to know if he would give you a place among the midshipmen of the navy; but my hope of seeing you a leading lawyer, and finally a judge on the bench, together with the possibility that your mother would not consent, and the possibility that you would not wish to go, deterred me: although I think I commenced a letter.

Among the books which profoundly influenced my mode of life and thought during the period embraced in the foregoing extracts were Fowler's "Phrenology" and Combe's "Constitution of Man." It may appear strange to the reader if a system so completely exploded as that of phrenology should have any value as a mental discipline. Its real value consisted, not in what it taught about the position of the "organs," but in presenting a study of human nature which, if not scientific in form, was truly so in spirit. I acquired the habit of looking on the characters and capabilities of men as the result of their organism. A hot and impulsive temper was checked by the reflection that it was beneath the dignity of human nature to allow a rush of blood to the organs of "combativeness" and "destructiveness" to upset one's mental equilibrium.

That I have gotten along in life almost without making (so far as I am aware) a personal enemy may be attributed to this early discipline, which led me into the habit of dealing with antagonism and personal opposition as I would deal with any physical opposition — evade it, avoid it, or overcome it. It goes without saying, however, that no discipline of this sort will avail to keep the passions of a youth always in check, and my own were no exception. When about fifteen I once made a great scandal by taking out my knife in prayer meeting and assaulting a young man who, while I was kneeling down during the prayer, stood above me and squeezed my neck. He escaped with a couple of severe though not serious cuts in his hand. He announced his intention of thrashing me when we should meet again; so for several days thereafter I tried, so far as possible, in going afield to keep a pitchfork within reach, determined that if he tried the job and I failed to kill him, it would be because I was unable to do so. Fortunately for both of us he never made the attempt.

I read Combe's "Constitution of Man" when between ten and twelve years of age. Though based on the ideas of phrenology and not, I believe, of high repute as a system of philosophy, it was as good a moral tonic as I can imagine to be placed in the hands of a youth, however fallacious may have been its general doctrines. So far as I can recall, it taught that all individual and social ills were due to men's disregard of the laws of Nature, which were classified as physical and moral. Obey the laws of health and we and our posterity will all reach the age of one hundred years. Obey the moral law and social evils will disappear. Its reading was accompanied by some qualms of conscience, arising from the non-accordance of many of its tenets with those of the "Catechism" and the "New England Primer." The combination of the two, however, led to the optimistic feeling that all wrongs would be righted, every act of injustice punished, and truth and righteousness eventually triumph through the regular processes of Nature and Society. I have been led to abandon this doctrine only by much experience, some of which will be found in the following pages.

In the direction of mathematical and physical science and reading generally, I may add something to what I have quoted from my father. My grandfather Simon had a small collection of books in the family. Among those purely literary were several volumes of "The Spectator" and "Roderick Random." Of the former I read a good deal. The latter was a story which a boy who had scarcely read any other would naturally follow with interest. Two circumstances connected with the reading, one negative and the other positive, I recall. Looking into the book after attaining years of maturity, I found it to contain many incidents of a character that would not be admitted into a modern work. Yet I read it through without ever noticing or retaining any impression of the indelicate side of the story. The other impression was a feeling of horror that a man fighting a duel and finding himself, as he supposed, mortally wounded by his opponent, should occupy his mind with avenging his own death instead of making his peace with Heaven.

Three mathematical books were in the collection, Hammond's Algebra, Simpson's Euclid, and Moore's Navigator, the latter the predecessor of Bowditch. The first was a miserable book, and I think its methods, which were crude in the extreme, though not incorrect, were rather more harmful than beneficial. The queer diagrams in Euclid had in my early years so little attraction for me that my curiosity never led me to examine its text. I at length did so in consequence of a passage in the algebra which referred to the 47th proposition of the First Book. It occurred to me to look into the book and see what this was. It was the first conception of mathematical proof that I had ever met with. I saw that the demonstration referred to a previous proposition, went back to that, and so on to the beginning. A new world of thought seemed to be opened. That principles so profound should be reached by methods so simple was astonishing. I was so enraptured that I explained to my brother Thomas while walking out of doors one day how the Pythagorean proposition, as it is now called, could be proved from first principles, drawing the necessary diagrams with a pencil on a piece of wood. I thought that even cattle might understand geometry could they only be communicated with and made to pay attention to it.

Some one at school had a copy of Mrs. Marcet's "Conversations on Natural Philosophy." With this book I was equally enraptured. Meagre and even erroneous though it was, it presented in a pleasing manner the first principles of physical science. I used to steal into the schoolhouse after hours to read a copy of the book, which belonged to one of the scholars, and literally devoured it in a few evenings.

My first undertaking in the way of scientific experiment was in the field of economics and psychology. When about fourteen I spent the winter in the house of an old farmer named Jefferson. He and his wife were a very kindly couple and took much interest in me. He was fond of his pipe, as most old farmers are. I questioned whether anything else would not do just as well as tobacco to smoke, and whether he was not wasting his money by buying that article when a cheap substitute could be found. So one day I took his pipe, removed the remains of the tobacco ashes, and stuffed the pipe with tea leaves that had been steeped, and which in color and general appearance looked much like tobacco. I took care to be around when he should again smoke. He lit the pipe as usual and smoked it with, seemingly, as much satisfaction as ever, only essaying the remark, " This tobacco tastes like tea." My conscience pricked me, but I could say nothing.

My father bought a copy of Lardner's " Popular Lectures on Science and Art." In this I first read of electricity. I recall an incident growing out of it. In Lardner's description of a Leyden jar, water is the only internal conductor. The wonders of the newly invented telegraph were then explained to the people in out of the way places by traveling lecturers. One of these came to Clements, where we then lived, with a lot of apparatus, amongst which was what I recognized as a Leyden jar. It was coated with tin-foil on the outside, but I did not see the inner coating, or anything which could serve as the necessary conductor. So with great diffidence I asked the lecturer while he was arranging his things, if he was not going to put water into the jar.

"No, my lad," was his reply, "I put lightning into it."

I wondered how the "lightning" was going to be conveyed to the interior surface of the glass without any conductor, such as water, but was too much abashed to ask the question.

Moore's "Navigator" taught not only a very crude sort of trigonometry, but a good deal about the warship of his time. To a boy living on the seacoast, who naturally thought a ship of war one of the greatest works of man, the book was of much interest.

Notwithstanding the intellectual pleasure which I have described, my boyhood was on the whole one of sadness. Occasionally my love of books brought a word of commendation from some visitor, perhaps a Methodist minister, who patted me on the head with a word of praise. Otherwise it caused only exclamations of wonder which were distasteful.

"You wouldn't believe what larnin' that boy has got. He has more larnin' than all the people around here put together," I heard one farmer say to another, looking at me, in my own view of the case, as if I were some monster misshapen in the womb. Instead of feeling that my bookish taste was something to be valued, I looked upon myself as a *lusus naturæ* whom Nature had cruelly formed to suffer from an abnormal constitution, and lamented that somehow I never could be like other boys.

The maladroitness described by my father, of which I was fully conscious, added to the feeling of my unfitness for the world around me. The skill required on a farm was above my reach, where efficiency in driving oxen was one of the most valued of accomplishments. I keenly felt my inability to acquire even respectable mediocrity in this branch of the agricultural profession. It was mortifying to watch the dexterous motions of the whip and listen to the torrent of imperatives with which a young farmer would set a team of these stolid animals in motion after they had failed to respond to my gentle requests, though conveyed in the best of ox language.

I had indeed gradually formed, from reading, a vague conception of a different kind of world, — a world of light, — where dwelt men who wrote books and people who knew the men who wrote books, — where lived boys who went to college and devoted themselves to learning, instead of driving oxen. I longed much to get into this world, but no possibility of doing so presented itself. I had no idea that it would be imbued with sympathy for a boy outside of it who wanted to learn. True, I had once read in some story, perhaps fictitious, how a nobleman had found a boy reading Newton's "Principia," and not only expressed his pleased surprise at the performance, but actually got the boy educated. But there was no nobleman in sight of the backwoods of Nova Scotia. I read in the autobiography of Franklin how he had made his way in life. But he was surrounded with opportunities from which I was cut off. It does seem a little singular that, well known as my tastes were to those around me, we never met a soul to say, "That boy ought to be educated." So far as I know, my father's idea of making me a lawyer met with nothing but ridicule from the neighbors. Did not a lawyer have to know Latin and have money to pursue his studies? In my own daydreams I was a farmer driving his own team; in my mother's a preacher, though she had regretfully to admit that I might never be good enough for this profession.

II

DOCTOR FOSHAY

In the summer of 1851, when I had passed the age of sixteen, we lived in a little school district a mile or two from the town of Yarmouth, N. S. Late in the summer we had a visit from a maternal uncle and aunt. As I had not seen Moncton since I was six years old, and as I wanted very much to visit my grandfather Prince once more, it was arranged that I should accompany them on their return home. An additional reason for this was that my mother's health had quite failed; there was no prospect of my doing anything where I was, and it was hoped that something might turn up at Moncton. There was but one difficulty; the visitors had driven to St. John in their own little carriage, which would hold only two people; so they could not take me back. I must therefore find my own way from St. John to Moncton.

We crossed the Bay of Fundy in a little sailing vessel. Among the passengers was an English ship captain who had just been wrecked off the coast of Newfoundland, and had the saved remnant of his crew with him. On the morning of our departure the weather was stormy, so that our vessel did not put to sea — a precaution for which the captain passenger expressed great contempt. He did not understand how a vessel should delay going to sea on account of a little storm.

The walk of one hundred miles from St. John to Moncton was for me, at that time, a much less formidable undertaking than it would appear in our times and latitude. A thirty-mile tramp was a bagatelle, and houses of entertainment — farmhouses where a traveler could rest or eat for a few pennies — were scattered along the road. But there was one great difficulty at the start. My instructions had been to follow the telegraph wires. I soon found that the line of telegraph came into the town from one direction, passed through it, and then left, not in the opposite direction, but perhaps at right angles to it. In which direction was the line to be followed? It was difficult to make known what I wanted. "Why, my boy, you can't walk to Moncton," was one answer. In a shop the clerks thought I wanted to ride on the telegraph, and, with much chuckling, directed me to the telegraph office where the man in charge would send me on. I tried in one direction which I thought could not be right, then I started off in the opposite one; but it soon became evident that that branch led up the river to Frederickton. So I had to retrace my steps and take the original line, which proved to be the right one.

The very first night I found that my grandfather's name was one to conjure with. I passed it with a hearty old farmer who, on learning who I was, entertained me with tales of Mr. Prince. The quality which most impressed the host was his enormous physical strength. He was rather below the usual stature and, as I remember him, very slightly built. Yet he could shoulder a barrel of flour and lift a hogshead of molasses on its end, feats of strength which only the most powerful men in the region were equal to.

On reaching my destination, I was not many days in learning that my grandfather was a believer in the maxims of "Poor Richard's Almanac," and disapproved of the aimless way in which I had been bred. He began to suggest the desirableness of my learning to do something to make a living. I thought of certain mechanical tastes which had moved me in former years to whittle and to make a reel on which to wind yarn, and to mend things generally. So I replied that I thought the trade of a carpenter was the one I could most easily learn. He approved of the idea, and expressed the intention of finding a carpenter who would want my services; but before he did so, I was started in a new and entirely different direction.

On her last visit to her birthplace, my mother brought back glowing reports of a wonderful physician who lived near Moncton and effected cures of the sick who had been given up by other doctors. I need hardly remark that physicians of wonderful proficiency — Diomeds of the medical profession, before whose shafts all forms of disease had to fall

— were then very generally supposed to be realities. The point which specially commended Dr. Foshay to us was that he practiced the botanic system of medicine, which threw mineral and all other poisons out of the materia medica and depended upon the healing powers of plants alone. People had seen so much of the evil effects of calomel, this being the favorite alterative of the profession, that they were quite ready to accept the new system. Among the remarkable cures which had given Dr. Foshay his great reputation was one of a young man with dyspepsia. He was reduced to a shadow, and the regular doctors had given him up as incurable. The new doctor took him to his home. The patient was addicted to two practices, both of which had been condemned by his former medical advisers. One was that of eating fat pork, which he would do at any hour of the day or night. The new doctor allowed him to eat all he wanted. Another was getting up in the night and practicing an ablution of the stomach by a method too heroic to be described in anything but a medical treatise.[1] He was now allowed to practice it to his heart's content. The outcome of the whole proceeding was that he was well in a few months, and, when I saw him, was as lusty a youth as one could desire to meet.

[1] I may remark, for the benefit of any medical reader, that it involved the use of two pails, one full of water, the other empty. When he got through the ablution, one pail was empty, and the other full. My authority for the actuality of this remarkable proceeding was some inmate of the house at the time, and I give credence to the story because it was not one likely to be invented.

Before Mr. Prince could see a carpenter, he was taken ill. I was intensely interested to learn that his physician was the great doctor I had heard of, who lived in the village of Salisbury, fifteen miles on the road to St. John.

One of my aunts had an impression that the doctor wanted a pupil or assistant of some kind, and suggested that a possible opening might here be offered me. She promised to present me to the doctor on his next visit, after she had broached the subject to him.

The time for which I waited impatiently at length arrived. Never before had I met so charming a man. He was decidedly what we should now call magnetic. There was an intellectual flavor in his talk which was quite new to me. What fascinated me most of all was his speaking of the difficulties he encountered in supplying himself with sufficient "reading matter." He said it as if mental food was as much a necessity as his daily bread. He was evidently a denizen of that world of light which I had so long wished to see. He said that my aunt was quite right in her impression, and our interview terminated in the following liberal proposition on his part: —

S. N. to live with the doctor, rendering him all the assistance in his power in preparing medicines, attending to business, and doing generally whatever might be required of him in the way of help.

The doctor, on his part, to supply S. N.'s bodily needs in food and clothing, and teach him medical botany and the botanic system of medicine. The contract to terminate when the other party should attain the age of twenty-one.

After mentioning the teaching clause, he corrected himself a moment, and added: " At least all I know about it."

All he knows about it! What more could heart desire or brain hold?

The brilliancy of the offer was dimmed by only a single consideration; I had never felt the slightest taste for studying medicine or caring for the sick. That my attainments in the line could ever equal those of my preceptor seemed a result too hopeless to expect. But, after all, something must be done, and this was better than being a carpenter.

Before entering upon the new arrangement, a ratification was required on both sides. The doctor had to make the necessary household arrangements, and secure the consent of his wife. I had to ask the approval of my father, which I did by letter. Like General Grant and many great men, he was a man of exceptional sagacity in matters outside the range of his daily concerns. He threw much cold water on the scheme, but consented to my accepting the arrangement temporarily, as there was nothing better to be done.

I awaited the doctor's next visit with glowing anticipation. In due course of time I stepped with him into his gig for the long drive, expecting nothing less on the journey than a complete outline of the botanic system of medicine and a programme of my future studies. But scarcely had we started when a chilling process commenced. The man erstwhile so effusive was silent, cold, impassive, — a marble statue of his former self. I scarcely got three sentences out of him during the journey, and these were of the most commonplace kind. Could it be the same man?

There was something almost frightful in being alongside a man who knew so much. When we reached our destination the horse had to be put away in the stable. I jumped up to the haymow to throw down the provender. It was a very peculiar feeling to do so under the eye of a man who, as he watched me, knew every muscle that I was setting in operation.

A new chill came on when we entered the house and I was presented to its mistress.

"So you're the boy that's come to work for the doctor, are you?"

"I have come to study with him, ma'am," was my interior reply, but I was too diffident to say it aloud. Naturally the remark made me very uncomfortable. The doctor did not correct her, and evidently must have told her something different from what he told me. Her tone was even more depressing than her words; it breathed a coldness, not to say harshness, to which I had not been accustomed in a woman. There was nothing in her appearance to lessen the unpleasant impression. Small in stature, with florid complexion, wide cheek bones that gave her face a triangular form, she had the eye and look of a well-trained vixen.

As if fate were determined to see how rapid my downfall should be before the close of the day, it continued to pursue me. I was left alone for a few minutes. A child some four years old entered and made a very critical inspection of my person. The result was clearly unfavorable, for she soon asked me to go away. Finding me indisposed to obey the order, she proceeded to the use of force and tried to expel me with a few strong pushes. When I had had enough of this, I stepped aside as she was making a push. She fell to the floor, then picked herself up and ran off crying, " Mamma." The latter soon appeared with added ire infused into her countenance.

" What did you hit the child for? "

" I did n't hit her. What should I want to strike a child like that for? "

" But she says you hit her and knocked her down."

" I did n't, though — she was trying to push me and fell and hurt herself."

A long piercing look of doubt and incredulity followed.

" Strange, very strange. I never knew that child to tell a lie, and she says you struck her."

It was a new experience — the first time I had ever known my word to be questioned.

During the day one thought dominated all

others: where are those treasures of literature which, rich though they are, fail to satisfy their owner's voracious intellectual appetite? As houses were then built, the living and sleeping rooms were all on one main floor. Here they comprised a kitchen, dining room, medicine room, a little parlor, and two small sleeping rooms, one for the doctor and one for myself. Before many hours I had managed to see the interior of every one except the doctor's bedroom, and there was not a sign of a book unless such common ones as a dictionary or a Bible. What could it all mean?

Next day the darkness was illuminated, at least temporarily, by a ray of light. The doctor had been absent most of the day before on a visit to some distant patient. Now he came to me and told me he wanted to show me how to make bilious powders. Several trays of dried herbs had been drying under the kitchen stove until their leaves were quite brittle. He took these and I followed him to the narrow stairway, which we slowly ascended, he going ahead. As I mounted I looked for a solution of the difficulty. Here upstairs must be where the doctor kept his books. At each step I peered eagerly ahead until my head was on a level with the floor. Rafters and a window at the other end had successively come into view and now the whole interior was visible. Nothing was there but a loft, at the further end of which was a bed for the housemaid. The floor was strewn with dried plants. Nothing else was visible. The disillusion seemed complete. My heart sank within me.

On one side of the stairway at a level with the floor was screwed a large coffee mill. The doctor spread a sheet of paper out on the floor on the other side, and laid a fine sieve upon it. Then he showed me how to grind the dry and brittle leaves in the coffee mill, put them into the sieve, and sift them on the paper. This work had a scientific and professional look which infused a glimmer of light into the Cimmerian darkness. The bilious powders were made of the leaves of four plants familiarly known as spearmint, sunflower, smartweed, and yarrow. In his practice a heaping teaspoonful of the pulverized leaves was stirred in a cup of warm water and the grosser parts were allowed to settle, while the patient took the finer parts with the infusion. This was one of Dr. Foshay's staple remedies. Another was a pill of which the principal active ingredient was aloes. The art of making these pills seemed yet more scientific than the other, and I was much pleased to find how soon I could master it. Beside these a number of minor remedies were kept in the medicine room. Among them were tinctures of lobelia, myrrh, and capsicum. There was also a pill box containing a substance which, from its narcotic odor, I correctly inferred to be opium. This drug being prohibited by the Botanic School I could not but feel that Dr. Foshay's orthodoxy was painfully open to question.

Determined to fathom the mystery in which the doctor's plans for my improvement were involved, I announced my readiness to commence the study of the botanic system. He disappeared in the direction of his bedroom, and soon returned with — could my eyes believe it? — a big book. It was one which, at the time of its publication, some thirty or forty years before, was well known to the profession, — Miner and Tully on the " Fevers of the Connecticut Valley." He explained bringing me this book.

"Before beginning the regular study of the botanic system, you must understand something of the old system. You can do so by reading this book."

A duller book I never read. There was every sort of detail about different forms of fever, which needed different treatment; yet calomel and, I think, opium were its main prescriptions. In due time I got through it and reported to my preceptor.

"Well, what do you think of the book?"

"It praises calomel and opium too much. But I infer from reading it that there are so many kinds of fever and other diseases that an immense amount of study will be required to distinguish and treat them."

"Oh, you will find that all these minute distinctions are not necessary when we treat the sick on the botanic system."

"What is the next thing for me? Can I not

now go on with the study of the botanic system?"

"You are not quite ready for it yet. You must first understand something about phrenology. One great difference between us and doctors of the old school is that they take no account of difference of temperament, but treat the lymphatic and bilious in the same way. But we treat according to the temperament of the patient and must therefore be expert in distinguishing temperaments."

"But I studied phrenology long ago and think I understand it quite well."

He was evidently surprised at this statement, but after a little consideration said it was very necessary to be expert in the subject, and thought I had better learn it more thoroughly. He returned to his bedroom and brought a copy of Fowler's "Phrenology," the very book so familiar to me. I had to go over it again, and did so very carefully, paying special attention to the study of the four temperaments, — nervous, bilious, lymphatic, and sanguine.

Before many days I again reported progress. The doctor seemed a little impatient, but asked me some questions about the position of the organs and other matters pertaining to the subject, which I answered promptly and correctly by putting my fingers on them on my own head. But though satisfied with the answers, it was easy to see that he was not satisfied with me. He had, on one or two previous occasions, intimated that I was not wise and prudent in worldly matters. Now he expressed himself more plainly.

"This world is all a humbug, and the biggest humbug is the best man. That's the Yankee doctrine, and that's the reason the Yankees get along so well. You have no organ of secretiveness. You have a window in your breast that every one can look into and see what you are thinking about. You must shut that window up, like I do. No one can tell from my talk or looks what I am thinking about."

It may seem incredible to the reader that I marveled much at the hidden meaning of this allegorical speech, and never for one moment supposed it to mean: "I, Dr. Foshay, with my botanic system of medicine, am the biggest humbug in these parts, and if you are going to succeed with me you must be another." But I had already recognized the truth of his last sentence. Probably neither of us had heard of Talleyrand, but from this time I saw that his hearty laugh and lively talk were those of a manikin.

His demeanor toward me now became one of complete gravity, formality, and silence. He was always kindly, but never said an unnecessary word, and avoided all reference to reading or study. The mystery which enveloped him became deeper month after month. In his presence I felt a certain awe which prevented my asking any questions as to his intentions toward me.

It must, of course, be a matter of lifelong regret that two years so important in one's education should have been passed in such a way, — still, they were not wholly misspent. From a teacher named Monroe,[1] who then lived near Salisbury, I borrowed Draper's Chemistry, little thinking that I would one day count the author among my friends. A book peddler going his rounds offered a collection of miscellaneous books at auction. I bought, among others, a Latin and a Greek grammar, and assiduously commenced their study. With the first I was as successful as could be expected under the circumstances, but failed with the Greek, owing to the unfamiliarity of the alphabet, which seemed to be an obstacle to memory of the words and forms.

But perhaps the greatest event of my stay was the advent of a botanic druggist of Boston, who passed through the region with a large wagonload of medicines and some books. He was a pleasant, elderly gentleman, and seemed much interested on learning that I was a student of the botanic system. He had a botanic medical college in or near Boston, and strongly urged me to go thither as soon as I could get ready to complete my studies. From him the doctor, willing to do me a favor, bought some books, among them the "Eclectic Medical Dispensary," published in Cincinnati. Of this

---

[1] Rev. Alexander H. Monroe, who, I have understood, afterward lived in Montreal. I have often wished to find a trace of him, but do not know whether he is still living.

book the doctor spoke approvingly, as founded on the true system which he himself practiced, and though I never saw him read it, he was very ready to accept the knowledge which I derived from it. The result was quite an enlargement of his materia medica, both in the direction of native plants and medicines purchased from his druggist.

On one occasion this advance came near having serious consequences. I had compounded some pills containing a minute quantity of elaterium. The doctor gave them to a neighboring youth affected with a slight indisposition in which some such remedy was indicated. The directions were very explicit, — one pill every hour until the desired effect was produced.

"Pshaw," said the patient's brother, "there's nothin' but weeds in them pills, and a dozen of them won't hurt you."

The idea of taking weed pills one at a time seemed too ridiculous, and so the whole number were swallowed at a dose. The result was, happily, not fatal, though impressive enough to greatly increase the respect of the young man's family for our medicines.

The intellectual life was not wholly wanting in the village. A lodge of a temperance organization, having its headquarters in Maine, was formed at a neighboring village. It was modeled somewhat after the fashion of the Sons of Temperance. The presiding officer, with a high sounding title, was my mother's cousin, Tommy Nixon. He was the most popular young man of the neighborhood. The rudiments of a classical education gained at a reputable academy in Sackville had not detracted from his qualities as a healthy, rollicking young farmer. The lodge had an imposing ritual of which I well remember one feature. At stated intervals a password which admitted a member of any one lodge to a meeting of any other was received from the central authority — in Maine, I believe. It was never to be pronounced except to secure admission, and was communicated to the members by being written on a piece of paper in letters so large that all could read. After being held up to view for a few moments, the paper was held in the flame of a candle with these words: "This paper containing our secret password I commit to the devouring element in token that it no longer exists save in the minds of the faithful brethren." The fine sonorous voice of the speaker and his manly front, seen in the lurid light of the burning paper, made the whole scene very impressive.

There was also a society for the discussion of scientific questions, of which the founder and leading spirit was a youth named Isaac Steves, who was beginning the study of medicine. The president was a "Worthy Archon." Our discussions strayed into the field of physiological mysteries, and got us into such bad odor with Mrs. Foshay and, perhaps, other ladies of the community, that the meetings were abandoned.

A soil like that of the Provinces at this time was

fertile in odd characters including, possibly, here and there, a "heart pregnant with celestial fire." One case quite out of the common line was that of two or three brothers employed in a sawmill somewhere up the river Petticodiac. According to common report they had invented a new language in order to enable them to talk together without their companions knowing what they were saying. I knew one of them well and, after some time, ventured to inquire about this supposed tongue. He was quite ready to explain it. The words were constructed out of English by the very simple process of reversing the syllables or the spelling. Everything was pronounced backward. Those who heard it, and knew the key, had no difficulty in construing the words; to those who did not, the words were quite foreign.

The family of the neighborhood in which I was most intimate was that of a Scotch farmer named Parkin. Father, mother, and children were very attractive, both socially and intellectually, and in later years I wondered whether any of them were still living. Fifty years later I had one of the greatest and most agreeable surprises of my life in suddenly meeting the little boy of the family in the person of Dr. George R. Parkin, the well-known promoter of imperial federation in Australia and the agent in arranging for the Rhodes scholarships at Oxford which are assigned to America.

My duties were of the most varied character. I

composed a little couplet designating my professions as those of

> Physician, apothecary, chemist, and druggist,
> Girl about house and boy in the barn.

I cared for the horse, cut wood for the fire, searched field and forest for medicinal herbs, ordered other medicines from a druggist[1] in St. John, kept the doctor's accounts, made his pills, and mixed his powders. This left little time for reading and study, and such exercises were still farther limited by the necessity of pursuing them out of sight of the housewife.

As time passed on, the consciousness that I was wasting my growing years increased. I long cherished a vague hope that the doctor could and would do something to promote my growth into a physician, especially by taking me out to see his patients. This was the recognized method of commencing the study of medicine. But he never proposed such a course to me, and never told me how he expected me to become a physician. Every month showed my prospects in a less hopeful light. I had rushed into my position in blind confidence in the man, and without any appreciation of the requirements of a medical practitioner. But these requirements now presented themselves to my mind with constantly increasing force. Foremost among them was a knowledge of anatomy, and how could that be acquired except at a medical school? It

---

[1] Our druggist was Mr. S. L. Tilley, afterward Sir Leonard Tilley, the well-known Canadian Minister of Finance.

was every day more evident that if I continued in my position I should reach my majority without being trained for any life but that of a quack.

While in this state of perplexity, an event happened which suggested a way out. One day the neighborhood was stirred by the news that Tommy Nixon had run away — left his home without the consent of his parents, and sailed for the gold fields of Australia. I was struck by the absence of any word of reprobation for his act. The young men at least seemed to admire the enterprising spirit he had displayed. A few weeks after his departure a letter which he wrote from London, detailing his adventures in the great metropolis, was read in my presence to a circle of admiring friends with expressions of wonder and surprise. This little circumstance made it clear to me that the easiest way out of my difficulty was to cut the Gordian knot, run away from Dr. Foshay, and join my father in New England.

No doubt the uppermost question in the mind of the reader will be: Why did you wait so long without having a clear understanding with the doctor? Why not ask him to his face how he expected you to remain with him when he had failed in his pledges, and demand that he should either keep them or let you go?

One answer, perhaps the first, must be lack of moral courage to face him with such a demand. I have already spoken of the mystery which seemed to enshroud his personality, and of the fascination which, through it, he seemed to exercise over me. But behind this was the conviction that he could not do anything for me were he ever so well disposed. That he was himself uneducated in many essentials of his profession had gradually become plain enough; but what he knew or possibly might know remained a mystery. I had heard occasional allusions, perhaps from Mrs. Foshay rather than from himself, to an institution supposed to be in Maine, where he had studied medicine, but its name and exact location were never mentioned. Altogether, if I told him of my intention, it could not possibly do any good, and he might be able to prevent my carrying it out, or in some other way to do much harm. And so I kept silent.

Tuesday, September 13, 1853, was the day on which I fixed for the execution of my plan. The day previous I was so abstracted as to excite remarks both from Mrs. Foshay and her girl help, the latter more than once declaring me crazy when I made some queer blunder. The fact is I was oppressed by the feeling that the step about to be taken was the most momentous of my life. I packed a few books and clothes, including some mementoes of my mother, and took the box to the stage and post-office in the evening, to be forwarded to an assumed name in St. John the next afternoon. This box I never saw again; it was probably stopped by Foshay before being dispatched. My plan was to start early in the morning, walk as far as I could during the day, and, in the evening, take the mail stage when it should overtake me. This course was necessitated by the fact that the little money that I had in my pocket was insufficient to pay my way to Boston, even when traveling in the cheapest way.

I thought it only right that the doctor should be made acquainted with my proceeding and my reason for taking it, so I indited a short letter, which I tried to reproduce from memory ten years later with the following result: —

Dear Doctor, — I write this to let you know of the step I am about to take. When I came to live with you, it was agreed that you should make a physician of me. This agreement you have never shown the slightest intention of fulfilling since the first month I was with you. You have never taken me to see a patient, you have never given me any instruction or advice whatever. Beside this, you must know that your wife treats me in a manner that is no longer bearable.

I therefore consider the agreement annulled from your failure to fulfill your part of it, and I am going off to make my own way in the world. When you read this, I shall be far away, and it is not likely that we shall ever meet again.

If my memory serves me right, the doctor was absent on a visit to some distant patient on the night in question, and I did not think it likely that he would return until at least noon on the following day. By this time my box would have been safely off in the stage, and I would be far out of reach. To delay his receiving the letter as much as possible, I did not leave it about the house, but put it in the window of a shop across the way, which served the neighbors as a little branch post-office.

But he must have returned sooner than I expected, for, to my great regret, I never again saw or heard of the box, which contained, not only the entire outfit for my journey, but all the books of my childhood which I had, as well as the little mementoes of my mother. The postmaster who took charge of the goods was a Mr. Pitman. When I again passed through Salisbury, as I did ten years later, he had moved away, no one could tell me exactly where.

I was on the road before daybreak, and walked till late at night, occasionally stopping to bathe my feet in a brook, or to rest for a few minutes in the shadow of a tree. The possibility of my being pursued by the doctor was ever present to my mind, and led me to keep a sharp lookout for coming vehicles. Toward sunset a horse and buggy appeared, coming over a hill, and very soon the resemblance of vehicle and driver to the turnout of the doctor became so striking that I concealed myself in the shrubbery by the wayside until the sound of the wheels told me he was well past. The probability that my pursuer was in front of me was an added source of discomfort which led me to avoid the road and walk in the woods wherever the former was not visible to some distance ahead. But I neither saw nor heard anything more of the supposed pursuer, though, from what

I afterward learned, there can be little doubt that it was actually Foshay himself.

The advent of darkness soon relieved me of the threatened danger, but added new causes of solicitude. The evening advanced, and the lights in the windows of the houses were becoming fewer and fewer, and yet the stage had not appeared. I slackened my pace, and made many stops, beginning to doubt whether I might not as well give up the stage and look for an inn. It was, I think, after ten o'clock when the rattling of wheels announced its approach. It was on a descending grade, and passed me like a meteor, in the darkness, quite heedless of my calls and gesticulations. Fortunately a house was in sight where I was hospitably entertained, and I was very soon sound asleep, as became one who had walked fifty miles or more since daylight.

Thus ended a day to which I have always looked back as the most memorable of my life. I felt its importance at the time. As I walked and walked, the question in my mind was, what am I doing and whither am I going? Am I doing right or wrong? Am I going forward to success in life, or to failure and degradation? Vainly, vainly, I tried to peer into the thick darkness of the future. No definite idea of what success might mean could find a place in my mind. I had sometimes indulged in daydreams, but these come not to a mind occupied as mine on that day. And if they had, and if fancy had been allowed its wildest flight in portraying a future, it is safe to say that the figure of an honorary academician of France, seated in the chair of Newton and Franklin in the palace of the Institute, would not have been found in the picture.

As years passed away I have formed the habit of looking back upon that former self as upon another person, the remembrance of whose emotions has been a solace in adversity and added zest to the enjoyment of prosperity. If depressed by trial, I think how light would this have appeared to that boy had a sight of the future been opened up to him. When, in the halls of learning, I have gone through the ceremonies which made me a citizen of yet another commonwealth in the world of letters, my thoughts have gone back to that day; and I have wished that the inexorable law of Nature could then have been suspended, if only for one moment, to show the scene that Providence held in reserve.

Next morning I was on my way betimes, having still more than thirty miles before me. And the miles seemed much longer than they did the day before, for my feet were sore and my limbs stiff. Quite welcome, therefore, was a lift offered by a young farmer, who, driving a cart, overtook me early in the forenoon. He was very sociable, and we soon got into an interesting conversation.

I knew that Dr. Foshay hailed from somewhere in this region, where his father still lived, so I asked my companion whether he knew a family of that name. He knew them quite well.

" Do you know anything of one of the sons who is a doctor ? "

" Yes indeed ; I know all about him, but he ain't no doctor. He tried to set up for one in Salisbury, but the people there must a' found him out before this, and I don't know where he is now."

" But I thought he studied medicine in Fredericton or Maine or somewhere on the border."

" Oh, he went off to the States and pretended to study, but he never did it. I tell you he ain't no more a doctor nor I am. He ain't smart enough to be a doctor."

I fell into a fit of musing long enough to hear, in my mind's ear, with startling distinctness, the words of two years before : " This world is all a humbug, and the biggest humbug is the best man. . . . You have a window in your breast and you must close that window before you can succeed in life." Now I grasped their full meaning.

Ten years later I went through the province by rail on my wedding journey. At Dorchester, the next village beyond Moncton, I was shown a place where insolvent debtors were kept " on the limits."

" By stopping there," said my informant, " you can see Dr. Foshay."

I suggested the question whether it was worth while to break our journey for the sake of seeing him. The reply of my informant deterred me.

" It can hardly be worth while to do so. He will be a painful object to see, — a bloated sot, drinking himself to death as fast as he can."

The next I heard of him was that he had succeeded.

I reached St. John on the evening that a great celebration of the commencement of work on the first railway in the province was in progress. When things are undecided, small matters turn the scale. The choice of my day for starting out on my adventurous journey was partly fixed by the desire to reach St. John and see something of the celebration. Darkness came on when I was yet a mile or two from the city; then the first rocket I had ever beheld rose before me in the sky. Two of what seemed like unfortunate incidents at the time were most fortunate. Subsequent and disappointing experience showed that had I succeeded in getting the ride I wished in the stage, the resulting depletion of my purse would have been almost fatal to my reaching my journey's end. Arriving at the city, I naturally found all the hotels filled. At length a kindly landlady said that, although she had no bed to give me, I was quite welcome to lie on a soft carpeted floor, in the midst of people who could not find any other sleeping place. No charge was made for this accommodation. My hope of finding something to do which would enable me to earn a little money in St. John over and above the cost of a bed and a daily loaf of bread was disappointed. The efforts of the next week are so painful to recall that I will not harrow the feelings of the reader by describing them. Suffice it to say that the adventure was wound up by an interview at

Calais, a town on the Maine border, a few miles from Eastport, with the captain of a small sailing vessel, hardly more than a boat. He was bound for Salem. I asked him the price of a passage.

"How much money have you?" he replied.

I told him; whether it was one or two dollars I do not recall.

"I will take you for that if you will help us on the voyage."

The offer was gladly accepted. The little craft was about as near the opposite of a clipper ship as one can imagine, never intended to run in any but fair winds, and even with that her progress was very slow. There was a constant succession of west winds, and the result was that we were about three weeks reaching Salem. Here I met my father, who, after the death of my mother, had come to seek his fortune in the "States." He had reached the conclusion, on what grounds I do not know, that the eastern part of Maryland was a most desirable region, both in the character of its people and in the advantages which it offered us. The result was that, at the beginning of 1854, I found myself teacher of a country school at a place called Massey's Cross Roads in Kent County. After teaching here one year, I got a somewhat better school at the pleasant little village of Sudlersville, a few miles away.

Of my abilities as a manager and teacher of youth the reader can judge. Suffice it to say that, looking back at those two years, I am deeply impressed with the good nature of the people in tolerating me at all.

My most pleasant recollection is that of two of my best pupils of Sudlersville, nearly my own age. One was Arthur E. Sudler, for whose special benefit some chemical apparatus was obtained from Philadelphia. He afterwards studied medicine at the University of Pennsylvania and delighted me by writing that what I had taught him placed him among the best in his class in chemistry. The other was B. S. Elliott, who afterward became an engineer or surveyor.

One of my most vivid recollections at Massey's relates to a subject which by no means forms a part of one's intellectual development, and yet is at the bottom of all human progress, that of digestion. The staple food of the inhabitants of a Southern farming region was much heartier than any to which I had been accustomed. "Pork and pone" were the staples, the latter being a rather coarse cake with little or no seasoning, baked from cornmeal. This was varied by a compound called "shortcake," a mixture of flour and lard, rapidly baked in a pan, and eaten hot. Though not distasteful, I thought it as villainous a compound as a civilized man would put into his stomach.

Quite near my school lived a young bachelor farmer who might be designated as William Bowler, Esq., though he was better known as Billy Bowler. He had been educated partly at Delaware College, Newark, and was therefore an interesting young man to know. In describing his experiences at the college, he once informed me that they were all very pleasant except in a single point; that was the miserably poor food that the students got to eat. He could not, he declared, get along without good eating. This naturally suggested that my friend was something of a gourmand. Great, therefore, was my delight when, a few weeks later, he expressed a desire to have me board with him. I accepted the offer as soon as possible. Much to my disappointment, shortcake was on the table at the first meal and again at the second. It proved to be the principal dish twice, and I am not sure but three times a day. The other staple was fried meat. On the whole this was worse than pork and pone, which, if not toothsome, was at least wholesome. As the days grew into weeks, I wondered what Delaware College could give its students to eat. To increase the perplexity, there were plenty of chickens in the yard and vegetables in the garden. I asked the cook if she could not boil some vegetables and bring them on the table.

"Mas'er Bowler don't like wegetable."

Then I found that the chickens were being consumed in the kitchen and asked for one.

"Mas'er Bowler don't like chicken," was the reply, with an added intimation that the chickens belonged to the denizens of the kitchen.

The mystery was now so dark and deep that I determined to fathom it. I drew Mr. Bowler into conversation once more about Delaware College, and asked him what the students had to eat when there.

He had evidently forgotten his former remark and described what seemed to me a fairly well provided students' table. Now I came down on him with my crusher.

"You told me once that the table was miserably poor, so that you could hardly stand it. What fault had you to find with it?"

He reflected a moment, apparently recalling his impression, then replied: "Oh, they had no short-cake there!"

In 1854 I availed myself of my summer vacation to pay my first visit to the national capital, little dreaming that it would ever be my home. I went as far as the gate of the observatory, and looked wistfully in, but feared to enter, as I did not know what the rules might be regarding visitors. I speculated upon the possible object of a queer red sandstone building, which seemed so different from anything else, and heard for the first time of the Smithsonian Institution.

On the very beginning of my work at Massey's the improvement in my position was so remarkable that I felt my rash step of a few months before fully justified. I wrote in triumph to my favorite aunt, Rebecca Prince, that leaving Dr. Foshay was the best thing I had ever done. I was no longer "that boy," but a respectable young man with a handle to my name.

Just what object I should pursue in life was still

doubtful; the avenues of the preferment I would have liked seemed to be closed through my not being a college graduate. I had no one to advise me as to the subjects I should pursue or the books I should study. On such books as I could get, I passed every spare hour. My father sent me Cobbett's English Grammar, which I found amusing and interesting, especially the criticisms upon the grammar found here and there in royal addresses to Parliament and other state papers. On the whole I am not sure but that the book justified my father's good opinion, although I cannot but think that it was rather hypercritical. I had been taught the rudiments of French in Wallace when quite a child by a Mr. Oldright, of whose methods and pronunciation my memory gives me a most favorable impression. I now got Cobbett's French Grammar, probably a much less commendable book than his English one. I had never yet fathomed the mysteries of analytic geometry or the calculus, and so got Davies' books on those subjects. That on the calculus was perhaps the worst that could be put into the hands of a person situated as I was. Two volumes of Bezout's Mathematics, in French, about a century old, were, I think, rather better. Say's Political Economy was the first book I read on that subject, and it was quite a delight to see human affairs treated by scientific methods.

I finally reached the conclusion that mathematics was the study I was best fitted to follow, though I did not clearly see in what way I should turn the subject to account. I knew that Newton's "Principia" was a celebrated book, so I got a copy of the English translation. The path through it was rather thorny, but I at least caught the spirit here and there. No teacher at the present time would think of using it as a text-book, yet as a mental discipline, and for the purpose of enabling one to form a mental image of the subject, its methods at least are excellent. I got a copy of the "American Journal of Science," hoping it might enlighten me, but was frightened by its big words, and found nothing that I could understand.

During the year at Sudlersville I made several efforts which, though they were insignificant so far as immediate results were concerned, were in some respects of importance for my future work. With no knowledge of algebra except what was derived from the meagre text-books I could pick up, — not having heard even the name of Abel, or knowing what view of the subject was taken by professional mathematicians, — I made my first attempt at a scientific article, " A New Demonstration of the Binomial Theorem." This I sent to Professor Henry, secretary of the Smithsonian Institution, to see if he deemed it suitable for publication. He promptly replied in the negative, but offered to submit it to a professional mathematician for an opinion of its merits. I gladly accepted this proposal, which was just what I wanted. In due course a copy of the report was sent me. One part of the work was praised for its elegance, but a lack of completeness and rigor was pointed out. It was accompanied by a pleasant note from Professor Henry remarking that, while not so favorable as I might have expected, it was sufficiently so to encourage me in persevering.

The other effort to which I refer was of quite a different character. A copy of the "National Intelligencer," intended for some subscriber who had left Sudlersville, came to the post-office for several months, and, there being no claimant, I frequently had an opportunity to read it. One of its features was frequent letters from volunteer writers on scientific subjects. Among these was a long letter from one G. W. Eveleth, the object of which was to refute the accepted theory of the universe, especially the view of Copernicus. For aught I knew Mr. Eveleth held as high a position as any one else in the world of science and letters, so I read his article carefully. It was evidently wholly fallacious, yet so plausible that I feared the belief of the world in the doctrine of Copernicus might suffer a severe shock, and hastened to the rescue by writing a letter over my own name, pointing out the fallacies. This was published in the "National Intelligencer" — if my memory serves me right — in 1855. My full name, printed in large capitals, in a newspaper, at the bottom of a letter, filled me with a sense of my temerity in appearing so prominently in print, as if I were intruding into company where I might not be wanted.

My letter had two most unexpected and gratifying results. One was a presentation of a copy of Lee's "Tables and Formulæ," which came to me a few days later through the mail with the compliments of Colonel Abert. Not long afterward came a letter from Professor J. Lawrence Smith, afterward a member of the National Academy of Sciences, transmitting a copy of a pamphlet by him on the theory that meteorites were masses thrown up from the volcanoes of the moon, and asking my opinion on the subject.

I had not yet gotten into the world of light. But I felt as one who, standing outside, could knock against the wall and hear an answering knock from within.

The beginning of 1856 found me teaching in the family of a planter named Bryan, residing in Prince George County, Md., some fifteen or twenty miles from Washington. This opened up new opportunities. I could ride into Washington whenever I wished, leave my horse at a livery stable, and see whatever sights the city offered. The Smithsonian Library was one of the greatest attractions. Sometime in May, 1856, I got permission from the attendant in charge to climb into the gallery and see the mathematical books. Here I was delighted to find the greatest treasure that my imagination had ever pictured, — a work that I had thought of almost as belonging to fairyland. And here it was right before my eyes — four enormous volumes, — "Mécanique Céleste, by the

Marquis de Laplace, Peer of France; translated by Nathaniel Bowditch, LL. D., Member of the Royal Societies of London, Edinburg, and Dublin." I inquired as to the possibility of my borrowing the first volume, and was told that this could be done only by special authority of Professor Henry. I soon got the necessary authority through Mr. Rhees, the chief clerk, whose kindness in the matter deeply impressed me, signed a promise to return it within one month, and carried it in triumph to my little schoolhouse. I dipped into it here and there, but at every step was met by formulæ and methods quite beyond the power of one who knew so little of mathematics. In due time I brought the book back as promised.

Up to this time I think I had never looked upon a real live professor; certainly not upon one of eminence in the scientific world. I wondered whether there was any possibility of my making the acquaintance of so great a man as Professor Henry. Some time previous a little incident had occurred which caused me some uneasiness on the subject. I had started out very early on a visit to Washington, or possibly I had stayed there all night. At any rate, I reached the Smithsonian Building quite early, opened the main door, stepped cautiously into the vestibule, and looked around. Here I was met by a short, stout, and exceedingly gruff sort of a man, who looked upon my entrance with evident displeasure. He said scarcely a word, but motioned me out of the door, and showed me a paper or something in the entrance which intimated that the Institution would be open at nine o'clock. It was some three minutes before that hour, so I was an intruder. The man looked so respectable and so commanding in his appearance that I wondered if he could be Professor Henry, yet sincerely hoped he was not. I afterward found that he was only "Old Peake," the janitor.[1] When I found the real Professor Henry he received me with characteristic urbanity, told me something of his own studies, and suggested that I might find something to do in the Coast Survey, but took no further steps at that time.

The question whether I was fitted for any such employment now became of great interest. The principal question was whether one must know celestial mechanics in order to secure such a position, so, after leaving Professor Henry, I made my way to the Coast Survey office, and was shown to the chief clerk, as the authority for the information. I modestly asked him whether a knowledge of physical astronomy was necessary to a position in that office. Instead of frankly telling me that he did not know what physical astronomy was, he answered in the affirmative. So I left with the impression that I must master the "Mécanique Céleste" or some similar treatise before finding any opening there.

[1] Peake, notwithstanding his official title, would seem to have been more than an ordinary janitor, as he was the author of a *Guide* to the Smithsonian Institution.

I could not, of course, be satisfied with a single visit to such a man, and so called several times during the year. One thing I wondered about was whether he would remember me when he again saw me. On one occasion I presented him with a plan for improving the Cavendish method of determining the density of the earth, which he took very kindly. I subsequently learned that he was much interested in this problem. On another occasion he gave me a letter to Mr. J. E. Hilgard, assistant in charge of the Coast Survey office. My reception by the latter was as delightful as that by Professor Henry. I found from my first interview with him that the denizens of the world of light were up to the most sanguine conceptions I ever could have formed.

At this time, or probably some time before, I bought a copy of the "American Ephemeris" for 1858, and amused myself by computing on a slate the occultations visible at San Francisco during the first few months of the year. At this time I had learned nothing definite from Mr. Hilgard as to employment in his office. But about December, 1856, I received a note from him stating that he had been talking about me to Professor Winlock, superintendent of the "Nautical Almanac," and that I might possibly get employment on that work. When I saw him again I told him that I had not yet acquired such a knowledge of physical astronomy as would be necessary for the calculations in question; but he assured me that this was no drawback, as formulæ for all the computations would be supplied me. I was far from satisfied at the prospect of doing nothing more than making routine calculations with formulæ prepared by others; indeed, it was almost a disappointment to find that I was considered qualified for such a place. I could only console myself by the reflection that the ease of the work would not hinder me from working my way up. Shortly afterward I understood that it was at least worth while to present myself at Cambridge, and so started out on a journey thither about the last day of the year 1856.

At that time even a railroad journey was quite different from what it is now. The cars were drawn through Baltimore by horses. At Havre de Grace the train had to stop and the passengers were taken across the river in a ferryboat to another train. At Philadelphia the city had to be traversed by transfer coaches. Looking around for this conveyance, I met a man who said he had it. He shoved me into it and drove off. I remarked with suspicion that no other coaches were accompanying us. After a pretty long drive the speed of the horses gradually began to slacken. At length it came to a complete stop in front of a large building, and I got out. But it was only a freight station, locked up and dark throughout. The driver mumbled something about his fare, then rolled back on his seat, seemingly dead drunk. The nearest sign of life was at a tavern a block or two away. There I found that I was only a short distance from the station of departure, and reached my train barely in time.

Landing in New York at the first glimmer of dawn, near the end of the line of passengers I was momentarily alarmed to see a man pick up what seemed to be a leather purse from right between my feet. It was brown and, so far as I could see, just like my own. I immediately felt the breast pocket of my coat and found that my own was quite safe. The man who picked up the purse inquired in the politest tone possible if it was mine, to which I replied in the negative. He retreated a short distance and then a bystander came up and chided me in a whisper for my folly in not claiming the purse. The only reply he got was, "Oh, I'm up to all your tricks." On a repetition of this assurance the pair sneaked away.

Arriving at Cambridge, I sought out Professor Winlock and was informed that no immediate employment was open at his office. It would be necessary for him to get authority from Washington. After this was obtained some hope might be held out, so I appeared in the office from time to time as a visitor, my first visit being that described in the opening chapter.

III

THE WORLD OF SWEETNESS AND LIGHT

The term " Nautical Almanac " is an unfortunate misnomer for what is, properly speaking, the "Astronomical Ephemeris." It is quite a large volume, from which the world draws all its knowledge of times and seasons, the motions of the heavenly bodies, the past and future positions of the stars and planets, eclipses, and celestial phenomena generally which admit of prediction. It is the basis on which the family almanac is to rest. It also contains the special data needed to enable the astronomer and navigator to determine their position on land or sea. The first British publication of the sort, prepared by Maskelyne, Astronomer Royal, a century ago, was intended especially for the use of navigators; hence the familiar appellation, which I call unfortunate because it leads to the impression that the work is simply an enlargement and improvement of the household almanac.

The leading nations publish ephemerides of this sort. The introductions and explanations are, of course, in the languages of the respective countries; but the contents of the volume are now so much alike that the duplication of work involved in preparing them seems quite unnecessary. Yet national pride and emulation will probably continue it for some time to come.

The first appropriation for an American ephemeris and nautical almanac was made by Congress in 1849. Lieutenant Charles Henry Davis, as a leader and moving spirit in securing the appropriation, was naturally made the first superintendent of the work. At that time astronomical science in our country was so far from being reduced to a system that it seemed necessary to have the work prepared at some seat of learning. So, instead of founding the office in Washington, it was established at Cambridge, the seat of Harvard University, where it could have the benefit of the technical knowledge of experts, and especially of Professor Benjamin Peirce, who was recognized as the leading mathematician of America. Here it remained until 1866, when conditions had so far changed that the office was removed to Washington, where it has since remained.

To this work I was especially attracted because its preparation seemed to me to embody the highest intellectual power to which man had ever attained. The matter used to present itself to my mind somewhat in this way: Supply any man with the fundamental data of astronomy, the times at which stars and planets cross the meridian of a place, and other matters of this kind. He is informed that each of these bodies whose observations he is to use is attracted by all the others with a force which varies as the inverse square of their distance apart. From these data he is to weigh the bodies, predict their motion in all future time, compute their orbits, determine what changes of form and position these orbits will undergo through thousands of ages, and make maps showing exactly over what cities and towns on the surface of the earth an eclipse of the sun will pass fifty years hence, or over what regions it did pass thousands of years ago. A more hopeless problem than this could not be presented to the ordinary human intellect. There are tens of thousands of men who could be successful in all the ordinary walks of life, hundreds who could wield empires, thousands who could gain wealth, for one who could take up this astronomical problem with any hope of success. The men who have done it are therefore in intellect the select few of the human race, — an aristocracy ranking above all others in the scale of being. The astronomical ephemeris is the last practical outcome of their productive genius.

On the question whether the world generally reasoned in this way, I do not remember having any distinct idea. This was certainly not because I was indifferent to the question, but because it never strongly presented itself to my mind. From my point of view it would not have been an important one, because I had already formed the conviction that one should choose that sphere in life to which he was most strongly attracted, or for which his faculties best fitted him.

A few months previous to my advent Commander Davis had been detached from the superintendency and ordered to command the sloop St. Mary's. He was succeeded by Professor Joseph Winlock, who afterward succeeded George P. Bond as director of the Harvard Observatory. Most companionable in the society of his friends, Winlock was as silent as General Grant with the ordinary run of men. Withal, he had a way of putting his words into exact official form. The following anecdote of him used to be current. While he was attached to the Naval Academy, he was introduced one evening at a reception to a visiting lady. He looked at the lady for a decorous length of time, and she looked at him; then they parted without saying a word. His introducer watched the scene, and asked him, "Why did you not talk to that lady?"

"I had no statement to make to her," was the reply.

Dr. Gould told me this story was founded on fact, but when, after Winlock's death, it was put off on me with some alterations, I felt less sure.

The following I believe to be authentic. It occurred several years later. Hilgard, in charge of the Coast Survey office, was struck by the official terseness of the communications he occasionally received from Winlock, and resolved to be his rival. They were expecting additions to their families about the same time, and had doubtless spoken of the subject. When Hilgard's arrived, he addressed a communication to Winlock in these terms:—

"Mine's a boy. What's yours?"

In due course of time the following letter was received in reply: —

Dear Hilgard: —
    Boy.
        Yours, etc., J. Winlock.

When some time afterward I spoke to Winlock on the subject, and told him what Hilgard's motive was, he replied, "It was not fair in Hilgard to try and take me unawares in that way. Had I known what he was driving at, I might have made my letter still shorter." I did not ask him how he would have done it. It is of interest that the "boy" afterward became one of the assistant secretaries of the Smithsonian Institution.

One of the most remarkable features of the history of the "Nautical Almanac" is the number of its early assistants who have gained prominence or distinction in the various walks of life. It would be difficult to find so modest a public work to exceed it in this respect.

John D. Runkle, who lived till 1902, was, as I have said, the senior and leading assistant in the office. He afterward became a professor in the Institute of Technology, and succeeded Rogers as its president. In 1876 he started the school of manual training, which has since been one of the great features of the Institute. He afterward resigned the presidency, but remained its principal professor of mathematics. He was the editor and founder of the "Mathematical Monthly," of which I shall presently have more to say.

The most wonderful genius in the office, and the one who would have been the most interesting subject of study to a psychologist, was Truman Henry Safford. In early childhood he had excited attention by his precocity as what is now sometimes called a "lightning calculator." A committee of the American Academy of Arts and Science was appointed to examine him. It very justly and wisely reported that his arithmetical powers were not in themselves equal to those of some others on record, especially Zerah Colburn, but that they seemed to be the outcome of a remarkable development of the reasoning power. When nine years old, he computed almanacs, and some of his work at this age is still preserved in the Harvard University Library. He graduated at Harvard in 1854, and was soon afterward taken into the Nautical Almanac Office, while he also worked from time to time at the Cambridge observatory. It was found, however, that the power of continuous work was no greater in him than in others, nor did he succeed in doing more than others in the course of a year.

The mental process by which certain gifted arithmetical computers reach almost in an instant the results of the most complicated calculations is a psychological problem of great interest, which has never been investigated. No more promising subject for the investigation could ever have been found than Safford, and I greatly regret having lost all opportunities to solve the problem. What was of interest in Safford's case was the connection of this faculty with other remarkable mental powers of an analogous but yet different kind. He had a remarkable faculty for acquiring, using, and reading languages, and would have been an accomplished linguist had he turned his attention in that direction. He was a walking bibliography of astronomy, which one had only to consult in order to learn in a moment what great astronomers of recent times had written on almost any subject, where their work was published, and on what shelf of the Harvard Library the book could be found. But the faculty most closely connected with calculation was a quickness and apprehension of vision, of which the following is an example: —

About 1876 he visited the Naval Observatory in Washington for the first time in his life. We wanted a certain catalogue of stars and went together into the library. The required catalogue was on one of a tier of shelves containing altogether a hundred, or perhaps several hundred volumes. "I do not know whether we have the book," said I, "but if we have, it is on one of these shelves." I began to go through the slow process of glancing at the books one by one until my eyes should strike the right title. He stood back six or eight feet and took in all the shelves seemingly at one glance, then stepped forward and said, "Here it is." I might have supposed this an accident, but that he subsequently did practically the same thing in my office, selecting in a moment a book we wanted to see, after throwing a rapid glance over shelves containing perhaps a hundred volumes.

An example of his apprehension and memory for numbers was narrated by Mr. Alvan Clark. When the latter had completed one of his great telescopes for the University of Chicago, Safford had been named as director, and accompanied the three members of the firm to the city when they carried the object glass thither. On leaving the train all four took their seats in a hotel omnibus, Safford near the door. Then they found that they had forgotten to give their baggage checks to the expressman; so the other three men passed their checks to Safford, who added his own and handed all four to the conductor of the omnibus.

When it was time for the baggage to come to the hotel, there was such a crowd of new arrivals that the attendants could not find it. The hotel clerk remarked on inquiry, "If I only knew the numbers of your checks, I would have no difficulty in tracing your trunks." Safford at once told off the four numbers, which he had read as he was passing the checks to the conductor.

The great fire practically put an end to the activity of the Chicago Observatory and forced its director to pursue his work in other fields. That he failed to attain that commanding position due to his genius is to be ascribed to a cause prevalent among us during all the middle part of the century; perhaps that from which most brilliant intellects fail to reach eminence: lack of the power of continuous work necessary to bring important researches to a completion.

Another great intellect of the office was Chauncey Wright. If Wright had systematically applied his powers, he might have preceded or supplanted Herbert Spencer as the great exponent of the theory of evolution. He had graduated at Harvard in 1853, and was a profound student of philosophy from that time forward, though I am not aware that he was a writer. When in 1858 Sir William Hamilton's "Lectures on Metaphysics" appeared, he took to them with avidity. In 1859 appeared Darwin's "Origin of Species," and a series of meetings was held by the American Academy, the special order of which was the discussion of this book. Wright and myself, not yet members, were invited to be present. To judge of the interest it is only necessary to remark that Agassiz and Gray were the two leading disputants, the first taking ground against Darwin, the other in his favor. Wright was a Darwinist from the very beginning, explaining the theory in private conversation from a master's point of view, and soon writing upon it in the "North American Review" and in other publications. Of one of his articles Darwin has been quoted as saying that it was the best exposition of his theory that had then appeared. After his untimely death in 1875, Wright's papers were collected and published under the title of "Philosophical Discussions."[1] Their style is clear-cut and faultless in logical form, yet requiring such close attention to every word as to be less attractive to the general reader of to-day than that of Spencer. In a more leisurely age, when men wanted to think profoundly as they went along in a book, and had little to disturb the current of their thoughts, it would have commanded wide attention among thinking men.

A singular peculiarity which I have sometimes noticed among men of intelligence is that those who are best informed on the subject may be most reckless as regards the laws of health. Wright did all of his office work in two or three months of the year. During those months he worked at his computations far into the hours of the morning, stimulating his strength with cigars, and dropping his work only to take it up when he had had the necessary sleep. A strong constitution might stand this for a few years, as his did. But the ultimate result hardly needs to be told.

Besides the volume I have mentioned, Wright's letters were collected and printed after his death by the subscription of his friends. In these his philosophic views are from time to time brought out in a light, easy way, much more charming than the style of his elaborate discussions. It was in one of his letters that I first found the apothegm, "Men are born either Platonists or Aristotelians," a happy drawing of the line which separates the

[1] Henry Holt & Co.: New York, 1877.

hard-headed scientific thinker of to-day from the thinkers of all other classes.

William Ferrell, a much older man than myself, entered the office about the same time as I did. He published papers on the motions of fluids on the earth's surface in the "Mathematical Monthly," and became one of the great authorities on dynamic meteorology, including the mathematical theory of winds and tides. He was, I believe, the first to publish a correct theory of the retardation produced in the rotation of the earth by the action of the tides, and the consequent slow lengthening of the day.

James Edward Oliver might have been one of the great mathematicians of his time had he not been absolutely wanting in the power of continuous work. It was scarcely possible to get even his year's office work out of him. Yet when I once wrote him a question on certain mathematical forms which arise in the theory of "least squares," he replied in a letter which, with some developments and change of form, would have made a worthy memoir in any mathematical journal. As a matter of fact, the same thoughts did appear some years after, in an elaborate paper by Professor J. W. L. Glaisher, of England, published by the Royal Astronomical Society.

Oliver, who afterward became professor of higher mathematics at Cornell University, was noted for what I think should be considered the valuable quality of absent-mindedness. It was said of him that he was once walking on the seashore with a small but valuable gold watch loose in his pocket. While deep in thought he started a kind of distraction by picking up flat stones and skipping them on the water. Taking his watch from his pocket he skipped it as a stone. When I became well acquainted with him I took the liberty of asking him as to the correctness of this story. He could not positively say whether it was true or not. The facts were simply that he had the watch, that he had walked on the seashore, had skipped stones, missed the watch at some subsequent time, and never saw it again.

More definite was an observation made on his movements one afternoon by a looker-out from a window of the Nautical Almanac Office. Across the way the road was bounded by no fence, simply passing along the side of an open field. As Oliver got near the office, his chin on his breast, deep in thought, he was seen gradually to deviate from the sidewalk, and direct his steps along the field. He continued on this erratic course until he ran almost against the fence at the other end. This awoke him from his reverie, and he started up, looked around, and made his way back to the road.

I have spoken only of the men who were employed at the office at the time I entered. Previous to my time were several who left to accept professorships in various parts of the country. Among them were Professors Van Vleck, of Middletown, and Hedrick and Kerr, of North Carolina. Not desiring to leave upon the mind of the reader the impression that all of whom I have not spoken remained in obscurity, I will remark that Mr. Isaac Bradford rose to the position of mayor of the city of Cambridge, and that fugitive pieces in prose and poetry by Mr. E. J. Loomis were collected in a volume.[1]

The discipline of the public service was less rigid in the office at that time than at any government institution I ever heard of. In theory there was an understanding that each assistant was "expected" to be in the office five hours a day. The hours might be selected by himself, and they generally extended from nine until two, the latter being at that time the college and family dinner hour. As a matter of fact, however, the work was done pretty much where and when the assistant chose, all that was really necessary being to have it done on time.

It will be seen that the excellent opportunities offered by this system were well improved by those who enjoyed them — improved in a way that I fear would not be possible in any other surroundings. I took advantage of them by enrolling myself as a student of mathematics in the Lawrence Scientific School. On this occasion I well remember my pleasant reception by Charles W. Eliot, tutor in mathematics, and E. N. Horsford, professor of chemistry, and, I believe, dean of the school. As a newcomer into the world of light, it was pleasant to feel the spirit with which they welcomed

[1] *Wayside Sketches*, by E. J. Loomis. Roberts: Boston.

me. The departments of chemistry and engineering were about the only ones which, at that time, had any distinct organization. As a student of mathematics it could hardly be said that anything was required of me either in the way of attendance on lectures or examinations until I came up for the degree of Bachelor of Science. I was supposed, however, to pursue my studies under the direction of Professor Peirce.

So slight a connection with the university does not warrant me in assuming an authoritative position as an observer of its men or its workings. Yet there are many features associated with it which I have not seen in print, which have probably disappeared with the progress of the age, and to which, therefore, allusion may be made. One, as it presents itself to my memory, is the great variety and picturesqueness of character which the university then presented. I would like to know whether the changes in men which one fancies he sees during his passage from youth to age are real, or only relative to his point of view. If my impressions are correct, our educational planing mill cuts down all the knots of genius, and reduces the best of the men who go through it to much the same standard. Does not the Harvard professor of to-day always dine in a dress coat? Is he not free from every eccentricity? Do the students ever call him "Benny" or "Tobie"? Is any "Old Soph"[1] now

---

[1] Evangelinus Apostolides Sophocles, a native Greek and a learned professor of the literature of his country.

ambulant on the college green? Is not the administration of the library a combination of liberality and correctness? Is such a librarian as John Langdon Sibley possible?

Mr. Sibley, under a rough exterior, was one of the best-hearted and most admirable of men, with whom I ultimately formed an intimate friendship. But our first acquaintance was of a very unfavorable kind. It came about in this way: not many days after being taken into the Nautical Almanac Office I wanted a book from the university library, and asked a not over-bright old gentleman in the office what formalities were necessary in order to borrow it.

"Just go over and tell them you want it for the Nautical Almanac."

"But they don't know me at the library, and surely will not give a book to any stray caller because he says he wants it for the Nautical Almanac."

"You have only to say 'Nautical Almanac' and you will get the book."

I argued the matter as stoutly as courtesy admitted, but at length, concluding that I was new to the rules and regulations of the place, accepted the supposedly superior knowledge of my informer and went over to the library with a due measure of assurance. The first attendant whom I addressed referred me to the assistant librarian, and he again to the librarian. After these formalities, conducted with impressive gravity, my assurance wilted when

I was ushered into the august presence of the chief librarian.

As the mental picture of the ensuing scene has shaped itself through more than forty years it shows a personage of imposing presence, gigantic features, and forbidding countenance, standing on a dais behind a desk, expounding the law governing the borrowing of books from the library of Harvard College to an abashed youth standing before him. I left without the book, but with a valuable addition to my knowledge of library management. We both remembered this interview, and exchanged impressions about it long years after.

" I thought you the most crusty and disobliging old man I had ever seen."

" And I thought *you* the most presumptuous youth that had ever appeared in the library."

One of Mr. Sibley's professional doctrines was that at least one copy of everything printed was worth preserving. I strove to refute him, but long failed. Half in derision, I offered the library the stub of my wash-book. Instead of throwing it into the wastebasket he kept it, with the remark that the wash-book of a nineteenth century student would at some future time be of interest to the antiquarian. In due time I received a finely engraved acknowledgment of the gift. But I forced him from his position at last. He had to admit that copies of the theatre posters need not all be preserved. It would suffice to keep a few specimens.

Professor Peirce was much more than a mathematician. Like many men of the time, he was a warm lover and a cordial hater. It could not always be guessed which side of a disputed question he would take; but one might be fairly sure that he would be at one extreme or the other. As a speaker and lecturer he was very pleasing, neither impressive nor eloquent, and yet interesting from his earnestness and vivacity. For this reason it is said that he was once chosen to enforce the views of the university professors at a town meeting, where some subject of interest to them was coming up for discussion. Several of the professors attended the meeting, and Peirce made his speech. Then a townsman rose and took the opposite side, expressing the hope that the meeting would not allow itself to be dictated to by these nabobs of Harvard College. When he sat down, Peirce remained in placid silence, making no reply. When the meeting broke up, some one asked Peirce why he had not replied to the man.

"Why! did you not hear what he called us? He said we were nabobs! I so enjoyed sitting up there and seeing all that crowd look up to me as a nabob that I could not say one word against the fellow."

The first of the leading astronomers whose acquaintance I made was Dr. Benjamin Apthorp Gould. Knowing his eminence, I was quite surprised by his youthful vivacity. His history, had I time to recount it, might be made to serve well the purpose of a grave lesson upon the conditions required, even by the educated public, of a scientific investigator, capable of doing the highest and best work in his branch. The soul of generosity and the pink of honor, ever ready to lend a hand to a struggling youth whom he found deserving of help, enthusiastically devoted to his favorite science, pursuing it in the most exalted spirit, animated by not a single mean motive, it might have been supposed that all the facilities the world could offer would have been open to him in his career. If such was not the case to the extent one might have wished, I do not mean to intimate that his life can be regarded as a failure. In whatever respect the results may have fallen off from his high ideal, it is more to be regretted on the score of science than on his own.

Scorning pretense and charlatanry of all kinds, believing that only the best were to be encouraged, he was far from being a man of the people. Only a select few enjoyed his favor, but these few well deserved it. That no others would have deserved it I should be far from intimating. The undisguised way in which he expressed his sentiments for any one, no matter how influential, who did not come up to the high standard he set, was not adapted to secure the favor even of the most educated community. Of worldly wisdom in this matter he seemed, at least in his early days, to know nothing.

He graduated at Harvard in 1845, in one of the very distinguished classes. Being fond of astronomy, he was struck with the backward condition of that science in our country. He resolved to devote his life to building up the science in America. He went to Germany, then the only country in which astronomy was pursued in its most advanced form, studied under Gauss and Argelander, and took his degree at Göttingen in 1848. Soon after his return he founded the "Astronomical Journal," and also took a position as Chief of the Longitude Department in the Coast Survey.

The great misfortune of his life, and temporarily at least, a severe blow to American astronomy, were associated with his directorship of the Dudley Observatory at Albany. This institution was founded by the munificence of a wealthy widow of Albany. The men to whom she intrusted the administration of her gift were among the most prominent and highly respected citizens of the place. The trustees went wisely to work. They began by forming an advisory scientific council, consisting of Bache, Henry, and Peirce. Under the direction of this council the observatory was built and equipped with instruments. When ready for active work in 1857, Gould moved thither and took personal charge. Very soon rumors of dissension were heard. The affair gradually grew into a contest between the director and the trustees, exceeding in bitterness any I have ever known in the world of learning or even of politics. It doubtless had its origin in very small beginnings. The policy of the director recognized no end but scientific efficiency. The trustees, as the responsible administrators of the trust, felt that they had certain rights in the matter, especially that of introducing visitors to inspect the institution and look through the telescope. How fatal the granting of such courtesies is to continuous work with an instrument only astronomers know; and one of the most embarrassing difficulties the director of such an institution meets with is to effect a prudent compromise between the scientific efficiency of his institution and the wishes of the public. But Gould knew no such word as compromise. It was humiliating to one in the position of a trustee to send some visitor with a permit to see the observatory, and have the visitor return with the report that he had not been received with the most distinguished courtesy, and, perhaps, had not seen the director at all, but had only been informed by an assistant of the rules of the place and the impossibility of securing admission.

This spark was enough to kindle a fire. When the fire gathered strength, the director, instead of yielding, called on the scientific council for aid. It is quite likely that, had these wise and prudent men been consulted at each step, and their advice been followed, he would have emphasized his protest by resigning. But before they were called in, the affair had gone so far that, believing the director to be technically right in the ground he had taken and the work he had done, the council felt bound to defend him. The result was a war in which the shots were pamphlets containing charges, defenses, and rejoinders. The animosity excited may be shown by the fact that the attacks were not confined to Gould and his administration, but extended to every institution with which he and the president of the council were supposed to be connected. Bache's administration of the Coast Survey was held up to scorn and ridicule. It was supposed that Gould, as a Cambridge astronomer, was, as a matter of course, connected with the Nautical Almanac Office, and paid a high salary. This being assumed, the office was included in the scope of attack, and with such success that the item for its support for the year 1859, on motion of Mr. Dawes, was stricken out of the naval bill. How far the fire spread may be judged by the fact that a whole edition of the "Astronomical Journal," supposed to have some mention of the affair in the same cover, was duly sent off from the observatory, but never reached its destination through the mails. Gould knew nothing of this fact until, some weeks later, I expressed my surprise to him at not receiving No. 121. How or by whom it was intercepted, I do not know that he ever seriously attempted to inquire. The outcome of the matter was that the trustees asserted their right by taking forcible possession of the observatory.

During my first year at Cambridge I made the acquaintance of a senior in the college whose untimely death seven years later I have never ceased to deplore. This was William P. G. Bartlett, son of a highly esteemed Boston physician, Dr. George Bartlett. The latter was a brother of Sidney Bartlett, long the leader of the Boston bar. Bartlett was my junior in years, but his nature and the surrounding circumstances were such that he exercised a powerful influence upon me. His virile and aggressive honesty could not be exceeded. His mathematical abilities were of a high order, and he had no ambition except to become a mathematician. Had he entered public life at Washington, and any one had told me that he was guilty of a dishonest act, I should have replied, " You might as well tell me that he picked up the Capitol last night and carried it off on his back." The fact that one could say so much of any man, I have always looked upon as illustrating one of the greatest advantages of having a youth go through college. The really important results I should look for are not culture or training alone, but include the acquaintance of a body of men, many of whom are to take leading positions in the world, of a completeness and intimacy that can never be acquired under other circumstances. The student sees his fellow students through and through as he can never see through a man in future years.

It was, and I suppose still is, the custom for the members of a graduating class at Harvard to add to their class biographies a motto expressing their aspirations or views of life. Bartlett's was, " I love mathematics and hate humbug." What the latter clause would have led to in his case, had he gone out into the world, one can hardly guess.

"I have had a long talk with my Uncle Sidney," he said to me one day. "He wants me to study law, maintaining that the wealth one can thereby acquire, and the prominence he may assume, will give him a higher position in society and public esteem than mere learning ever can. But I told him that if I could stand high in the esteem of twenty such men as Cayley, Sylvester, and Peirce, I cared nothing to be prominent in the eyes of the rest of the world." Such an expression from an eminent member of the Boston bar, himself a Harvard graduate, was the first striking evidence I met with that my views of the exalted nature of astronomical investigation were not shared by society at large. One of the greatest advantages I enjoyed through Bartlett was an intimate acquaintance with a cultured and refined Boston family.

In 1858 Mr. Runkle founded the "Mathematical Monthly," having secured, in advance, the coöperation of the leading professors of the subject in the country. The journal was continued, under many difficulties, for three years. As a vehicle for publishing researches in advanced mathematics, it could not be of a high order, owing to the necessity of a subscription list. Its design was therefore to interest students and professors in the subject, and thus prepare the way for the future growth of mathematical study among us. Its principal feature was the offer of prize problems to students as well as prizes for essays on mathematical subjects. The first to win a prize for an essay was George W. Hill, a graduate of Rutgers just out of college, who presented a memoir in which the hand of the future master was evident throughout.

In the general conduct of the journal Bartlett and myself, though not ostensibly associate editors, were at least assistants. Simple though the affair was, some of our experiences were of an interesting and, perhaps, instructive nature.

Soon after the first number appeared, a contribution was offered by a professor in a distant State. An important part of the article was found to be copied bodily from Walton's "Problems in Mechanics," an English book which, it might be supposed, was not much known in this country. Runkle did not want to run the risk of injuring his subscription list by offending one occupying an influential position if he could help it with honor to the journal. Of course it was not a question of publishing the paper, but only of letting the author know why he did not do so, — "letting him down easy."

Bartlett's advice was characteristic. "Just write to the fellow that we don't publish stolen articles. That's all you need say."

I suggested that we might inflict on him all necessary humiliation by letting him know in the gentlest manner possible that we saw the fraud. Of course Runkle preferred this course, and wrote him, calling his attention to a similarity between his treatment of the subject and that of Walton, which materially detracted from the novelty of the former. I think it was suggested that he get the book, if possible, and assure himself on the subject.

A vigorous answer came by return of mail. He was a possessor of Walton's book, knew all about the similar treatment of the subject by Walton, and did not see that that should be any bar to the publication of the article. I think it was he who wound up his letter with the statement that, while he admitted the right of the editor to publish what he pleased, he, the writer, was too busy to spend his time in writing rejected articles.

An eminent would-be contributor was a prominent Pennsylvania politician, who had read a long and elaborate article, before some teachers' association, on an arithmetical problem about oxen eating grass, the power to solve which was taken as the highest mark of mathematical ability, among school teachers during the first half of the century. The association referred the paper to the editor of the "Mathematical Monthly," by whom it was, I believe, consigned to the wastebasket. The result was a good deal of correspondence, such a proceeding being rather humiliating to a man of eminence who had addressed so distinguished an assembly. The outcome of the matter was that the paper, which was much more in the nature of a legal document than of a mathematical investigation, was greatly reduced in length by its author, and then still further shorn by the editor, until it would fill only two or three pages of the journal; thus reduced, it was published.

The time was not yet ripe for the growth of mathematical science among us, and any development that might have taken place in that direction was rudely stopped by the civil war. Perhaps this may account for the curious fact that, so far as I have ever remarked, none of the student contributors to the journal, Hill excepted, has made himself known as a mathematical investigator. Not only the state of mathematical learning, but the conditions of success at that time in a mathematical text-book, are strikingly illustrated by one of our experiences.

One of the leading publishing houses of educational text-books in the country issued a very complete and advanced series, from the pen of a former teacher of the subject. They were being extensively introduced, and were sent to the " Mathematical Monthly " for review. They were distinguished by quite apt illustrations, well fitted, perhaps, to start the poorly equipped student in the lower branches of the work, but the advanced works, at least, were simply ridiculous. A notice appeared in which the character of the books was pointed out. The evidence of the worthlessness of the entire series was so strong that the publishers had it entirely rewritten by more competent authors. Now came the oddest part of the whole affair. The new series was issued under the name of the same author as the old one, just as if the acknowledgment of his total failure did not detract from the value of his name as an author.

In 1860 a total eclipse of the sun was visible in British America. The shadow of the moon, starting from near Vancouver's Island, crossed the continent in a northeast direction, passed through the central part of the Hudson Bay region, crossed Hudson Bay itself and Greenland, then inclining southward, swept over the Atlantic to Spain. As this was the first eclipse of the kind which had recently been visible, much interest was taken in its observation. On the part of the Nautical Almanac Office, I computed the path of the shadow and the times of crossing certain points in it. The results were laid down on a map which was published by the office. One party, fitted out in connection with the American Association for the Advancement of Science, was sent to Greenland. Admiral Davis desired to send another, on behalf of his own office, into the central regions of the continent. As members of this party Mr. Ferrel and myself were chosen. At the request of Professor Agassiz one of the assistants in the Museum of Comparative Zoölogy, Mr. Samuel H. Scudder, accompanied us. More than twenty years later Mr. Scudder published a little book describing some of our adventures, which was illustrated with sketches showing the experiences of a party in the wild West at that time.

Our course lay from St. Paul across Minnesota to the Red River of the North, thence north to Fort Garry near the southern end of Lake Winnipeg, then over the lake and some distance up the Saskatchewan River. At St. Paul we paid our respects to Governor Ramsey, afterward Senator from Minnesota and Secretary of War. We were much surprised at the extraordinary deference paid by the community to a Mr. Burbank, a leading citizen of the town, and owner of the stages which we had to engage for our journey across the country. He seemed to be a man whom every one was afraid to offend. Even the local newspapers were careful what they printed about matters in which he was interested.

The two or three days which we passed in getting things ready to start were rather dull. The morning after our arrival I saw, during a morning walk, on a hill just outside the town, a large new building, on which the word "Athenæum" was conspicuously shown. The Boston Athenæum had a very fine library; is it not possible that this may have a beginning of something of the same sort? Animated by this hope, I went up the hill and entered the building, which seemed to be entirely vacant. The first words that met my eyes were "Bar Room" painted over a door. It was simply a theatre, and I left it much disappointed.

Here we were joined by a young Methodist clergyman, — Edward Eggleston, — and the four of us, with our instruments and appliances, set out on our journey of five days over the plains. On the first day we followed partly the line of a projected railway, of which the embankments had been completed, but on which work had, for some reason, been stopped to await a more prosperous season. Here was our first experience of towns on paper. From the tone in which the drivers talked of the places where we were to stop over night one might have supposed that villages, if not cities, were plentiful along our track. One example of a town at that time will be enough. The principal place on our route, judging from the talk, was Breckenridge. We would reach it at the end of the fourth day, where we anticipated a pleasant change after camping out in our tent for three nights. It was after dark before we arrived, and we looked eagerly for signs of the town we were approaching.

The team at length stopped in front of an object which, on careful examination in the darkness, appeared to be the most primitive structure imaginable. It had no foundations, and if it had a wall at all, it was not more than two or three feet in height. Imagine the roof taken off a house forty feet long and twenty feet wide and laid down on the ground, and you have the hotel and only building, unless perhaps a stable, in Breckenridge at that time. The entrance was at one end. Going in, a chimney was seen in the middle of the building. The floor was little more than the bare ground. On each side of the door, by the flickering light of a fire, we saw what looked like two immense boxes. A second glance showed that these boxes seemed to be filled with human heads and legs. They were, in fact, the beds of the inhabitants of Breckenridge. Beds for the arriving travelers, if they existed at all, which I do not distinctly remember, were in the back of the house. I think the other members of the party occupied that portion. I simply spread my blanket out on the hearth in front of the fire, wrapped up, and slept as soundly as if the bed was the softest of a regal palace.

At Fort Garry we were received by Governor McTavish, with whom Captain Davis had had some correspondence on the subject of our expedition, and who gave us letters to the "factors" of the Hudson Bay Company scattered along our route. We found that the rest of our journey would have to be made in a birch bark canoe. One of the finest craft of this class was loaned us by the governor. It had been, at some former time, the special yacht of himself or some visiting notable. It was manned by eight half-breeds, men whose physical endurance I have never seen equaled.

It took three or four days to get everything ready, and this interval was, of course, utilized by Scudder in making his collections. He let the fishermen of the region know that he wanted specimens of every kind of fish that could be found in the lake. A very small reward stirred them into activity, and, in due time, the fish were brought to the naturalist, — but lo! all nicely dressed and fit for cooking. They were much surprised when told that all their pains in dressing their catch had spoiled it for the purposes of the visiting naturalist, who wanted everything just as it was taken from the water.

Slow indeed was progress through the lake. A canoe can be paddled only in almost smooth water, and we were frequently stormbound on some desolate island or point of land for two or three days at a time. When, after many adventures, some of which looked like hairbreadth escapes, we reached the Saskatchewan River, the eclipse was only three or four days ahead, and it became doubtful whether we should reach our station in time for the observation. It was to come off on the morning of July 18, and, by dint of paddling for twenty-four hours at a stretch, our men brought us to the place on the evening before.

Now a new difficulty occurred. In the wet season the Saskatchewan inundates the low flat region through which it flows, much like the Nile. The country was practically under water. We found the most elevated spot we could, took out our instruments, mounted them on boxes or anything else in the shallow puddles of water, and slept in the canoe. Next morning the weather was hopelessly cloudy. We saw the darkness of the eclipse and nothing more.

Astronomers are greatly disappointed when, having traveled halfway around the world to see an eclipse, clouds prevent a sight of it; and yet a sense of relief accompanies the disappointment. You are not responsible for the mishap; perhaps something would have broken down when you were making your observations, so that they would have failed in the best of weather; but now you are relieved from all responsibility. It was much easier to go back and tell of the clouds than it would have been to say that the telescope got disarranged at the critical moment so that the observations failed.

On our return across Minnesota we had an experience which I have always remembered as illustrative of the fallacy of all human testimony about ghosts, rappings, and other phenomena of that character. We spent two nights and a day at Fort Snelling. Some of the officers were greatly surprised by a celestial phenomenon of a very extraordinary character which had been observed for several nights past. A star had been seen, night after night, rising in the east as usual, and starting on its course toward the south. But instead of continuing that course across the meridian, as stars invariably had done from the remotest antiquity, it took a turn toward the north, sunk toward the horizon, and finally set near the north point of the horizon. Of course an explanation was wanted.

My assurance that there must be some mistake in the observation could not be accepted, because this erratic course of the heavenly body had been seen by all of them so plainly that no doubt could exist on the subject. The men who saw it were not of the ordinary untrained kind, but graduates of West Point, who, if any one, ought to be free from optical deceptions. I was confidently invited to look out that night and see for myself. We all watched with the greatest interest.

In due time the planet Mars was seen in the east making its way toward the south. "There it is!" was the exclamation.

"Yes, there it is," said I. "Now that planet is going to keep right on its course toward the south."

"No, it is not," said they; "you will see it turn around and go down towards the north."

Hour after hour passed, and as the planet went on its regular course, the other watchers began to get a little nervous. It showed no signs of deviating from its course. We went out from time to time to look at the sky.

"There it is," said one of the observers at length, pointing to Capella, which was now just rising a little to the east of north; "there is the star setting."

"No, it is n't," said I; "there is the star we have been looking at, now quite inconspicuous near the meridian, and that star which you think is setting is really rising and will soon be higher up."

A very little additional watching showed that no deviation of the general laws of Nature had occurred, but that the observers of previous nights had jumped at the conclusion that two objects, widely apart in the heavens, were the same.

I passed more than four years in such life, surroundings, and activities as I have described. In 1858 I received the degree of D. S. from the Lawrence Scientific School, and thereafter remained on the rolls of the university as a resident graduate. Life in the new atmosphere was in such pleasant and striking contrast to that of my former world that I intensely enjoyed it. I had no very well marked object in view beyond continuing studies and researches in mathematical astronomy. Not long after my arrival in Cambridge some one, in speaking of Professor Peirce, remarked to me that he had a European reputation as a mathematician. It seemed to me that this was one of the most exalted positions that a man could attain, and I intensely longed for it. Yet there was no hurry. Reputation would come to him who deserved it by his works; works of the first class were the result of careful thought and study, and not of hurry. A suggestion had been made to me looking toward a professorship in some Western college, but after due consideration, I declined to consider the matter. Yet the necessity of being on the alert for some opening must have seemed quite strong, because in 1860 I became a serious candidate for the professorship of physics in the newly founded Washington University at St. Louis. I was invited to visit the university, and did so on my way to observe the eclipse of 1860. My competitor was

Lieutenant J. M. Schofield of the United States Army, then an instructor at West Point. It will not surprise the reader to know that the man who was afterward to command the army of the United States received the preference, so I patiently waited more than another year.

IV

LIFE AND WORK AT AN OBSERVATORY

IN August, 1861, while I was passing my vacation on Cape Ann, I received a letter from Dr. Gould, then in Washington, informing me that a vacancy was to be filled in the corps of professors of mathematics attached to the Naval Observatory, and suggesting that I might like the place. I was at first indisposed to consider the proposition. Cambridge was to me the focus of the science and learning of our country. I feared that, so far as the world of learning was concerned, I should be burying myself by moving to Washington. The drudgery of night work at the observatory would also interfere with carrying on any regular investigation. But, on second thought, having nothing in view at the time, and the position being one from which I could escape should it prove uncongenial, I decided to try, and indited the following letter: —

NAUTICAL ALMANAC OFFICE,
CAMBRIDGE, MASS., August 22, 1861.

SIR, — I have the honor to apply to you for my appointment to the office of Professor of Mathematics in the United States Navy.

I would respectfully refer you to Commander Charles

Henry Davis, U. S. N., Professor Benjamin Peirce, of Harvard University, Dr. Benjamin A. Gould, of Cambridge, and Professor Joseph Henry, Secretary of the Smithsonian Institution, for any information respecting me which will enable you to judge of the propriety of my appointment.

With high respect,
Your obedient servant,
SIMON NEWCOMB,
Assistant, Nautical Almanac.

HON. GIDEON WELLES,
    Secretary of the Navy,
        Washington, D. C.

I also wrote to Captain Davis, who was then on duty in the Navy Department, telling him what I had done, but made no further effort. Great was my surprise when, a month later, I found in the post-office, without the slightest premonition, a very large official envelope, containing my commission duly signed by Abraham Lincoln, President of the United States. The confidence in the valor, abilities, etc., of the appointee, expressed in the commission, was very assuring. Accompanying it was a letter from the Secretary of the Navy directing me to report to the Bureau of Ordnance and Hydrography, in Washington, for such duty as it might assign me. I arrived on October 6, and immediately called on Professor J. S. Hubbard, who was the leading astronomer of the observatory. On the day following I reported as directed, and was sent to Captain Gilliss, the recently appointed Superintendent of the Naval Observatory, before whom I stood with much trepidation. In reply to his questions I had to confess my entire inexperience in observatory work or the making of astronomical observations. A coast survey observer had once let me look through his transit instrument and try to observe the passage of a star. On the eclipse expedition mentioned in the last chapter I had used a sextant. This was about all the experience in practical astronomy which I could claim. In fact I had never been inside of an observatory, except on two or three occasions at Cambridge as a visitor. The captain reassured me by saying that no great experience was expected of a newcomer, and told me that I should go to work on the transit instrument under Professor Yarnall, to whose care I was then confided.

As the existence of a corps of professors of mathematics is peculiar to our navy, as well as an apparent, perhaps a real, anomaly, some account of it may be of interest. Early in the century — one hardly knows when the practice began — the Secretary of the Navy, in virtue of his general powers, used to appoint men as professors of mathematics in the navy, to go to sea and teach the midshipmen the art of navigation. In 1844, when work at the observatory was about to begin, no provision for astronomers was made by Congress. The most convenient way of supplying this want was to have the Secretary appoint professors of mathematics, and send them to the observatory on duty.

A few years later the Naval Academy was

founded at Annapolis, and a similar course was pursued to provide it with a corps of instructors. Up to this time the professors had no form of appointment except a warrant from the Secretary of the Navy. Early in the history of the academy the midshipmen burned a professor in effigy. They were brought before a court-martial on the charge of disrespect to a superior officer, but pleaded that the professor, not holding a commission, was not their superior officer, and on this plea were acquitted. Congress thereupon took the matter up, provided that the number of professors should not exceed twelve, and that they should be commissioned by the President by and with the advice and consent of the Senate. This raised their rank to that of a commissioned corps in the navy. They were to perform such duty as the Secretary of the Navy might direct, and were, for the most part, divided between the Naval Academy and the Observatory.

During the civil war some complaint was made that the midshipmen coming from the academy were not well trained in the duties of a seagoing officer; and it was supposed that this was due to too much of their time being given to scientific studies. This was attributed to the professors, with the result that nearly all those attached to the academy were detached during the four years following the close of the civil war and ordered elsewhere, mostly to the observatory. Their places were taken by line officers who, in the intervals between their turns of sea duty, were made heads of departments and teachers of the midshipmen in nearly every branch.

This state of things led to the enactment of a law (in 1869, I think), " that hereafter no vacancy in the grade of professors of mathematics in the navy shall be filled."

In 1873 this provision was annulled by a law, again providing for a corps of twelve professors, three of whom should have the relative rank of captain, four of commander, and the remainder of lieutenant-commander or lieutenant.

Up to 1878 the Secretary of the Navy was placed under no restrictions as to his choice of a professor. He could appoint any citizen whom he supposed to possess the necessary qualifications. Then it was enacted that, before appointment, a candidate should pass a medical and a professional examination.

I have said that the main cause of hesitation in making my application arose from my aversion to very late night work. It soon became evident that there was less ground than I had supposed for apprehension on this point. There was a free and easy way of carrying on work which was surprising to one who had supposed it all arranged on strict plans, and done according to rule and discipline. Professor Yarnall, whose assistant I was, was an extremely pleasant gentleman to be associated with. Although one of the most industrious workers at the observatory, there was nothing of the martinet about him. He showed me how to handle the instrument and record my observations. There was a Nautical Almanac and a Catalogue of Stars. Out of these each of us could select what he thought best to observe.

The custom was that one of us should come on every clear evening, make observations as long as he chose, and then go home. The transit instrument was at one end of the building and the mural circle, in charge of Professor Hubbard, at the other. He was weak in health, and unable to do much continuous work of any kind, especially the hard work of observing. He and I arranged to observe on the same nights; but I soon found that there was no concerted plan between the two sets of observers. The instruments were old-fashioned ones, of which mine could determine only the right ascension of a star and his only its declination; hence to completely determine the position of a celestial body, observations must be made on the same object with both instruments. But I soon found that there was no concert of action of this kind. Hubbard, on the mural circle, had his plan of work; Yarnall and myself, on the transit, had ours. When either Hubbard or myself got tired, we could "vote it cloudy" and go out for a plate of oysters at a neighboring restaurant.

In justice to Captain Gilliss it must be said that he was not in any way responsible for this lack of system. It grew out of the origin and history of the establishment and the inaction of Congress. The desirableness of our having a national observatory of the same rank as those of other countries was pointed out from time to time by eminent statesmen from the first quarter of the century. John Quincy Adams had, both while he filled the presidential office and afterward, made active efforts in this direction; but there were grave doubts whether Congress had any constitutional authority to erect such an institution, and the project got mixed up with parties and politics. So strong was the feeling on the subject that, when the Coast Survey was organized, it was expressly provided that it should not establish an astronomical observatory.

The outcome of the matter was that, in 1842, when Congress at length decided that we should have our national observatory, it was not called such, but was designated as a " house " to serve as a depot for charts and instruments for the navy. But every one knew that an observatory was meant. Gilliss was charged with its erection, and paid a visit to Europe to consult with astronomers there on its design, and to order the necessary instruments. When he got through with this work and reported it as completed he was relieved, and Lieutenant Matthew F. Maury was appointed superintendent of the new institution.

Maury, although (as he wrote a few years later) quite without experience in the use of astronomical instruments, went at his work with great energy and efficiency, so that, for two or three years, the institution bade fair to take a high place in science. Then he branched off into what was, from a practical standpoint, the vastly more important work of studying the winds and currents of the ocean. The epoch-making character of his investigations in this line, and their importance to navigation when ships depended on sails for their motive power, were soon acknowledged by all maritime nations, and the fame which he acquired in pursuing them added greatly to the standing of the institution at which the work was done, though in reality an astronomical outfit was in no way necessary to it. The new work was so absorbing that he seemed to have lost interest in the astronomical side of the establishment, which he left to his assistants. The results were that on this side things fell into the condition I have described, and stayed there until Maury resigned his commission and cast his fortunes with the Confederacy. Then Gilliss took charge and had to see what could be done under the circumstances.

It soon became evident to him that no system of work of the first order of importance could be initiated until the instrumental equipment was greatly improved. The clocks, perfection in which is almost at the bottom of good work, were quite unfit for use. The astronomical clock with which Yarnall and I made our observations kept worse time than a high-class pocket watch does to-day. The instruments were antiquated and defective in several particulars. Before real work could be commenced new ones must be procured. But the civil war was in progress, and the times were not favorable to immediately securing them. That the work of the observatory was kept up was due to a feeling of pride on the part of our authorities in continuing it without interruption through the conflict. The personnel was as insufficient as the instruments. On it devolved not only the making of the astronomical observations, but the issue of charts and chronometers to the temporarily immense navy. In fact the observatory was still a depot of charts for the naval service, and continued to be such until the Hydrographic Office was established in 1866.

In 1863 Gilliss obtained authority to have the most pressing wants supplied by the construction of a great transit circle by Pistor and Martins in Berlin. He had a comprehensive plan of work with this instrument when it should arrive, but deferred putting any such plan in operation until its actual reception.

Somehow the work of editing, explaining, and preparing for the press the new series of observations made by Yarnall and myself with our old transit instrument devolved on me. To do this in the most satisfactory way, it was necessary to make a careful study of the methods and system at the leading observatories of other countries in the line we were pursuing, especially Greenwich. Here I was struck by the superiority of their system to ours. Everything was there done on an exact and uniform plan, and one which seemed to me better adapted to get the best results than ours was. For the non-astronomical reader it may be remarked that after an astronomer has made and recorded his observations, a large amount of calculation is necessary to obtain the result to which they lead. Making such calculations is called "reducing" the observations. Now in the previous history of the observatory, the astronomers fell into the habit of every one not only making his observations in his own way, but reducing them for himself. Thus it happened that Yarnall had been making and reducing his observations in his own way, and I, on alternate nights, had been making and reducing mine in my way, which was modeled after the Greenwich fashion, and therefore quite different from his. Now I suddenly found myself face to face with the problem of putting these two heterogeneous things together so as to make them look like a homogeneous whole. I was extremely mortified to see how poor a showing would be made in the eyes of foreign astronomers. But I could do nothing more than to describe the work and methods in such a way as to keep in the background the want of system that characterized them.

Notwithstanding all these drawbacks of the present, the prospect of future success seemed brilliant. Gilliss had the unlimited confidence of the Secretary of the Navy, had a family very popular in Washington society, was enthusiastically devoted to building up the work of the observatory, and was drawing around him the best young men that could be found to do that work. He made it a point that his relations with his scientific subordinates should be not only official, but of the most friendly social character. All were constantly invited to his charming family circle. It was from the occasional talks thus arising that I learned the details of his plan of work with the coming instrument.

In 1862 Gilliss had the working force increased by the appointment of four " aides," as they were then called, — a number that was afterwards reduced to three. This was the beginning of the corps of three assistant astronomers, which is still maintained. It will be of interest to know that the first aide was Asaph Hall; but before his appointment was made, an impediment, which for a time looked serious, had to be overcome. Gilliss desired that the aide should hold a good social and family position. The salary being only $1000, this required that he should not be married. Hall being married, with a growing family, his appointment was long objected to, and it was only through much persuasion on the part of Hubbard and myself that Gilliss was at length induced to withdraw his objections. Among other early appointees were William Harkness and John A. Eastman, whose subsequent careers in connection with the observatory are well known.

The death of Professor Hubbard in 1863 led to my taking his place, in charge of the mural circle, early in September of that year. This gave me an opportunity of attempting a little improvement in the arrangements. I soon became conscious of the fact, which no one had previously taken much account of, that upon the plan of each man reducing his own observations, not only was there an entire lack of homogeneity in the work, but the more work one did at night the more he had to do by day. It was with some trepidation that I presented the case to Gilliss, who speedily saw that work done with the instruments should be regarded as that of the observatory, and reduced on a uniform plan, instead of being considered as the property of the individual who happened to make it. Thus was introduced the first step toward a proper official system.

In February, 1865, the observatory sustained the greatest loss it had ever suffered, in the sudden death of its superintendent. What it would have grown to had he lived it is useless to guess, but there is little doubt that its history would have been quite different from what it is.

Soon afterward Admiral Davis left his position as Chief of the Bureau of Navigation to take the subordinate one of Superintendent of the Observatory. This step was very gratifying to me. Davis had not only a great interest in scientific work, especially astronomy, but a genuine admiration of scientific men which I have never seen exceeded, accompanied with a corresponding love of association with them and their work.

In October, 1865, occurred what was, in my eyes, the greatest event in the history of the observatory.

LIFE AND WORK AT AN OBSERVATORY

The new transit circle arrived from Berlin in its boxes. Now for the first time in its history, the observatory would have a meridian instrument worthy of it, and would, it was hoped, be able to do the finest work in at least one branch of astronomy. To my great delight, Davis placed me in charge of it. The last three months of the year were taken up with mounting it in position and making those investigations of its peculiarities which are necessary before an instrument of the kind is put into regular use. On the 1st day of January, 1866, this was all done, and we were ready to begin operations. An opportunity thus arose of seeing what we could do in the way of a regular and well-planned piece of work. In the greater clearness of our sky, and the more southern latitude of our observatory, we had two great advantages over Greenwich. Looking back at his first two or three years of work at the observatory, Maury wrote to a friend, " We have beaten Greenwich hollow." It may be that I felt like trying to do the same thing over again. At any rate, I mapped out a plan of work the execution of which would require four years.

It was a piece of what, in astronomy, is called " fundamental work," in which results are to be obtained independent of any previously obtained by other observers. It had become evident to me from our own observations, as well as from a study of those made at European observatories, that an error in the right ascension of stars, so that stars in opposite quarters of the heavens would not agree, might very possibly have crept into nearly all the modern observations at Greenwich, Paris, and Washington. The determination of this error was no easy matter. It was necessary that, whenever possible, observations should be continued through the greater part of the twenty-four hours. One observer must be at work with comparative steadiness from nine o'clock in the morning until midnight or even dawn of the morning following. This requirement was, however, less exacting than might appear when stated. One half the nights would, as a general rule, be cloudy, and an observer was not expected to work on Sunday. Hence no one of the four observers would probably have to do such a day's work as this more than thirty or forty times in a year.

All this was hard work enough in itself, but conditions existed which made it yet harder. No houses were then provided for astronomers, and the observatory itself was situated in one of the most unhealthy parts of the city. On two sides it was bounded by the Potomac, then pregnant with malaria, and on the other two, for nearly half a mile, was found little but frame buildings filled with quartermaster's stores, with here and there a few negro huts. Most of the observers lived a mile or more from the observatory; during most of the time I was two miles away. It was not considered safe to take even an hour's sleep at the observatory. The result was that, if it happened to clear off after a cloudy evening, I frequently arose from my bed at any hour of the night or morning and walked two miles to the observatory to make some observation included in the programme.

This was certainly a new departure from the free and easy way in which we had been proceeding, and it was one which might be unwelcome to any but a zealous astronomer. As I should get the lion's share of credit for its results, whether I wanted to or not, my interest in the work was natural. But it was unreasonable to expect my assistants, one or two of whom had been raised to the rank of professor, to feel the same interest, and it is very creditable to their zeal that we pursued it for some time as well as we did. If there was any serious dissatisfaction with the duty, I was not informed of that fact.

During the second year of this work Admiral Davis was detached and ordered to sea. The question of a successor interested many besides ourselves. Secretary Welles considered the question what policy should be pursued in the appointment. Professor Henry took part in the matter by writing the secretary a letter, in which he urged the appointment of an astronomer as head of the institution. His position prevented his supporting any particular candidate; so he submitted a list of four names, any one of which would be satisfactory. These were: Professor William Chauvenet, Dr. B. A. Gould, Professor J. H. C. Coffin, U. S. N., and Mr. James Ferguson. The latter held a civil position at the observatory, under the title of "assistant astronomer," and was at the time the longest in service of any of its force.

A different view was urged upon the secretary in terms substantially these: " Professors so able as those of the observatory require no one to direct their work. All that the observatory really needs is an administrative head who shall preserve order, look after its business generally, and see that everything goes smoothly." Such a head the navy can easily supply.

The secretary allowed it to be given out that he would be glad to hear from the professors upon the subject. I thereupon went to him and expressed my preference for Professor Coffin. He asked me, "How would it do to have a purely administrative head?"

I replied that we might get along for a time if he did not interfere with our work.

"No," said the secretary, "he shall not interfere. That shall be understood."

As I left him there was, to my inexperienced mind, something very odd in this function, or absence of function, of the head of an establishment; but of course I had to bow to superior wisdom and could say nothing.

The policy of Commodore (afterward Rear-Admiral) Sands, the incoming superintendent, toward the professors was liberal in the last degree. Each was to receive due credit for what he did, and was in every way stimulated to do his best at any piece of scientific work he might undertake with the approval of the superintendent. Whether he wanted to observe an eclipse, determine the longitude of a town or interior station, or undertake some abstruse investigation, every facility for doing it and every encouragement to go on with it was granted him.

Under this policy the observatory soon reached the zenith of its fame and popularity. Whenever a total eclipse of the sun was visible in an accessible region parties were sent out to observe it. In 1869 three professors, I being one, were sent to Des Moines, Iowa, to observe the solar eclipse which passed across the country in June of that year. As a part of this work, I prepared and the observatory issued a detailed set of instructions to observers in towns at each edge of the shadow-path to note the short duration of totality. The object was to determine the exact point to which the shadow extended. At this same eclipse Professor Harkness shared with Professor Young of Princeton the honor of discovering the brightest line in the spectrum of the sun's corona. The year following parties were sent to the Mediterranean to observe an eclipse which occurred in December, 1870. I went to Gibraltar, although the observation of the eclipse was to me only a minor object. Some incidents connected with this European trip will be described in a subsequent chapter.

The reports of the eclipse parties not only described the scientific observations in great detail, but also the travels and experiences, and were sometimes marked by a piquancy not common in official documents. These reports, others pertaining to longitude, and investigations of various kinds were published in full and distributed with great liberality. All this activity grew out of the stimulating power and careful attention to business of the head of the observatory and the ability of the young professors of his staff. It was very pleasant to the latter to wear the brilliant uniform of their rank, enjoy the protection of the Navy Department, and be looked upon, one and all, as able official astronomers. The voice of one of our scientific men who returned from a visit abroad declaring that one of our eclipse reports was the laughing-stock of Europe was drowned in the general applause.

In the latter part of 1869 I had carried forward the work with the transit circle as far as it could be profitably pursued under existing conditions. On working up my observations, the error which I had suspected in the adopted positions of the stars was proved to be real. But the discovery of this error was due more to the system of observation, especially the pursuit of the latter through the day and night, than it was to any excellence of the instrument. The latter proved to have serious defects which were exaggerated by the unstable character of the clayey soil of the hill on which the observatory was situated. Other defects also existed, which seemed to preclude the likelihood that the future work of the instrument would be of a high class. I had also found that very difficult mathematical investigations were urgently needed to unravel one of the greatest mysteries of astronomy, that of the moon's motion. This was a much more important work than making observations, and I wished to try my hand at it. So in the autumn I made a formal application to the Secretary of the Navy to be transferred from the observatory to the Nautical Almanac Office for the purpose of engaging in researches on the motion of the moon. On handing this application to the superintendent he suggested that the work in question might just as well be done at the observatory. I replied that I thought that the business of the observatory was to make and reduce astronomical observations with its instruments, and that the making of investigations of the kind I had in view had always been considered to belong to the Nautical Almanac Office. He replied that he deemed it equally appropriate for the observatory to undertake it. As my objection was founded altogether on a principle which he refused to accept, and as by doing the work at the observatory I should have ready access to its library, I consented to the arrangement he proposed. Accordingly, in forwarding my application, he asked that my order should be so worded as not to detach me from the observatory, but to add the duty I asked for to that which I was already performing.

So far as I was personally concerned, this change was fortunate rather than otherwise. As things go in Washington, the man who does his work in a fine public building can gain consideration for it much more readily than if he does it in a hired office like that which the "Nautical Almanac" then occupied. My continued presence on the observatory staff led to my taking part in two of the great movements of the next ten years, the construction and inauguration of the great telescope and the observations of the transit of Venus. But for the time being my connection with the regular work of the observatory ceased.

On the retirement of Admiral Sands in 1874, Admiral Davis returned to the observatory, and continued in charge until his death in February, 1877. The principal event of this second administration was the dispatch of parties to observe the transit of Venus. Of this I shall speak in full in a subsequent chapter.

One incident, although of no public importance, was of some interest at the time. This was a visit of the only emperor who, I believe, had ever set foot on our shores,— Dom Pedro of Brazil. He had chosen the occasion of our Centennial for a visit to this country, and excited great interest during his stay, not only by throwing off all imperial reserve during his travels, but by the curiosity and vigor with which he went from place to place examining and studying everything he could find, and by the singular extent of his knowledge on almost every subject of a scientific or technical character. A Philadelphia engineer with whom he talked was quoted as saying that his knowledge of engineering was not merely of the ordinary kind to be expected in an intelligent man, but extended to the minutest details and latest improvements in the building of bridges, which was the specialty of the engineer in question.

Almost as soon as he arrived in Washington I received the following letter by a messenger from the Arlington Hotel: —

MR.:

En arrivant à Washington j'ai tout-de-suite songé à votre observatoire, où vous avez acquis tant de droit à l'estime de tout ceux qui achèvent la science. Je m'y rendrai donc aujourd'hui à 7 heures du soir, et je compte vous y trouver, surtout pour vous remercier de votre beau mémoire que j'ai reçu peu avant mon départ de mon pays, et que je n'ai pas pu, par conséquent, apprécier autant que je l'aurais voulu. En me plaisant de l'espoir de vous connaître personnellement je vous prie de me compter parmi vos affectionnés.

D. PEDRO D'ALCANTARA.

7 Mai, 1876.

Like other notes which I subsequently received from him, it was in his own autograph throughout: if he brought any secretary with him on his travels I never heard of it.

The letter placed me in an embarrassing position, because its being addressed to me was in contravention of all official propriety. Of course I lost no time in calling on him and trying to explain the situation. I told him that Admiral Davis, whom he well knew from his being in command of the Brazilian station a few years before, was the head of the observatory, and hinted as plainly as I could that a notification of the coming of such a visitor as he should be sent to the head of the institution. But he refused to take the hint, and indicated that he expected me to arrange the whole matter for him. This I did by going to the observatory and frankly explaining the matter to Admiral Davis. Happily the latter was not a stickler for official forms, and was cast in too large a mould to take offense where none was intended. At his invitation I acted as one of the receiving party. The carriage drove up at the appointed hour, and its occupant was welcomed by the admiral at the door with courtly dignity. The visitor had no time to spend in preliminaries; he wished to look through the establishment immediately.

The first object to meet his view was a large marble-cased clock which, thirty years before, had acquired some celebrity from being supposed to embody the first attempt to apply electricity to the recording of astronomical observations. It was said to have cost a large sum, paid partly as a reward to its inventor. Its only drawbacks were that it would not keep time and had never, so far as I am aware, served any purpose but that of an ornament. The first surprise came when the visitor got down on his hands and knees in front of the clock, reached his hands under it, and proceeded to examine its supports. We all wondered what it could mean. When he arose, it was explained. He did not see how a clock supported in this way

LIFE AND WORK AT AN OBSERVATORY

could keep the exact time necessary in the work of an astronomer. So we had to tell him that the clock was not used for this purpose, and that he must wait until we visited the observing rooms to see our clocks properly supported.

The only evidence of the imperial will came out when he reached the great telescope. The moon, near first quarter, was then shining, but the night was more than half cloudy, and there was no hope of obtaining more than a chance glimpse at it through the clouds. But he wished to see the moon through the telescope. I replied that the sky was now covered, and it was very doubtful whether we should get a view of the moon. But he required that the telescope should be at once pointed at it. This was done, and at that moment a clear space appeared between the clouds. I remarked upon the fact, but he seemed to take it as a matter of course that the cloud would get out of the way when he wanted to look.

I made some remark about the "vernier" of one of the circles on the telescope.

"Why do you call it a vernier?" said he. "Its proper term is a nonius, because Nonius was its inventor and Vernier took the idea from him."

In this the national spirit showed itself. Nonius, a Portuguese, had invented something on a similar principle and yet essentially different from the modern vernier, invented by a Frenchman of that name.

Accompanying the party was a little girl, ten or

twelve years old, who, though an interested spectator, modestly kept in the background and said nothing. On her arrival home, however, she broke her silence by running upstairs with the exclamation, —

"Oh, Mamma, he's the funniest emperor you ever did see!"

My connection with the observatory ceased September 15, 1877, when I was placed in charge of the Nautical Almanac Office. It may not, however, be out of place to summarize the measures which have since been taken both by the Navy Department and by eminent officers of the service to place the work of the institution on a sound basis. One great difficulty in doing this arises from the fact that neither Congress nor the Navy Department has ever stated the object which the government had in view in erecting the observatory, or assigned to it any well-defined public functions. The superintendent and his staff have therefore been left to solve the question what to do from time to time as best they could.

In the spring of 1877 Rear-Admiral John Rodgers became the superintendent of the observatory. As a cool and determined fighter during the civil war he was scarcely second even to Farragut, and he was at the same time one of the ablest officers and most estimable men that our navy ever included in its ranks. "I would rather be John Rodgers dead than any other man I know living," was said by one of the observatory assistants after his death. Not many months after his accession he began to consider the question whether the wide liberty which had been allowed the professors in choosing their work was adapted to attain success. The Navy Department also desired to obtain some expressions of opinion on the subject. The result was a discussion and an official paper, not emanating from the admiral, however, in which the duty of the head of the observatory was defined in the following terms: —

" The superintendent of the observatory should be a line officer of the navy, of high rank, who should attend to the business affairs of the institution, thus leaving the professors leisure for their proper work."

Although he did not entirely commit himself to this view, he was under the impression that to get the best work out of the professors their hearts must be in it; and this would not be the case if any serious restraint was placed upon them as to the work they should undertake.

After Rodgers's death Vice-Admiral Rowan was appointed superintendent. About this time it would seem that the department was again disposed to inquire into the results of the liberal policy heretofore pursued. Commander (since Rear-Admiral) William T. Sampson was ordered to the observatory, not as its head, but as assistant to the superintendent. He was one of the most proficient men in practical physics that the navy has ever produced. I believe that one reason for choosing so able and energetic an officer for the place was to see if any improvement could be made on the system. As I was absent at the Cape of Good Hope to observe the transit of Venus during the most eventful occasion of his administration, I have very little personal knowledge of it. It seems, however, that newspaper attacks were made on him, in which he was charged with taking possession of all the instruments of the observatory but two, and placing them in charge of naval officers who were not proficient in astronomical science. In reply he wrote an elaborate defense of his action to the "New York Herald," which appeared in the number for February 13, 1883. The following extract is all that need find a place in the present connection.

When I came here on duty a little more than a year since, I found these instruments disused. The transit instrument had not been used since 1878, and then only at intervals for several years previous; the mural circle had not been used since 1877; the prime vertical had not been used since 1867. These instruments had been shamefully neglected and much injured thereby. . . . The small equatorial and comet seeker were in the same disgraceful condition, and were unfit for any real work.

Admiral Franklin was made superintendent sometime in 1883, I believe, and issued an order providing that the work of the observatory should be planned by a board consisting of the superintendent, the senior line officer, and the senior professor. Professors or officers in charge of instruments were required to prepare a programme for their proposed work each year in advance, which programme would be examined by the board. Of the work of this board or its proceedings, no clear knowledge can be gleaned from the published reports, nor do I know how long it continued.

In 1885 Secretary Whitney referred to the National Academy of Sciences the question of the advisability of proceeding promptly with the erection of a new naval observatory upon the site purchased in 1880. The report of the academy was in the affirmative, but it was added that the observatory should be erected and named as a national one, and placed under civilian administration. The year following Congress made the preliminary appropriation for the commencement of the new building, but no notice was taken of the recommendation of the academy.

In 1891 the new buildings were approaching completion, and Secretary Tracy entered upon the question of the proper administration of the observatory. He discussed the subject quite fully in his annual report for that year, stating his conclusion in the following terms: —

I therefore recommend the adoption of legislation which shall instruct the President to appoint, at a sufficient salary, without restriction, from persons either within or outside the naval service, the ablest and most accomplished astronomer who can be found for the position of superintendent.

At the following session of Congress Senator Hale introduced an amendment to the naval appropriation bill, providing for the expenses of a commission to be appointed by the Secretary of the Navy, to consider and report upon the organization of the observatory. The House non-concurred in this amendment, and it was dropped from the bill.

At the same session, all the leading astronomers of the country united in a petition to Congress, asking that the recommendation of the Secretary of the Navy should be carried into effect. After a very patient hearing of arguments on the subject by Professor Boss and others, the House Naval Committee reported unanimously against the measure, claiming that the navy had plenty of officers able to administer the observatory in a satisfactory way, and that there was therefore no necessity for a civilian head.

Two years later, Senator Morrill offered an amendment to the legislative appropriation bill, providing that the superintendent of the observatory should be selected from civil life, and be learned in the science of astronomy. He supported his amendment by letters from a number of leading astronomers of the country in reply to questions which he had addressed to them.

This amendment, after being approved by the Senate Naval Committee, was referred by the Committee on Appropriations to the Secretary of the Navy. He recommended a modification of the measure so as to provide for the appointment of a "Director of Astronomy," to have charge of the astronomical work of the observatory, which should, however, remain under a naval officer as superintendent. This arrangement was severely criticised in the House by Mr. Thomas B. Reed, of Maine, and the whole measure was defeated in conference.

In 1892, when the new observatory was being occupied, the superintendent promulgated regulations for its work. These set forth in great detail what the observatory should do. Its work was divided into nine departments, each with its chief, besides which there was a chief astronomical assistant and a chief nautical assistant to the superintendent, making eleven chiefs in all. The duties of each chief were comprehensively described. As the entire scientific force of the observatory numbered some ten or twelve naval officers, professors, and assistant astronomers, with six computers, it may be feared that some of the nine departments were short-handed.

In September, 1894, new regulations were established by the Secretary of the Navy, which provided for an "Astronomical Director," who was to "have charge of and to be responsible for the direction, scope, character, and preparation for publication of all work purely astronomical, which is performed at the Naval Observatory." As there was no law for this office, it was filled first by the detail of Professor Harkness, who served until his retirement in 1899, then by the detail of Professor Brown, who served until March, 1901.

In 1899 the Secretary of the Navy appointed a Board of Visitors to the observatory, comprising

Senator Chandler, of New Hampshire, Hon. A. G. Dayton, House of Representatives, and Professors Pickering, Comstock, and Hale. This board, "in order to obviate a criticism that the astronomical work of the observatory has not been prosecuted with that vigor and continuity of purpose which should be shown in a national observatory," recommended that the Astronomical Director and the Director of the Nautical Almanac should be civil officers, with sufficient salaries. A bill to this effect was introduced into each House of Congress at the next session, and referred to the respective naval committees, but never reported.

In 1901 Congress, in an amendment to the naval appropriation bill, provided a permanent Board of Visitors to the observatory, in whom were vested full powers to report upon its condition and expenditures, and to prescribe its plan of work. It was also provided in the same law that the superintendent of the observatory should, until further legislation by Congress, be a line officer of the navy of a rank not below that of captain. In the first annual report of this board is the following clause: —

"We wish to record our deliberate and unanimous judgment that the law should be changed so as to provide that the official head of the observatory — perhaps styled simply the Director — should be an eminent astronomer appointed by the President by and with the consent of the Senate."

Although the board still has a legal existence,

Congress, in 1902, practically suspended its functions by declining to make any appropriation for its expenses. Moreover, since the detachment of Professor Brown, Astronomical Director, no one has been appointed to fill the vacancy thus arising. At the time of the present writing, therefore, the entire responsibility for planning and directing the work of the observatory is officially vested in the naval superintendent, as it was at the old observatory.

V

GREAT TELESCOPES AND THEIR WORK

ONE hardly knows where, in the history of science, to look for an important movement that had its effective start in so pure and simple an accident as that which led to the building of the great Washington telescope, and went on to the discovery of the satellites of Mars. Very different might have been a chapter of astronomical history, but for the accident of Mr. Cyrus Field, of Atlantic cable fame, having a small dinner party at the Arlington Hotel, Washington, in the winter of 1870. Among the guests were Senators Hamlin and Casserly, Mr. J. E. Hilgard of the Coast Survey, and a young son of Mr. Field, who had spent the day in seeing the sights of Washington. Being called upon for a recital of his experiences, the youth described his visit to the observatory, and expressed his surprise at finding no large telescope. The only instrument they could show him was much smaller and more antiquated than that of Mr. Rutherfurd in New York.

The guests listened to this statement with incredulity, and applied to Mr. Hilgard to know whether the visitor was not mistaken, through a failure to find the great telescope of the observatory. Mr. Hilgard replied that the statement was quite correct, the observatory having been equipped at a time when the construction of great refracting telescopes had not been commenced, and even their possibility was doubted.

"This ought not to be," said one of the senators. "Why is it so?"

Mr. Hilgard mentioned the reluctance of Congress to appropriate money for a telescope.

"It must be done," replied the senator. "You have the case properly represented to Congress, and we will see that an appropriation goes through the Senate at least."

It chanced that this suggestion had an official basis which was not known to the guests. Although Mr. Alvan Clark had already risen into prominence as a maker of telescopes, his genius in this direction had not been recognized outside of a limited scientific circle. The civil war had commenced just as he had completed the largest refracting telescope ever made, and the excitement of the contest, as well as the absorbing character of the questions growing out of the reconstruction of the Union, did not leave our public men much time to think about the making of telescopes. Mr. Clark had, however, been engaged by Captain Gilliss only a year or two after the latter had taken charge of the observatory, to come to Washington, inspect our instruments, and regrind their glasses. The result of his work was so striking to the observers using the instruments before and after his work on them, that no doubt of his ability could be felt. Accordingly, in preparing items for the annual reports of the observatory for the years 1868 and 1869, I submitted one to the superintendent setting forth the great deficiency of the observatory in respect to the power of its telescope, and the ability of Mr. Clark to make good that deficiency. These were embodied in the reports. It was recommended that authority be given to order a telescope of the largest size from Mr. Clark.

It happened, however, that Secretary Welles had announced in his annual reports as his policy that he would recommend no estimates for the enlargement and improvement of public works in his department, but would leave all matters of this kind to be acted on by Congress as the latter might deem best. As the telescope was thrown out of the regular estimates by this rule, this subject had failed to be considered by Congress.

Now, however, the fact of the recommendation appearing in the annual report, furnished a basis of action. Mr. Hilgard did not lose a day in setting the ball in motion.

He called upon me immediately, and I told him of the recommendations in the last two reports of the superintendent of the observatory. Together we went to see Admiral Sands, who of course took the warmest interest in the movement, and earnestly promoted it on the official side. Mr. Hilgard telegraphed immediately to some leading men of science, who authorized their signatures to a petition. In this paper attention was called to the wants of the observatory, as set forth by the superintendent, and to the eminent ability of the celebrated firm of the Clarks to supply them. The petition was printed and put into the hands of Senator Hamlin for presentation to the Senate only three or four days after the dinner party. The appropriation measure was formally considered by the Committee on Naval Affairs and that on Appropriations, and was adopted in the Senate as an amendment to the naval appropriation bill without opposition. The question then was to get the amendment concurred in by the House of Representatives. The session was near its close, and there was no time to do much work.

Several members of the House Committee on Appropriations were consulted, and the general feeling seemed to be favorable to the amendment. Great, therefore, was our surprise to find the committee recommending that the amendment be not concurred in. To prevent a possible misapprehension, I may remark that the present system of non-concurring in all amendments to an appropriation bill, in order to bring the whole subject into conference, had not then been introduced, so that this action showed a real opposition to the movement. One of the most curious features of the case is that the leader in the opposition was said to be Mr. Washburn, the chairman of the committee, who, not many years later, founded the Washburn Observatory of the University of Wisconsin. There is, I believe, no doubt that his munificence in this direction arose from what he learned about astronomy and telescopes in the present case.

It happened, most fortunately, that the joint committee of conference included Drake of the Senate and Niblack of the House, both earnestly in favor of the measure. The committee recommended concurrence, and the clause authorizing the construction became a law. The price was limited to $50,000, and a sum of $10,000 was appropriated for the first payment.

No sooner were the Clarks consulted than difficulties were found which, for a time, threatened to complicate matters, and perhaps delay the construction. In the first place, our currency was then still on a paper basis. Gold was at a premium of some ten or fifteen per cent., and the Clarks were unwilling to take the contract on any but a gold basis. This, of course, the Government could not do. But the difficulty was obviated through the action of a second one, which equally threatened delay. Mr. L. J. McCormick, of reaping-machine fame, had conceived the idea of getting the largest telescope that could be made. He had commenced negotiations with the firm of Alvan Clark & Sons before we had moved, and entered into a contract while the appropriation was still pending in Congress. If the making of one great telescope was a tedious job, requiring many years for its completion, how could two be made?

I was charged with the duty of negotiating the government contract with the Clarks. I found that the fact of Mr. McCormick's contract being on a gold basis made them willing to accept one from the Government on a currency basis; still they considered that Mr. McCormick had the right of way in the matter of construction, and refused to give precedence to our instrument. On mature consideration, however, the firm reached the conclusion that two instruments could be made almost simultaneously, and Mr. McCormick very generously waived any right he might have had to precedence in the matter.

The question how large an instrument they would undertake was, of course, one of the first to arise. Progress in the size of telescopes had to be made step by step, because it could never be foreseen how soon the limit might be met; and if an attempt were made to exceed it, the result would be not only failure for the instrument, but loss of labor and money by the constructors. The largest refracting telescope which the Clarks had yet constructed was one for the University of Mississippi, which, on the outbreak of the civil war, had come into the possession of the Astronomical Society of Chicago. This would have been the last step, beyond which the firm would not have been willing to go to any great extent, had it not happened that, at this very time, a great telescope had been mounted in England. This was made by Thomas Cooke & Sons of York, for Mr. R. S. Newall of

Gateshead on Tyne, England. The Clarks could not, of course, allow themselves to be surpassed or even equalled by a foreign constructor; yet they were averse to going much beyond the Cooke telescope in size. Twenty-six inches aperture was the largest they would undertake. I contended as strongly as I could for a larger telescope than Mr. McCormick's, but they would agree to nothing of the sort, — the supposed right of that gentleman to an instrument of equal size being guarded as completely as if he had been a party to the negotiations. So the contract was duly made for a telescope of twenty-six inches clear aperture.

At that time Cooke and Clark were the only two men who had ever succeeded in making refracting telescopes of the largest size. But in order to exercise their skill, an art equally rare and difficult had to be perfected, that of the glassmaker. Ordinary glass, even ordinary optical glass, would not answer the purpose at all. The two disks, one of crown glass and the other of flint, must be not only of perfect transparency, but absolutely homogeneous through and through, to avoid inequality of refraction, and thus cause all rays passing through them to meet in the same focus. It was only about the beginning of the century that flint disks of more than two or three inches diameter could be made. Even after that, the art was supposed to be a secret in the hands of a Swiss named Guinand, and his family. Looking over the field, the Clarks concluded that the only firm that could be relied on to furnish the glass was that of Chance & Co., of Birmingham, England. So, as soon as the contracts were completed, one of the Clark firm visited England and arranged with Chance & Co. to supply the glass for the two telescopes. The firm failed in a number of trials, but by repeated efforts finally reached success at the end of a year. The glasses were received in December, 1871, and tested in the following month. A year and a half more was required to get the object glasses into perfect shape; then, in the spring or summer of 1873, I visited Cambridge for the purpose of testing the glasses. They were mounted in the yard of the Clark establishment in a temporary tube, so arranged that the glass could be directed to any part of the heavens.

I have had few duties which interested me more than this. The astronomer, in pursuing his work, is not often filled with those emotions which the layman feels when he hears of the wonderful power of the telescope. Not to say anything so harsh as that "familiarity breeds contempt," we must admit that when an operation of any sort becomes a matter of daily business, the sentiments associated with it necessarily become dulled. Now, however, I was filled with the consciousness that I was looking at the stars through the most powerful telescope that had ever been pointed at the heavens, and wondered what mysteries might be unfolded. The night was of the finest, and I remember, sweeping at random, I ran upon what seemed to be a little cluster of stars, so small and faint that it could scarcely have been seen in a smaller instrument, yet so distant that the individual stars eluded even the power of this instrument. What cluster it might have been it was impossible to determine, because the telescope had not the circles and other appliances necessary for fixing the exact location of an object. I could not help the vain longing which one must sometimes feel under such circumstances, to know what beings might live on planets belonging to what, from an earthly point of view, seemed to be a little colony on the border of creation itself.

In his report dated October 9, 1873, Admiral Sands reported the telescope as "nearly completed." The volume of Washington observations showed that the first serious observations made with it, those on the satellites of Neptune, were commenced on November 10 of the same year. Thus, scarcely more than a month elapsed from the time that the telescope was reported still incomplete in the shop of its makers until it was in regular nightly use.

Associated with the early history of the instrument is a chapter of astronomical history which may not only instruct and amuse the public, but relieve the embarrassment of some astronomer of a future generation who, reading the published records, will wonder what became of an important discovery. If the faith of the public in the absolute certainty of all astronomical investigation is thereby impaired, what I have to say will be in the interest of truth;

and I have no fear that our science will not stand the shock of the revelation. Of our leading astronomical observers of the present day — of such men as Burnham and Barnard — it may be safely said that when they see a thing it is there. But this cannot always be said of every eminent observer, and here is a most striking example of this fact.

When the telescope was approaching completion I wrote to the head of one of the greatest European observatories, possessing one of the best telescopes of the time, that the first thing I should attempt with the telescope would be the discovery of the companion of Procyon. This first magnitude star, which may be well seen in the winter evenings above Orion, had been found to move in an exceedingly small orbit, one too small to be detected except through the most refined observations of modern precision. The same thing had been found in the case of Sirius, and had been traced to the action of a minute companion revolving around it, which was discovered by the Clarks a dozen years before. There could be no doubt that the motion of Procyon was due to the same cause, but no one had ever seen the planet that produced it, though its direction from the star at any time could be estimated.

Now, it happened that my European friend, as was very natural, had frequently looked for this object without seeing it. Whether my letter set him to looking again, or whether he did not receive it until a later day, I do not know. What is certain is that, in the course of the summer, he published the discovery of the long-looked-for companion, supplemented by an excellent series of observations upon it, made in March and April.

Of course I was a little disappointed that the honor of first finding this object did not belong to our own telescope. Still I was naturally very curious to see it. So, on the very first night on which the telescope could be used, I sat up until midnight to take a look at Procyon, not doubting that, with the greater power of our telescope, it would be seen at the first glance. To my great concern, nothing of the sort was visible. But the night was far from good, the air being somewhat thick with moisture, which gave objects seen through it a blurred appearance; so I had to await a better night and more favorable conditions. Better nights came and passed, and still not a trace of the object could be seen. Supposing that the light of the bright star might be too dazzling, I cut it off with a piece of green glass in the focus. Still no companion showed itself. Could it be that our instrument, in a more favorable location, would fail to show what had been seen with one so much smaller? This question I could not answer, but wrote to my European friend of my unavailing attempts.

He replied expressing his perplexity and surprise at the occurrence, which was all the greater that the object had again been seen and measured in April, 1874. A fine-looking series of observations was published, similar to those of the preceding year. What made the matter all the more certain was that there was a change in the direction of the object which corresponded very closely to the motion as it had been predicted by Auwers. The latter published a revision of his work, based on the new observations.

A year later, the parties that had been observing the transit of Venus returned home. The head of one of them, Professor C. H. F. Peters of Clinton, stopped a day or two at Washington. It happened that a letter from my European friend arrived at the same time. I found that Peters was somewhat skeptical as to the reality of the object. Sitting before the fire in my room at the observatory, I read to him and some others extracts from the letter, which cited much new evidence to show the reality of the discovery. Not only had several of his own observers seen the object, but it had been seen and measured on several different nights by a certain Professor Blank, with a telescope only ten or twelve inches aperture.

"What," said Peters, "has Blank seen it?"

"Yes, so the letter says."

"Then it is n't there!"

And it really was not there. The maker of the discovery took it all back, and explained how he had been deceived. He found that the telescope through which the observations were made seemed to show a little companion of the same sort alongside of every very bright star. Everything was explained by this discovery. Even the seeming motion of the imaginary star during the twelve months was accounted for by the fact that in 1873 Procyon was much nearer the horizon when the observations were made than it was the year following.[1]

There is a sequel to the history, which may cause its revision by some astronomer not many years hence. When the great telescope was mounted at the Lick Observatory, it is understood that Burnham and Barnard, whose eyes are of the keenest, looked in vain for the companion of Procyon. Yet, in 1895, it was found with the same instrument by Schaeberle, and has since been observed with the great Yerkes telescope, as well as by the observers at Mount Hamilton, so that the reality of the discovery is beyond a doubt. The explanation of the failure of Burnham and Barnard to see it is very simple: the object moves in an eccentric orbit, so that it is nearer the planet at some points of its orbit than at others. It was therefore lost in the rays of the bright star during the years 1887–94. Is it possible that it could have been far enough away to be visible in 1873–74? I need scarcely add that this question must be answered in the negative, yet it may be worthy of consideration, when the exact orbit of the body is worked out twenty or thirty years hence.

In my work with the telescope I had a more

[1] In justice to Mr. Blank, I must say that there seems to have been some misunderstanding as to his observations. What he had really seen and observed was a star long well known, much more distant from Procyon than the companion in question.

definite end in view than merely the possession of a great instrument. The work of reconstructing the tables of the planets, which I had long before mapped out as the greatest one in which I should engage, required as exact a knowledge as could be obtained of the masses of all the planets. In the case of Uranus and Neptune, the two outer planets, this knowledge could best be obtained by observations on their satellites. To the latter my attention was therefore directed. In the case of Neptune, which has only one satellite yet revealed to human vision, and that one so close to the planet that the observations are necessarily affected by some uncertainty, it was very desirable that a more distant one should be found if it existed. I therefore during the summer and autumn of 1874 made most careful search under the most favorable conditions. But no second satellite was found. I was not surprised to learn that the observers with the great Lick telescope were equally unsuccessful. My observations with the instrument during two years were worked up and published, and I turned the instrument over to Professor Hall in 1875.

The discovery of the satellites of Mars was made two years later, in August, 1877. As no statement that I took any interest in the discovery has ever been made in any official publication, I venture, with the discoverer's permission, to mention the part that I took in verifying it.

One morning Professor Hall confidentially showed me his first observations of an object near Mars, and asked me what I thought of them. I remarked, "Why, that looks very much like a satellite."

Yet he seemed very incredulous on the subject; so incredulous that I feared he might make no further attempt to see the object. I afterward learned, however, that this was entirely a misapprehension on my part. He had been making a careful search for some time, and had no intention of abandoning it until the matter was cleared up one way or the other.

The possibility of the object being an asteroid suggested itself. I volunteered to test this question by looking at the ephemerides of all the small planets in the neighborhood of Mars. A very little searching disproved the possibility of the object belonging to this class. One such object was in the neighborhood, but its motion was incompatible with the measures.

Then I remarked that, if the object were really a satellite, the measures already made upon it, and the approximately known mass of the planet, would enable the motion of the satellite to be determined for a day or two. Thus I found that on that night the satellite would be hidden in the early evening by the planet, but would emerge after midnight. I therefore suggested to Professor Hall that, if it was not seen in the early evening, he should wait until after midnight. The result was in accordance with the prediction, — the satellite was not visible in the early evening, but came out after midnight. No further doubt was possible, and the discovery was published. The labor of searching and observing was so exhausting that Professor Hall let me compute the preliminary orbit of the satellites from his early observations.

My calculations and suggestions lost an importance they might otherwise have claimed, for the reason that several clear nights followed. Had cloudy weather intervened, a knowledge of when to look for the object might have greatly facilitated its recognition.

It is still an open question, perhaps, whether a great refracting telescope will last unimpaired for an indefinite length of time. I am not aware that the twin instruments of Harvard and Pulkowa, mounted in 1843, have suffered from age, nor am I aware that any of Alvan Clark's instruments are less perfect to-day than when they left the hands of their makers. But not long after the discovery of the satellites of Mars, doubts began to spread in some quarters as to whether the great Washington telescope had not suffered deterioration. These doubts were strengthened in the following way: When hundreds of curious objects were being discovered in the heavens here and there, observers with small instruments naturally sought to find them. The result was several discoveries belonging to the same class as that of the satellite of Procyon. They were found with very insignificant instruments, but could not be seen in the large ones. Professor Hall published a letter in a European journal, remarking upon the curious fact that several objects were being discovered with very small instruments, which were invisible in the Washington telescope. This met the eye of Professor Wolf, a professor at the Sorbonne in Paris, as well as astronomer at the Paris Observatory. In a public lecture, which he delivered shortly afterward, he lamented the fact that the deterioration of the Washington telescope had gone so far as that, and quoted Professor Hall as his authority.

The success of the Washington telescope excited such interest the world over as to give a new impetus to the construction of such instruments. Its glass showed not the slightest drawbacks from its great size. It had been feared that, after a certain limit, the slight bending of the glass under its own weight would be injurious to its performance. Nothing of the kind being seen, the Clarks were quite ready to undertake much larger instruments. A 30-inch telescope for the Pulkova Observatory in Russia, the 36-inch telescope of the Lick Observatory in California, and, finally, the 40-inch of the Yerkes Observatory in Chicago, were the outcome of the movement.

Of most interest to us in the present connection is the history of the 30-inch telescope of the Pulkova Observatory, the object glass of which was made by Alvan Clark & Sons. It was, I think, sometime in 1878 that I received a letter from Otto Struve,[1] director of the Pulkova Observatory,

[1] Otto Struve was a brother of the very popular Russian minister to Washington during the years 1882–92. He retired from the

stating that he was arranging with his government for a grant of money to build one of the largest refracting telescopes. In answering him I called his attention to the ability of Alvan Clark & Sons to make at least the object glass, the most delicate and difficult part of the instrument. The result was that, after fruitless negotiations with European artists, Struve himself came to America in the summer of 1879 to see what the American firm could do. He first went to Washington and carefully examined the telescope there. Then he proceeded to Cambridge and visited the workshop of the Clarks. He expressed some surprise at its modest dimensions and fittings generally, but was so well pleased with what he saw that he decided to award them the contract for making the object glass. He was the guest of the Pickerings at the Cambridge Observatory, and invited me thither from where I was summering on the coast of Massachusetts to assist in negotiating the contract.

He requested that, for simplicity in conference, the preliminary terms should be made with but a single member of the firm to talk with. George B. Clark, the eldest member, was sent up to represent the firm. I was asked to take part in the negotiations as a mutual friend of both parties, and suggested the main conditions of the contract. A

---

direction of the Pulkowa Observatory about 1894. The official history of his negotiations and other proceedings for the construction of the telescope will be found in a work published in 1889 in honor of the jubilee of the observatory.

summary of these will be found in the publication to which I have already referred.

There was one provision the outcome of which was characteristic of Alvan Clark & Sons. Struve, in testing some object glasses which they had constructed and placed in their temporary tube, found so great physical exertion necessary in pointing so rough an instrument at any heavenly body with sufficient exactness, that he could not form a satisfactory opinion of the object glass. As he was to come over again when the glass was done, in order to test it preliminary to acceptance, he was determined that no such difficulty should arise. He therefore made a special provision that $1000 extra, to be repaid by him, should be expended in making a rough equatorial mounting in which he could test the instrument. George Clark demurred to this, on the ground that such a mounting as was necessary for this purpose could not possibly cost so much money. But Struve persistently maintained that one to cost $1000 should be made. The other party had to consent, but failed to carry out this provision. The tube was, indeed, made large enough to test not only Struve's glass but the larger one of the Lick Observatory, which, though not yet commenced, was expected to be ready not long afterward. Yet, notwithstanding this increase of size, I think the extra cost turned out to be much less than $1000, and the mounting was so rough that when Struve came over in 1883 to test the glass, he suffered much physical inconvenience and met, if my memory serves me aright, with a slight accident, in his efforts to use the rough instrument.

In points like this I do not believe that another such business firm as that of the Clarks ever existed in this country or any other. Here is an example. Shortly before the time of Struve's visit, I had arranged with them for the construction of a refined and complicated piece of apparatus to measure the velocity of light. As this apparatus was quite new in nearly all its details, it was impossible to estimate in advance what it might cost; so, of course, they desired that payment for it should be arranged on actual cost after the work was done. I assured them that the government would not enter into a contract on such terms. There must be some maximum or fixed price. This they fixed at $2500. I then arranged with them that this should be taken as a maximum and that, if it was found to cost less, they should accept actual cost. The contract was arranged on this basis. There were several extras, including two most delicate reflecting mirrors which would look flat to the eye, but were surfaces of a sphere of perhaps four miles diameter. The entire cost of the apparatus, as figured up by them after it was done, with these additions, was less than $1500, or about forty per cent. below the contract limit.

No set of men were ever so averse to advertising themselves. If anybody, in any part of the world, wanted them to make a telescope, he must write to them to know the price, etc. They could never be induced to prepare anything in the form of a price catalogue of the instruments they were prepared to furnish. The history of their early efforts and the indifference of our scientific public to their skill forms a mortifying chapter in our history of the middle of the century. When Mr. Clark had finished his first telescope, a small one of four inches aperture, which was, I have no reason to doubt, the best that human art could make, he took it to the Cambridge Observatory to be tested by one of the astronomers. The latter called his attention to a little tail which the glass showed as an appendage of a star, and which was, of course, non-existent. It was attributed to a defect in the glass, which was therefore considered a failure. Mr. Clark was quite sure that the tail was not shown when he had previously used the glass, but he could not account for it at the time. He afterwards traced it to the warm air collecting in the upper part of the tube and producing an irregular refraction of the light. When this cause was corrected the defect disappeared. But he got no further encouragement at home to pursue his work. The first recognition of his genius came from England, the agent being Rev. W. R. Dawes, an enthusiastic observer of double stars, who was greatly interested in having the best of telescopes. Mr. Clark wrote him a letter describing a number of objects which he had seen with telescopes of his own make. From this description Mr. Dawes saw that the instruments must be of great excellence, and the outcome of the matter was that he ordered one or more telescopes from the American maker. Not until then were the abilities of the latter recognized in his own country.

I have often speculated as to what the result might have been had Mr. Clark been a more enterprising man. If, when he first found himself able to make a large telescope, he had come to Washington, got permission to mount his instrument in the grounds of the capitol, showed it to members of Congress, and asked for legislation to promote this new industry, and, when he got it, advertised himself and his work in every way he could, would the firm which he founded have been so little known after the death of its members, as it now unhappily is? This is, perhaps, a rather academic question, yet not an unprofitable one to consider.

In recent years the firm was engaged only to make object glasses of telescopes, because the only mountings they could be induced to make were too rude to satisfy astronomers. The palm in this branch of the work went to the firm of Warner & Swasey, whose mounting of the great Yerkes telescope of the University of Chicago is the last word of art in this direction.

During the period when the reputation of the Cambridge family was at its zenith, I was slow to believe that any other artist could come up to their standard. My impression was strengthened by a curious circumstance. During a visit to the Strasburg Observatory in 1883 I was given permission to look through its great telescope, which was made by a renowned German artist. I was surprised to find the object glass affected by so serious a defect that it could not be expected to do any work of the first class. One could only wonder that European art was so backward. But, several years afterward, the astronomers discovered that, in putting the glasses together after being cleaned, somebody had placed one of them in the wrong position, the surface which should have been turned toward the star being now turned toward the observer. When the glass was simply turned over so as to have the right face outward, the defect disappeared.

VI

THE TRANSITS OF VENUS

It was long supposed that transits of Venus over the sun's disk afforded the only accurate method of determining the distance of the sun, one of the fundamental data of astronomy. Unfortunately, these phenomena are of the rarest. They come in pairs, with an interval of eight years between the transits of a pair. A pair occurred in 1761 and 1769, and again in 1874 and 1882. Now the whole of the twentieth century will pass without another recurrence of the phenomenon. Not until the years 2004 and 2012 will our posterity have the opportunity of witnessing it.

Much interesting history is associated with the adventures of the astronomers who took part in the expeditions to observe the transits of 1761 and 1769. In the almost chronic warfare which used to rage between France and England during that period, neither side was willing to regard as neutral even a scientific expedition sent out by the other. The French sent one of their astronomers, Le Gentil, to observe the transit at Pondicherry in the East Indies. As he was nearing his station, the presence of the enemy prevented him from making port, and he was still at sea on the day of the transit. When he at length landed, he determined to remain until the transit of 1769, and observe that. We must not suppose, however, that he was guilty of the eccentricity of doing this with no other object in view than that of making the observation. He found the field open for profitable mercantile enterprise, as well as interesting for scientific observations and inquiries. The eight long years passed away, and the morning of June 4, 1769, found him in readiness for his work. The season had been exceptionally fine. On the morning of the transit the sun shone in a cloudless sky, as it had done for several days previous. But, alas for all human hopes! Just before Venus reached the sun, the clouds gathered, and a storm burst upon the place. It lasted until the transit was over, and then cleared away again as if with the express object of showing the unfortunate astronomer how helpless he was in the hands of the elements.

The Royal Society of England procured a grant of £800 from King George II. for expeditions to observe the transit of 1761.[1] With this grant the Society sent the Rev. Nevil Maskelyne to the island of St. Helena, and, receiving another grant, it was

[1] For the incidents connected with the English observations of this transit, the author is indebted to Vice-Admiral W. H. Smyth's curious and rare book, *Speculum Hartwellianum*, London, 1860. It and other works of the same author may be described as queer and interesting jumbles of astronomical and other information, thrown into an interesting form; and, in the case of the present work, spread through a finely illustrated quarto volume of nearly five hundred pages.

used to dispatch Messrs. Mason and Dixon (those of our celebrated "line") to Bencoolen. The admiralty also supplied a ship for conveying the observers to their respective destinations. Maskelyne, however, would not avail himself of this conveyance, but made his voyage on a private vessel. Cloudy weather prevented his observations of the transit, but this did not prevent his expedition from leaving for posterity an interesting statement of the necessaries of an astronomer of that time. His itemized account of personal expenses was as follows: —

One year's board at St. Helena	£109	10s.	0d.
Liquors at 5s. per day	91	5	0
Washing at 9d. per day	13	13	9
Other expenses	27	7	6
Liquors on board ship for six months	50	0	0
	£291	16s.	3d.

Seven hundred dollars was the total cost of liquors during the eighteen months of his absence. Admiral Smyth concludes that Maskelyne "was not quite what is now ycleped a teetotaler." He was subsequently Astronomer Royal of England for nearly half a century, but his published observations give no indication of the cost of the drinks necessary to their production.

Mason and Dixon's expedition met with a mishap at the start. They had only got fairly into the English Channel when their ship fell in with a French frigate of superior force. An action ensued in which the English crew lost eleven killed and thirty-eight wounded. The Frenchman was driven off, but the victorious vessel had to return to Plymouth for repairs. This kind of a scientific expedition was more than the astronomers had bargained for, and they wrote from Plymouth to the Royal Society, describing their misfortune and resigning their mission. But the Council of the Society speedily let them know that they were unmoved by the misfortunes of their scientific missionaries, and pointed out to them in caustic terms that, having solemnly undertaken the expedition, and received money on account of it, their failure to proceed on the voyage would be a reproach to the nation in general, and to the Royal Society in particular. It would also bring an indelible scandal upon their character, and probably end in their utter ruin. They were assured that if they persisted in the refusal, they would be treated with the most inflexible resentment, and prosecuted with the utmost severity of the law.

Under such threats the unfortunate men could do nothing but accept the situation and sail again after their frigate had been refitted. When they got as far as the Cape of Good Hope, it was found very doubtful whether they would reach their destination in time for the transit; so, to make sure of some result from their mission, they made their observations at the Cape.

One of the interesting scraps of history connected with the transit of 1769 concerns the observations of Father Maximilian Hell, S. J., the leading astronomer of Vienna. He observed the transit at Wardhus, a point near the northern extremity of Norway, where the sun did not set at the season of the transit. Owing to the peculiar circumstances under which the transit was observed, — the ingress of the planet occurring two or three hours before the sun approached the northern horizon, and the end of the transit about as long afterward, — this station was the most favorable one on the globe. Hell, with two or three companions, one of them named Sajnovics, went on his mission to this isolated place under the auspices of the king of Denmark. The day was cloudless and the observations were made with entire success. He returned to Copenhagen, where he passed several months in preparing for the press a complete account of his expedition and the astronomical observations made at the station.

Astronomers were impatient to have the results for the distance of the sun worked out as soon as possible. Owing to the importance of Hell's observations, they were eagerly looked for. But he at first refused to make them known, on the ground that, having been made under the auspices of the king of Denmark, they ought not to be made known in advance of their official publication by the Danish Academy of Sciences. This reason, however, did not commend itself to the impatient astronomers; and suspicions were aroused that something besides official formalities was behind the delay. It was hinted that Hell was waiting for the observations made at other stations in order that he might so manipulate his own that they would fit in with those made elsewhere. Reports were even circulated that he had not seen the transit at all, owing to cloudy weather, and that he was manufacturing observations in Copenhagen. The book was, however, sent to the printer quite promptly, and the insinuations against its author remained a mere suspicion for more than sixty years. Then, about 1833, a little book was published on the subject by Littrow, Director of the Vienna Observatory, which excited much attention. Father Hell's original journal had been conveyed to Vienna on his return, and was still on deposit at the Austrian National Observatory. Littrow examined it and found, as he supposed, that the suspicions of alterations in observations were well founded; more especially that the originals of the all-important figures which recorded the critical moment of "contact" had been scraped out of the paper, and new ones inserted in their places. The same was said to be the case with many other important observations in the journal, and the conclusion to which his seemingly careful examination led was that no reliance could be placed on the genuineness of Hell's work. The doubts thus raised were not dispelled until another half-century had elapsed.

In 1883 I paid a visit to Vienna for the purpose of examining the great telescope which had just been mounted in the observatory there by Grubb, of Dublin. The weather was so unfavorable that it was necessary to remain two weeks, waiting for an opportunity to see the stars. One evening I visited the theatre to see Edwin Booth, in his celebrated tour over the Continent, play King Lear to the applauding Viennese. But evening amusements cannot be utilized to kill time during the day. Among the tasks I had projected was that of rediscussing all the observations made on the transits of Venus which had occurred in 1761 and 1769, by the light of modern science. As I have already remarked, Hell's observations were among the most important made, if they were only genuine. So, during my almost daily visits to the observatory, I asked permission of Director Weiss to study Hell's manuscript.

At first the task of discovering anything which would lead to a positive decision on one side or the other seemed hopeless. To a cursory glance, the descriptions given by Littrow seemed to cover the ground so completely that no future student could turn his doubt into certainty. But when one looks leisurely at an interesting object, day after day, he continually sees more and more. Thus it was in the present case. One of the first things to strike me as curious was that many of the alleged alterations had been made before the ink got dry. When the writer made a mistake, he had rubbed it out with his finger, and made a new entry.

The all-important point was a certain suspicious

record which Littrow affirmed had been scraped out so that the new insertion could be made. As I studied these doubtful figures, day by day, light continually increased. Evidently the heavily written figures, which were legible, had been written over some other figures which were concealed beneath them, and were, of course, completely illegible, though portions of them protruded here and there outside of the heavy figures. Then I began to doubt whether the paper had been scraped at all. To settle the question, I found a darkened room, into which the sun's rays could be admitted through an opening in the shutter, and held the paper in the sunlight in such a way that the only light which fell on it barely grazed the surface of the paper. Examining the sheet with a magnifying glass, I was able to see the original texture of the surface with all its hills and hollows. A single glance sufficed to show conclusively that no eraser had ever passed over the surface, which had remained untouched.

The true state of the case seemed to me almost beyond doubt. It frequently happened that the ink did not run freely from the pen, so that the words had sometimes to be written over again. When Hell first wrote down the little figures on which, as he might well suppose, future generations would have to base a very important astronomical element, he saw that they were not written with a distinctness corresponding to their importance. So he wrote them over again with the hand, and in the spirit of a man who was determined to leave no doubt on the subject, little weening that the act would give rise to a doubt which would endure for a century.

This, although the most important case of supposed alteration, was by no means the only one. Yet, to my eyes, all the seeming corrections in the journal were of the most innocent and commonplace kind, — such as any one may make in writing.

Then I began to compare the manuscript, page after page, with Littrow's printed description. It struck me as very curious that where the manuscript had been merely retouched with ink which was obviously the same as that used in the original writing, but looked a little darker than the original, Littrow described the ink as of a different color. In contrast with this, there was an important interlineation, which was evidently made with a different kind of ink, one that had almost a blue tinge by comparison; but in the description he makes no mention of this plain difference. I thought this so curious that I wrote in my notes as follows: —

"That Littrow, in arraying his proofs of Hell's forgery, should have failed to dwell upon the obvious difference between this ink and that with which the alterations were made leads me to suspect a defect in his sense of color."

Then it occurred to me to inquire whether, perhaps, such could have been the case. So I asked

Director Weiss whether anything was known as to the normal character of Littrow's power of distinguishing colors. His answer was prompt and decisive. "Oh, yes, Littrow was color blind to red. He could not distinguish between the color of Aldebaran and that of the whitest star." No further research was necessary. For half a century the astronomical world had based an impression on the innocent but mistaken evidence of a color-blind man respecting the tints of ink in a manuscript.

About the middle of the nineteenth century other methods of measuring the sun's distance began to be developed which, it was quite possible, might prove as good as the observation in question. But the relative value of these methods and of transits of Venus was a subject on which little light could be thrown; and the rarity of the latter phenomena naturally excited universal interest, both among the astronomers and among the public. For the purpose in question it was necessary to send expeditions to different and distant parts of the globe, because the result had to depend upon the times of the phases, as seen from widely separated stations.

In 1869 the question what stations should be occupied and what observations should be made was becoming the subject of discussion in Europe, and especially in England. But our country was still silent on the subject. The result of continued silence was not hard to foresee. Congress would, at the last moment, make a munificent appropriation for sending out parties to observe the transit. The plans and instruments would be made in a hurry, and the parties packed off without any well-considered ideas of what they were to do; and the whole thing would end in failure so far as results of any great scientific value were concerned.

I commenced the discussion by a little paper on the subject in the "American Journal of Science," but there was no one to follow it up. So, at the spring meeting of the National Academy of Sciences, in 1870, I introduced a resolution for the appointment of a committee to consider the subject and report upon the observations which should be made. This resolution was adopted, and a few days afterward Professor Henry invited me to call at his office in the evening to discuss with himself and Professor Peirce, then superintendent of the Coast Survey, the composition of the committee.

At the conference I began by suggesting Professor Peirce himself for chairman. Naturally this met with no opposition; then I waited for the others to go on. But they seemed determined to throw the whole onus of the matter on me. This was the more embarrassing, because I believe that, in parliamentary law and custom, the mover of a resolution of this sort has a prescribed right to be chairman of the committee which he proposes shall be appointed. If not chairman, it would seem that he ought at any rate to be a member. But I was determined not to suggest myself in any way, so I went on and suggested Admiral Davis. This nomination was, of course, accepted without hesitation. Then I remarked that the statutes of the academy permitted of persons who were not members being invited to serve on a committee, and as the Naval Observatory would naturally take a leading part in such observations as were to be made, I suggested that its superintendent, Admiral Sands, should be invited to serve as a member of the committee. "There," said Peirce, "we now have three names. Committees of three are always the most efficient. Why go farther?"

I suggested that the committee should have on it some one practiced in astronomical observation, but he deemed this entirely unnecessary, and so the committee of three was formed. I did not deem it advisable to make any opposition at the time, because it was easy to foresee what the result would be.

During the summer nothing was heard of the committee, and in the autumn I made my first trip to Europe. On my return, in May, 1871, I found that the committee had never even held a meeting, and that it had been enlarged by the addition of a number of astronomers, among them myself. But, before it went seriously to work, it was superseded by another organization, to be described presently.

At that time astronomical photography was in its infancy. Enough had been done by Rutherfurd to show that it might be made a valuable adjunct to astronomical investigation. Might we not then photograph Venus on the sun's disk, and by measurements of the plates obtain the desired result, perhaps better than it could be obtained by any kind of eye observation? This question had already suggested itself to Professor Winlock, who, at the Cambridge Observatory, had designed an instrument for taking the photographs. It consisted of a fixed horizontal telescope, into which the rays of the sun were to be thrown by a reflector. This kind of an instrument had its origin in France, but it was first practically applied to photographing the sun in this country. As whatever observations were to be made would have to be done at governmental expense, an appropriation of two thousand dollars was obtained from Congress for the expense of some preliminary instruments and investigations.

Admiral Sands, superintendent of the observatory, now took an active part in the official preparations. It was suggested to him, on the part of the academy committee, that it would be well to join hands with other organizations, so as to have the whole affair carried on with unity and harmony. To this he assented. The result was a provision that these and all other preparations for observing the transit of Venus should be made under the direction of a commission to be composed of the superintendent of the Naval Observatory, the superintendent of the United States Coast Survey, the president of the National Academy of Sciences, and two professors of mathematics attached to the Naval Observatory. Under this provision the commission was constituted as follows: Commodore B. F.

Sands, U. S. N., Professor Benjamin Peirce, Professor Joseph Henry, Professor Simon Newcomb, Professor William Harkness.

The academy committee now surrendered its functions to the commission, and the preparations were left entirely in the hands of the latter.

So far as scientific operations were concerned, the views of the commission were harmonious through the whole of their deliberations. It was agreed from the beginning that the photographic method offered the greatest promise of success. But how, with what sort of instruments, and on what plan, must the photographs be taken? Europeans had already begun to consider this question, and for the most part had decided on using photographic telescopes having no distinctive feature specially designed for the transit. In fact, one might almost say that the usual observations with the eye were to be made on the photograph instead of on the actual sun. The American commissioners were of opinion that this would lead to nothing but failure, and that some new system must be devised.

The result was a series of experiments and trials with Professor Winlock's instrument at the Cambridge Observatory. The outcome of the matter was the adoption of his plan, with three most important additions, which I shall mention, because they may possibly yet be adopted with success in other branches of exact astronomy if this telescope is used, as it seems likely it may be.

THE TRANSITS OF VENUS

The first feature was that the photographic telescope should be mounted exactly in the meridian, and that its direction should be tested by having the transit instrument mounted in front of it, in the same line with it. In this way the axis of the telescope was a horizontal north and south line.

The next feature was that, immediately in front of the photographic plate, in fact as nearly in contact with it as possible without touching it, a plumb line of which the thread was a very fine silver wire should be suspended, the bob of which passed down below, and was immersed in a vessel of water to prevent vibration. In this way the direction of the north and south line on the plate admitted of being calculated with the greatest exactness, and the plumb line being photographed across the disk of the sun, the position angle could be measured with the same precision that any other measure could be made.

The third feature was that the distance between the photographic plate and the object glass of the telescope should be measured by a long iron rod which was kept in position above the line of sight of the telescope itself. This afforded the means of determining to what angle a given measure on the plate would correspond. The whole arrangement would enable the position of the centre of Venus with respect to the centre of the sun to be determined by purely geometric methods. One reason for relying entirely on this was that the diameter of the sun, as photographed, would be greater the greater the intensity of the photographic impression, so that no reliance could be placed upon its uniformity.

Ours were the only parties whose photographic apparatus was fitted up in this way. The French used a similar system, but without the essentials of the plumb line and the measurement of the length of the telescope. The English and Germans used ordinary telescopes for the purpose.

One of the earliest works of the commission was the preparation and publication of several papers, which were published under the general title, "Papers relating to the Transit of Venus in 1874." The first of these papers was a discussion of our proposed plan of photographing, in which the difficulties of the problem, and the best way of surmounting them, were set forth. The next, called Part II., related to the circumstances of the transit, and was therefore entirely technical. Part III. related to the corrections of Hansen's table of the moon, and was published as a paper relating to the transit of Venus, because these corrections were essential in determining the longitudes of the stations by observations of the moon.

In England the preparations were left mostly in the hands of Professor Airy, Astronomer Royal, and, I believe, Captain Tupman, who at least took a leading part in the observations and their subsequent reduction. In France, Germany, and Russia, commissions were appointed to take charge of the work and plan the observations.

As coöperation among the parties from different countries would be generally helpful, I accepted an invitation to attend a meeting of the German commission, to be held at Hanover in August, 1873. Hansen was president of the commission, while Auwers was its executive officer. One of my main objects was to point out the impossibility of obtaining any valuable result by the system of photographing which had been proposed, but I was informed, in reply, that the preparations had advanced too far to admit of starting on a new plan and putting it in operation.

From the beginning of our preparations it began to be a question of getting from Congress the large appropriations necessary for sending out the expeditions and fitting them up with instruments. The sum of $50,000 was wanted for instruments and outfit. Hon. James A. Garfield was then chairman of the committee on appropriations. His principles and methods of arranging appropriations for the government were, in some features, so different from those generally in vogue that it will be of interest to describe them.

First of all, Garfield was rigidly economical in grants of money. This characteristic of a chairman of a committee on appropriations was almost a necessary one. But he possessed it in a different way from any other chairman before or since. The method of the "watch dogs of the treasury" who sometimes held this position was to grant most of the objects asked for, but to cut down the estimated amounts by one fourth or one third. This was a very easy method, and one well fitted to impress the public, but it was one that the executive officers of the government found no difficulty in evading, by the very simple process of increasing their estimate so as to allow for the prospective reduction.[1]

Garfield compared this system to ordering cloth for a coat, but economizing by reducing the quantity put into it. If a new proposition came before him, the question was whether it was advisable for the government to entertain it at all. He had to be thoroughly convinced before this would be done. If the question was decided favorably all the funds necessary for the project were voted.

When the proposition for the transit of Venus came before him, he proceeded in a manner which I never heard of the chairman of an appropriation committee adopting before or since. Instead of calling upon those who made the proposition to appear formally before the committee, he asked me to dinner with his family, where we could talk the matter over. One other guest was present, Judge Black of Pennsylvania. He was a dyed-in-

---

[1] "The War Department got ahead of us in the matter of furniture," said an officer of the Navy Department to me long afterwards, when the furniture for the new department building was being obtained. "They knew enough to ask for a third more than they wanted; we reduced our estimate to the lowest point. Both estimates were reduced one third by the Appropriations Committee. The result is that they have all the furniture they want, while we are greatly pinched."

the-wool Democrat, wielding as caustic a pen as was ever dipped into ink, but was, withal, a firm personal friend and admirer of Garfield. As may readily be supposed, the transit of Venus did not occupy much time at the table. I should not have been an enthusiastic advocate of the case against opposition, in any case, because my hopes of measuring the sun's distance satisfactorily by that method were not at all sanguine. My main interest lay in the fact that, apart from this, the transit would afford valuable astronomical data for the life work which I had mainly in view. So the main basis of my argument was that other nations were going to send out parties; that we should undoubtedly do the same, and that they must be equipped and organized in the best way.

It appears that Judge Black was an absent-minded man, as any man engaged in thought on very great subjects, whether of science, jurisprudence, or politics, has the right to be. Garfield asked him whether it was true that, on one occasion, when preparing an argument, and walking up and down the room, his hat chanced to drop on the floor at one end of the room, and was persistently used as a cuspidor until the argument was completed. Mr. Black neither affirmed nor denied the story, but told another which he said was true. While on his circuit as judge he had, on one occasion, tried a case of theft in which the principal evidence against the accused was the finding of the stolen article in his possession. He charged the jury that this fact was *prima facie* evidence that the man was actually the thief. When through his business and about to leave for home, he went into a jeweler's shop to purchase some little trinket for his wife. The jeweler showed him a number of little articles, but finding none to suit him, he stepped into his carriage and drove off. In the course of the day he called on a street urchin to water his horse. Reaching into his pocket for a reward, the first thing he got hold of was a diamond ring which must have been taken from the shop of the jeweller when he left that morning. "I wondered," said the judge, "how I should have come out had I been tried under my own law."

The outcome of the matter was that the appropriations were duly made; first, in 1872, $50,000 for instruments, then, the year following, $100,000 for the expeditions. In 1874, $25,000 more was appropriated to complete the work and return the parties to their homes.

The date of the great event was December 8–9, 1874. To have the parties thoroughly drilled in their work, they were brought together at Washington in the preceding spring for practice and rehearsal. In order that the observations to be made by the eye should not be wholly new, an apparatus representing the transit was mounted on the top of Winder's building, near the War Department, about two thirds of a mile from the observatory. When this was observed through the telescope from the roof of the observatory, an artificial black Venus was seen impinging upon an artificial sun, and entering upon its disk in the same way that the actual Venus would be seen. This was observed over and over until, as was supposed, the observers had gotten into good practice.

In order to insure the full understanding of the photographic apparatus, the instruments were mounted and the parties practiced setting them up and going through the processes of photographing the sun. To carry out this arrangement with success, it was advisable to have an expert in astronomical photography to take charge of the work. Dr. Henry Draper of New York was invited for this purpose, and gave his services to the commission for several weeks.

This transit was not visible in the United States. It did not begin until after the sun had set in San Francisco, and it was over before the rising sun next morning had reached western Europe. All the parties had therefore to be sent to the other side of the globe. Three northern stations were occupied, — in China, Japan, and Siberia; and five southern ones, at various points on the islands of the Pacific and Indian oceans. This unequal division was suggested by the fact that the chances of fair weather were much less in the southern hemisphere than in the northern.

The southern parties were taken to their destinations in the U. S. S. Swatara, Captain Ralph Chandler, U. S. N., commanding. In astronomical observations all work is at the mercy of the elements. Clear weather was, of course, a necessity to success at any station. In the present case the weather was on the whole unpropitious. While there was not a complete failure at any one station, the number or value of the observations was more or less impaired at all. Where the sky was nearly cloudless, the air was thick and hazy. This was especially the case at Nagasaki and Pekin, where from meteorological observations which the commission had collected through our consuls, the best of weather was confidently expected. What made this result more tantalizing was that the very pains we had taken to collect the data proved, by chance, to have made the choice worse. For some time it was deliberated whether the Japanese station should be in Nagasaki or Yokohama. Consultation with the best authorities and a study of the records showed that, while Yokohama was a favorable spot, the chances were somewhat better at Nagasaki. So to Nagasaki the party was sent. But when the transit came, while the sky was of the best at Yokohama, it was far from being so at Nagasaki.

Something of the same sort occurred at the most stormy of all the southern stations, that at Kerguelen Island. The British expeditions had, in the beginning, selected a station on this island known as Christmas Harbor. We learned that a firm of New London, Conn., had a whaling station on the island. It was therefore applied to to know what the weather chances were at various points in the island. Information was obtained from their men, and it was thus found that Molloy Point, bad though the weather there was, afforded better chances than Christmas Harbor; so it was chosen. But this was not all; the British parties, either in consequence of the information we had acquired, or through what was learned from the voyage of the Challenger, established their principal station near ours. But it happened that the day at Christmas Harbor was excellent, while the observations were greatly interfered with by passing clouds at Molloy Point.

After the return of the parties sent out by the various nations, it did not take long for the astronomers to find that the result was disappointing, so far, at least, as the determination of the sun's distance was concerned. It became quite clear that this important element could be better measured by determining the velocity of light and the time which it took to reach us from the sun than it could by any transit of Venus. It was therefore a question whether parties should be sent out to observe the transit of 1882. On this subject the astronomers of the country at large were consulted. As might have been expected, there was a large majority in favor of the proposition. The negative voices were only two in number, those of Pickering and myself. I took the ground that we should make ample provisions for observing it at various stations in our own country, where it would now be visible, but that, in view of the certain failure to get a valuable result for the distance of the sun by this method, it was not worth while for us to send parties to distant parts of the world. I supposed the committee on appropriations might make careful inquiry into the subject before making the appropriation, but a representation of the case was all they asked for, and $10,000 was voted for improving the instruments and $75,000 for sending out parties.

Expeditions being thus decided upon, I volunteered to take charge of that to the Cape of Good Hope. The scientific personnel of my party comprised an officer of the army engineers, one of the navy, and a photographer. The former were Lieutenant Thomas L. Casey, Jr., Corps of Engineers, U. S. A., and Lieutenant J. H. L. Holcombe, U. S. N. We took a Cunard steamer for Liverpool about the middle of September, 1882, and transported our instruments by rail to Southampton, there to have them put on the Cape steamship. At Liverpool I was guilty of a remissness which might have caused much trouble. Our apparatus and supplies, in a large number of boxes, were all gathered and piled in one place. I sent one of my assistants to the point to see that it was so collected that there should be no possibility of mistake in getting it into the freight car designed to carry it to Southampton, but did not require him to stay there and see that all was put on board. When the cases reached Southampton it was found that one was missing. It was one of the heaviest of the lot, containing the cast-iron pier on which the photoheliograph was to be mounted. While it was possible to replace this by something else, such a course would have been inconvenient and perhaps prejudicial. The steamer was about to sail, but would touch at Plymouth next day. Only one resource was possible. I telegraphed the mistake to Liverpool and asked that the missing box be sent immediately by express to Plymouth. We had the satisfaction of seeing it come on board with the mail just as the steamer was about to set sail.

We touched first at Madeira, and then at Ascension Island, the latter during the night. One of the odd things in nomenclature is that this island, a British naval station, was not called such officially, but was a "tender to Her Majesty's ship Flora," I believe. It had become astronomically famous a few years before by Gill's observations of the position of Mars to determine the solar parallax.

We touched six hours at St. Helena, enough to see the place, but scarcely enough to make a visit to the residence of Napoleon, even had we desired to see it. The little town is beautifully situated, and the rocks around are very imposing. My most vivid recollection is, however, of running down from the top of a rock some six hundred or eight hundred feet high, by a steep flight of steps, without stopping, or rather of the consequences of this imprudent gymnastic performance. I could scarcely move for the next three days.

Cape Town was then suffering from an epidemic of smallpox, mostly confined to the Malay population, but causing some disagreeable results to travelers. Our line of ships did not terminate their voyage at the Cape, but proceeded thence to other African ports east of the Cape. Here a rigid quarantine had been established, and it was necessary that the ships touching at the Cape of Good Hope should have had no communication with the shore. Thus it happened that we found, lying in the harbor, the ship of our line which had preceded us, waiting to get supplies from us, in order that it might proceed on its voyage. Looking at a row-boat after we had cast anchor, we were delighted to see two faces which I well knew: those of David Gill, astronomer of the Cape Observatory, and Dr. W. L. Elkin, now director of the Yale Observatory. The latter had gone to the Cape as a volunteer observer with Gill, their work being directed mostly to parallaxes of stars too far south to be well observed in our latitude. Our friends were not, however, even allowed to approach the ship, for fear of the smallpox, the idea appearing to be that the latter might be communicated by a sort of electric conduction, if the boat and the ship were allowed to come into contact, so we had to be put ashore without their aid.

We selected as our station the little town of Wellington, some forty miles northeast of Cape Town. The weather chances were excellent anywhere, but here they were even better than at the

Cape. The most interesting feature of the place was what we might call an American young ladies' school. The Dutch inhabitants of South Africa are imbued with admiration of our institutions, and one of their dreams is said to be a United States of South Africa modeled after our own republic. Desiring to give their daughters the best education possible, they secured the services of Miss Ferguson, a well-known New England teacher, to found a school on the American model. We established our station in the grounds of this school.

The sky on the day of the transit was simply perfect. Notwithstanding the intensity of the sun's rays, the atmosphere was so steady that I have never seen the sun to better advantage. So all our observations were successful.

On our departure we left two iron pillars, on which our apparatus for photographing the sun was mounted, firmly imbedded in the ground, as we had used them. Whether they will remain there until the transit of 2004, I do not know, but cannot help entertaining a sentimental wish that, when the time of that transit arrives, the phenomenon will be observed from the same station, and the pillars be found in such a condition that they can again be used.

All the governments, except our own, which observed the two transits of Venus on a large scale long ago completed the work of reduction, and published the observations in full. On our own part we have published a preliminary discussion of some observations of the transit of 1874. Of that of 1882 nothing has, I believe, been published except some brief statements of results of the photographs, which appeared in an annual report of the Naval Observatory. Having need in my tables of the planets of the best value of the solar parallax that could be obtained by every method, I worked up all the observations of contacts made by the parties of every country, but, of course, did not publish our own observations. Up to the present time, twenty-eight years after the first of the transits, and twenty years after the second, our observations have never been officially published except to the extent I have stated. The importance of the matter may be judged by the fact that the government expended $375,000 on these observations, not counting the salaries of its officers engaged in the work, or the cost of sailing a naval ship. As I was a member of the commission charged with the work, and must therefore bear my full share of the responsibility for this failure, I think it proper to state briefly how it happened, hoping thereby to enforce the urgent need of a better organization of some of our scientific work.

The work of reducing such observations, editing and preparing them for the press, involved much computation to be done by assistants, and I, being secretary of the commission, was charged with the execution of this part of the work. The appropriations made by Congress for the observations were considered available for the reduction also. There was a small balance left over, and I estimated that $3000 more would suffice to complete the work. This was obtained from Congress in the winter of 1875.

About the end of 1876 I was surprised to receive from the Treasury Department a notification that the appropriation for the transit of Venus was almost exhausted, when according to my accounts, more than $3000 still remained. On inquiry it was found that the sum appropriated about two years before had never been placed to the credit of the transit of Venus commission, having been, in fact, inserted in a different appropriation bill from that which contained the former grant.

I, as secretary of the commission, made an application to the Treasury Department to have the sum, late though it was, placed to our credit. But the money had been expended and nothing could be now done in the matter.[1] The computers had therefore to be discharged and the work stopped until a new appropriation could be obtained from Congress.

During the session of 1876–77, $5000 was therefore asked for for the reduction of the observations. It was refused by the House committee on appropriations. I explained the matter to Mr. Julius H.

[1] As this result would not be possible under our present system, which was introduced by the first Cleveland administration, I might remark that it resulted from a practice on the part of the Treasury of lumping appropriations on its books in order to simplify the keeping of the accounts.

Seelye, formerly president of Amherst College, who was serving a term in Congress. He took much interest in the subject, and moved the insertion of the item when the appropriation bill came up before the House. Mr. Atkins, chairman of the appropriations committee, opposed the motion, maintaining that the Navy Department had under its orders plenty of officers who could do the work, so there was no need of employing the help of computers. But the House took a different view, and inserted the item over the heads of the appropriations committee.

Now difficulties incident to the divided responsibility of the commission were met with. During the interim between the death of Admiral Davis, in February, 1877, and the coming of Admiral John Rodgers as his successor, a legal question arose as to the power of the commission over its members. The work had to stop until it was settled, and I had to discharge my computers a second time. After it was again started I discovered that I did not have complete control of the funds appropriated for reducing the observations. The result was that the computers had to be discharged and the work stopped for the third time. This occurred not long before I started out to observe the transit in 1882. For me the third hair was the one that broke the camel's back. I turned the papers and work over to Professor Harkness, by whom the subject was continued until he was made astronomical director of the Naval Observatory in 1894.

I do not know that the commission was ever formally dissolved. Practically, however, its functions may be said to have terminated in the year 1886, when a provision of law was enacted by which all its property was turned over to the Secretary of the Navy.

What the present condition of the work may be, and how much of it is ready for the press, I cannot say. My impression is that it is in that condition known in household language as "all done but finishing." Whether it will ever appear is a question for the future. All the men who took part in it or who understood its details are either dead or on the retired list, and it is difficult for one not familiar with it from the beginning to carry it to completion.

VII

THE LICK OBSERVATORY

In the wonderful development of astronomical research in our country during the past twenty years, no feature is more remarkable than the rise on an isolated mountain in California of an institution which, within that brief period, has become one of the foremost observatories of the world. As everything connected with the early history of such an institution must be of interest, it may not be amiss if I devote a few pages to it.

In 1874 the announcement reached the public eye that James Lick, an eccentric and wealthy Californian, had given his entire fortune to a board of trustees to be used for certain public purposes, one of which was the procuring of the greatest and most powerful telescope that had ever been made. There was nothing in the previous history of the donor that could explain his interest in a great telescope. I am sure he had never looked through a telescope in his life, and that if he had, and had been acquainted with the difficulties of an observation with it, it is quite likely the Lick Observatory would never have existed. From his point of view, as, indeed, from that of the public very generally, the question of telescopic vision is merely one of magnifying power. By making an instrument large and powerful enough we may hope even to discover rational beings on other planets.

The president of the first board of trustees was Mr. D. O. Mills, the well-known capitalist, who had been president of the Bank of California. Mr. Mills visited Washington in the summer or autumn of 1874, and conferred with the astronomers there, among others myself, on the question of the proposed telescope. I do not think that an observatory properly so called was, at first, in Mr. Lick's mind; all he wanted was an immense telescope.

The question was complicated by the result of some correspondence between Mr. Lick and the firm of Alvan Clark & Sons. The latter had been approached to know the cost of constructing the desired telescope. Without making any exact estimate, or deciding upon the size of the greatest telescope that could be constructed, they named a very large sum, $200,000 I believe, as the amount that could be put into the largest telescope it was possible to make. Mr. Lick deemed this estimate exorbitant, and refused to have anything more to do with the firm. The question now was whether any one else besides the Clarks could make what was wanted.

I suggested to Mr. Mills that this question was a difficult one to answer, as no European maker was known to rival the Clarks in skill in the desired direction. It was impossible to learn what could be done in Europe except by a personal visit to the great optical workshops and a few observatories where great telescopes had been mounted.

I also suggested that a director of the new establishment should be chosen in advance of beginning active work, so that everything should be done under his supervision. As such director I suggested that very likely Professor Holden, then my assistant on the great equatorial, might be well qualified. At least I could not, at the moment, name any one I thought would be decidedly preferable to him. I suggested another man as possibly available, but remarked that he had been unfortunate. "I don't want to have anything to do with unfortunate men," was the reply. The necessity of choosing a director was not, however, evident, but communication was opened with Professor Holden as well as myself to an extent that I did not become aware of until long afterward.

The outcome of Mr. Mills's visit was that in December, 1874, I was invited to visit the European workshops as an agent of the Lick trustees, with a view of determining whether there was any chance of getting the telescope made abroad. The most difficult and delicate question arose in the beginning; shall the telescope be a reflector or a refractor? The largest and most powerful one that could be made would be, undoubtedly, a reflector. And yet reflecting telescopes had not, as a rule, been successful in permanent practical work. The world's work in astronomy was done mainly with refracting telescopes. This was not due to any inherent superiority in the latter, but to the mechanical difficulties incident to so supporting the great mirror of a reflecting telescope that it should retain its figure in all positions. Assuming that the choice must fall upon a refractor, unless proper guarantees for one of the other kind should be offered, one of my first visits was to the glass firm of Chance & Co. in Birmingham, who had cast the glass disks for the Washington telescope. This firm and Feil of Paris were the only two successful makers of great optical disks in the world. Chance & Co. offered the best guarantees, while Feil had more enthusiasm than capital, although his skill was of the highest. Another Paris firm was quite willing to undertake the completion of the telescope, but it was also evident that its price was suggested by the supposed liberality of an eccentric California millionaire. I returned their first proposal with the assurance that it would be useless to submit it. A second was still too high to offer any inducement over the American firm. Besides, there was no guarantee of the skill necessary to success.

In Germany the case was still worse. The most renowned firm there, the successors of Fraunhofer, were not anxious to undertake such a contract. The outcome of the matter was that Howard Grubb, of Dublin, was the only man abroad with whom negotiations could be opened with any chance of success. He was evidently a genius who meant business. Yet he had not produced a work which would justify unlimited confidence in his ability to meet Mr. Lick's requirements. The great Vienna telescope which he afterward constructed was then only being projected.

Not long after my return with this not very encouraging report, Mr. Lick suddenly revoked his gift, through some dissatisfaction with the proceedings of his trustees, and appointed a new board to carry out his plans. This introduced legal complications, which were soon settled by a friendly suit on the part of the old trustees, asking authority to transfer their trust. The president of the new board was Mr. Richard S. Floyd, a member of the well-known Virginia family of that name, and a graduate, or at least a former cadet, of the United States Naval Academy. I received a visit from him on his first trip to the East in his official capacity, early in 1876, I believe. Some correspondence with Mr. Lick's home representative ensued, of which the most interesting feature was the donor's idea of a telescope. He did not see why so elaborate and expensive a mounting as that proposed was necessary, and thought that the object glass might be mounted on the simplest kind of a pole or tower which would admit of its having the requisite motions in connection with the eyepiece. Whether I succeeded in convincing him of the impracticability of his scheme, I do not know, as he died before the matter was settled.

This left the trustees at liberty to build and

organize the institution as they deemed best. It was speedily determined that the object glass should be shaped by the Clarks, who should also be responsible for getting the rough disks. This proved to be a very difficult task. Chance & Co. were unwilling to undertake the work and Feil had gone out of business, leaving the manufacture in the hands of his son. The latter also failed, and the father had to return. Ultimately the establishment was purchased by Mantois, whose success was remarkable. He soon showed himself able to make disks not only of much larger size than had ever before been produced, but of a purity and transparency which none before him had ever approached. He died in 1899 or 1900, and it is to be hoped that his successor will prove to be his equal.

The original plan of Mr. Lick had been to found the observatory on the borders of Lake Tahoe, but he grew dissatisfied with this site and, shortly before his death, made provisional arrangements for placing it on Mount Hamilton. In 1879 preparations had so far advanced that it became necessary to decide whether this was really a suitable location. I had grave doubts on the subject. A mountain side is liable to be heated by the rays of the sun during the day, and a current of warm air which would be fatal to the delicacy of astronomical vision is liable to rise up the sides and envelope the top of the mountain. I had even been informed that, on a summer evening, a piece of paper let loose on the mountain top would be carried up into the air by the current. But, after all, the proof of the pudding is in the eating, and Holden united with me in advising that an experienced astronomer with a telescope should be stationed for a few weeks on the mountain in order to determine, by actual trial, what the conditions of seeing were. The one best man for this duty was S. W. Burnham of Chicago, who had already attained a high position in the astronomical world by the remarkable skill shown in his observations of double stars. So, in August, 1879, huts were built on the mountain, and Burnham was transported thither with his telescope. I followed personally in September.

We passed three nights on the mountain with Captain Floyd, studying the skies by night and prospecting around in the daytime to see whether the mountain top or some point in the neighboring plateau offered the best location for the observatory. So far as the atmospheric conditions were concerned, the results were beyond our most sanguine expectations. What the astronomer wants is not merely a transparent atmosphere, but one of such steadiness that the image of a star, as seen in a telescope, may not be disturbed by movements of the air which are invisible to the naked eye.

Burnham found that there were forty-two first-class nights during his stay, and only seven which would be classed as low as medium. In the East the number of nights which he would call first-class are but few in a year, and even the medium night is by no means to be counted on. No further doubt could remain that the top of the mountain was one of the finest locations in the world for an astronomical observatory, and it was definitely selected without further delay.

Sometime after my return Mr. Floyd sent me a topographical sketch of the mountain, with a request to prepare preliminary plans for the observatory. As I had always looked on Professor Holden as probably the coming director, I took him into consultation, and the plans were made under our joint direction in my office. The position and general arrangement of the buildings remain, so far as I am aware, much as then planned; the principal change being the omission of a long colonnade extending over the whole length of the main front in order to secure an artistic and imposing aspect from the direction of San José.

In the summer of 1885, as I was in New York in order to sail next day to Europe, I was surprised by a visit from Judge Hagar, a prominent citizen of San Francisco, a member of the Board of Regents of the University of California, and an active politician, who soon afterward became collector of the port, to consult me on the question of choosing Professor Holden as president of the university. This was not to interfere with his becoming director of the Lick Observatory whenever that institution should be organized, but was simply a temporary arrangement to bridge over a difficulty.

In the autumn of 1887 I received an invitation

from Mr. Floyd to go with him to Cleveland, in order to inspect the telescope, which was now nearly ready for delivery. It was mounted in the year following, and then Holden stepped from the presidency of the university into the directorship of the observatory.

The institution made its mark almost from the beginning. I know of no example in the world in which young men, most of whom were beginners, attained such success as did those whom Holden collected around him. The names of Barnard, Campbell, and Schaeberle immediately became well known in astronomy, owing to the excellence of their work. Burnham was, of course, no beginner, being already well known, nor was Keeler, who was also on the staff.

In a few years commenced the epoch-making work of Campbell, in the most refined and difficult problem of observational astronomy, — that of the measurement of the motion of stars to or from us. Through the application of photography and minute attention to details, this work of the Lick Observatory almost immediately gained a position of preëminence, which it maintains to the present time. If any rival is to appear, it will probably be the Yerkes Observatory. The friendly competition which we are likely to see between these two establishments affords an excellent example of the spirit of the astronomy of the future. Notwithstanding their rivalry, each has done and will do all it can to promote the work of the other.

The smiles of fortune have been bestowed even upon efforts that seemed most unpromising. After work was well organized, Mr. Crossley, of England, presented the observatory with a reflecting telescope of large size, but which had never gained a commanding reputation. No member of the staff at first seemed ambitious to get hold of such an instrument, but, in time, Keeler gave it a trial in photographing nebulæ. Then it was found that a new field lay open. The newly acquired reflector proved far superior to other instruments for this purpose, the photographic plates showing countless nebulæ in every part of the sky, which the human eye was incapable of discerning in the most powerful of telescopes.

In 1892, only four years after the mounting of the telescope, came the surprising announcement that the work of Galileo on Jupiter had been continued by the discovery of a fifth satellite to that planet. This is the most difficult object in the solar system, only one or two observers besides Barnard having commanded the means of seeing it. The incident of my first acquaintance with the discoverer is not flattering to my pride, but may be worth recalling.

In 1877 I was president of the American Association for the Advancement of Science at the meeting held in Nashville. There I was told of a young man a little over twenty years of age, a photographer by profession, who was interested in astronomy, and who desired to see me. I was, of course, very glad to make his acquaintance. I found that with his scanty earnings he had managed either to purchase or to get together the materials for making a small telescope. He was desirous of doing something with it that might be useful in astronomy, and wished to know what suggestions I could make in that line. I did not for a moment suppose that there was a reasonable probability of the young man doing anything better than amuse himself. At the same time, feeling it a duty to encourage him, I suggested that there was only one thing open to an astronomical observer situated as he was, and that was the discovery of comets. I had never even looked for a comet myself, and knew little about the methods of exploring the heavens for one, except what had been told me by H. P. Tuttle. But I gave him the best directions I could, and we parted. It is now rather humiliating that I did not inquire more thoroughly into the case. It would have taken more prescience than I was gifted with to expect that I should live to see the bashful youth awarded the gold medal of the Royal Astronomical Society for his work.

The term of Holden's administration extended through some ten years. To me its most singular feature was the constantly growing unpopularity of the director. I call it singular because, if we confine ourselves to the record, it would be difficult to assign any obvious reason for it. One fact is indisputable, and that is the wonderful success of the director in selecting young men who were to make the institution famous by their abilities and industry. If the highest problem of administration is to select the right men, the new director certainly mastered it. So far as liberty of research and publication went, the administration had the appearance of being liberal in the extreme. Doubtless there was another side to the question. Nothing happens spontaneously, and the singular phenomenon of one who had done all this becoming a much hated man must have an adequate cause. I have several times, from pure curiosity, inquired about the matter of well-informed men. On one occasion an instance of maladroitness was cited in reply.

"True," said I, "it was not exactly the thing to do, but, after all, that is an exceedingly small matter."

"Yes," was the answer, "that was a small thing, but put a thousand small things like that together, and you have a big thing."

A powerful factor in the case may have been his proceeding, within a year of his appointment, to file an astounding claim for the sum of $12,000 on account of services rendered to the observatory in the capacity of general adviser before his appointment as director. These services extended from the beginning of preparations in 1874 up to the completion of the work. The trustees in replying to the claim maintained that I had been their principal adviser in preparing the plans. However true this may have been, it was quite evident, from Holden's statement, that they had been consulting him on a much larger scale than I had been aware of. This, however, was none of my concern. I ventured to express the opinion that the movement was made merely to place on record a statement of the director's services; and that no serious intention of forcing the matter to a legal decision was entertained. This surmise proved to be correct, as nothing more was heard of the claim.

Much has been said of the effect of the comparative isolation of such a community, which is apt to be provocative of internal dissension. But this cause has not operated in the case of Holden's successors. Keeler became the second director in 1897, and administered his office with, so far as I know, universal satisfaction till his lamented death in 1900. It would not be a gross overstatement to say that his successor was named by the practically unanimous voice of a number of the leading astronomers of the world who were consulted on the subject, and who cannot but be pleased to see how completely their advice has been justified by the result of Campbell's administration.

VIII

THE AUTHOR'S SCIENTIFIC WORK

PERHAPS an apology is due to the reader for my venturing to devote a chapter to my own efforts in the scientific line. If so, I scarcely know what apology to make, unless it is that one naturally feels interested in matters relating to his own work, and hopes to share that interest with his readers, and that it is easier for one to write such an account for himself than for any one else to do it for him.

Having determined to devote my life to the prosecution of exact astronomy, the first important problem which I took up, while at Cambridge, was that of the zone of minor planets, frequently called asteroids, revolving between the orbits of Mars and Jupiter. It was formerly supposed that these small bodies might be fragments of a large planet which had been shattered by a collision or explosion. If such were the case, the orbits would, for a time at least, all pass through the point at which the explosion occurred. When only three or four were known, it was supposed that they did pass nearly through the same point. When this was found not to be the case, the theory of an explosion was in no way weakened, because, owing to the gradual changes in the form and position of the orbits, produced by the attraction of the larger planets, these orbits would all move away from the point of intersection, and, in the course of thousands of years, be so mixed up that no connection could be seen between them. This result was that nothing could be said upon the subject except that, if the catastrophe ever did occur, it must have been many thousand years ago. The fact did not in any way militate against the theory because, in view of the age of the universe, the explosion might as well have occurred hundreds of thousands or even millions of years ago as yesterday. To settle the question, general formulæ must be found by which the positions of these orbits could be determined at any time in the past, even hundreds of thousands of years back. The general methods of doing this were known, but no one had applied them to the especial case of these little planets. Here, then, was an opportunity of tracing back the changes in these orbits through thousands of centuries in order to find whether, at a certain epoch in the past, so great a cataclysm had occurred as the explosion of a world. Were such the case, it would be possible almost to set the day of the occurrence. How great a feat would it be to bring such an event at such a time to light!

I soon found that the problem, in the form in which it had been attacked by previous mathematicians, involved no serious difficulty. At the Springfield meeting of the American Association for the Advancement of Science, in 1859, I read a paper explaining the method, and showed by a curve on the blackboard the changes in the orbit of one of the asteroids for a period, I think, of several hundred thousand years, — " beyond the memory of the oldest inhabitants " — said one of the local newspapers. A month later it was extended to three other asteroids, and the result published in the " Astronomical Journal." In the following spring, 1860, the final results of the completed work were communicated to the American Academy of Arts and Sciences in a paper " On the Secular Variations and Mutual Relations of the Orbits of the Asteroids." The question of the possible variations in the orbits and the various relations amongst them were here fully discussed. One conclusion was that, so far as our present theory could show, the orbits had never passed through any common point of intersection.

The whole trend of thought and research since that time has been toward the conclusion that no such cataclysm as that looked for ever occurred, and that the group of small planets has been composed of separate bodies since the solar system came into existence. It was, of course, a great disappointment not to discover the cataclysm, but next best to finding a thing is showing that it is not there. This, it may be remarked, was the first of my papers to attract especial notice in foreign scientific journals, though I had already published several short notes on various subjects in the " Astronomical Journal."

At this point I may say something of the problems of mathematical astronomy in the middle of the last century. It is well known that we shall at least come very near the truth when we say that the planets revolve around the sun, and the satellites around their primaries according to the law of gravitation. We may regard all these bodies as projected into space, and thus moving according to laws similar to that which governs the motion of a stone thrown from the hand. If two bodies alone were concerned, say the sun and a planet, the orbit of the lesser around the greater would be an ellipse, which would never change its form, size, or position. That the orbits of the planets and asteroids do change, and that they are not exact ellipses, is due to their attraction upon each other. The question is, do these mutual attractions completely explain all the motions down to the last degree of refinement? Does any world move otherwise than as it is attracted by other worlds?

Two different lines of research must be brought to bear on the question thus presented. We must first know by the most exact and refined observations that the astronomer can make exactly how a heavenly body does move. Its position, or, as we cannot directly measure distance, its direction from us, must be determined as precisely as possible from time to time. Its course has been mapped out for it in advance by tables which are published in the "Astronomical Ephemeris," and we may express its position by its deviation from these tables. Then comes in the mathematical problem how it ought to move under the attraction of all other heavenly bodies that can influence its motion. The results must then be compared, in order to see to what conclusion we may be led.

This mathematical side of the question is of a complexity beyond the powers of ordinary conception. I well remember that when, familiar only with equations of algebra, I first looked into a book on mechanics, I was struck by the complexity of the formulæ. But this was nothing to what one finds when he looks into a work on celestial mechanics, where a single formula may fill a whole chapter. The great difficulty arises from the fact that the constant action upon a planet exerted at every moment of time through days and years by another planet affects its motion in all subsequent time. The action of Jupiter upon our earth this morning changes its motion forever, just as a touch upon a ball thrown by a pitcher will change the direction of the ball through its whole flight.

The wondrous perfection of mathematical research is shown by the fact that we can now add up, as it were, all these momentary effects through years and centuries, with a view of determining the combined result at any one moment. It is true that this can be done only in an imperfect way, and at the expense of enormous labor; but, by putting more and more work into it, investigating deeper and deeper, taking into account smaller and smaller terms of our formulæ, and searching for the minutest effects, we may gradually approach, though we may never reach, absolute exactness. Here we see the first difficulty in reaching a definite conclusion. One cannot be quite sure that a deviation is not due to some imperfection in mathematical method until he and his fellows have exhausted the subject so thoroughly as to show that no error is possible. This is hard indeed to do.

Taking up the question on the observational side, a source of difficulty and confusion at once presented itself. The motions of a heavenly body from day to day and year to year are mapped out by comparative observations on it and on the stars. The question of the exact positions of the stars thus comes in. In determining these positions with the highest degree of precision, a great variety of data have to be used. The astronomer cannot reach a result by a single step, nor by a hundred steps. He is like a sculptor chiseling all the time, trying to get nearer and nearer the ideal form of his statue, and finding that with every new feature he chisels out, a defect is brought to light in other features. The astronomer, when he aims at the highest mathematical precision in his results, finds Nature warring with him at every step, just as if she wanted to make his task as difficult as possible. She alters his personal equation when he gets tired, makes him see a small star differently from a bright one, gives his instrument minute twists with heat and cold, sends currents of warm or cold air over his locality, which refract the rays of light, asks him to keep the temperature in which he works the same as that outside, in order to avoid refraction when the air enters his observing room, and still will not let him do it, because the walls and everything inside the room, being warmed up during the day, make the air warmer than it is outside. With all these obstacles which she throws in his way he must simply fight the best he can, exerting untiring industry to eliminate their effects by repeated observations under a variety of conditions.

A necessary conclusion from all this is that the work of all observing astronomers, so far as it could be used, must be combined into a single whole. But here again difficulties are met at every step. There has been, in times past, little or no concert of action among astronomers at different observatories. The astronomers of each nation, perhaps of each observatory, to a large extent, have gone to work in their own way, using discordant data, perhaps not always rigidly consistent, even in the data used in a single establishment. How combine all the astronomical observations, found scattered through hundreds of volumes, into a homogeneous whole?

What is the value of such an attempt? Certainly if we measure value by the actual expenditure of nations and institutions upon the work, it must be very great. Every civilized nation expends a large annual sum on a national observatory, while a still greater number of such institutions are supported at corporate expense. Considering that the highest value can be derived from their labors only by such a combination as I have described, we may say the result is worth an important fraction of what all the observatories of the world have cost during the past century.

Such was, in a general way, the great problem of exact astronomy forty or fifty years ago. Its solution required extended coöperation, and I do not wish to give the impression that I at once attacked it, or even considered it as a whole. I could only determine to do my part in carrying forward the work associated with it.

Perhaps the most interesting and important branch of the problem concerned the motion of the moon. This had been, ever since the foundation of the Greenwich Observatory, in 1670, a specialty of that institution. It is a curious fact, however, that while that observatory supplied all the observations of the moon, the investigations based upon these observations were made almost entirely by foreigners, who also constructed the tables by which the moon's motion was mapped out in advance. The most perfect tables made were those of Hansen, the greatest master of mathematical astronomy during the middle of the century, whose tables of the moon were published by the British government in 1857. They were based on a few of the Greenwich observations from 1750 to 1850. The period began with 1750, because that was the earliest at which observations of any exactness were made. Only a few observations were used, because Hansen, with the limited computing force at his command, — only a single assistant, I believe, — was not able to utilize a great number of the observations. The rapid motion of the moon, a circuit being completed in less than a month, made numerous observations necessary, while the very large deviations in the motion produced by the attraction of the sun made the problem of the mathematical theory of that motion the most complicated in astronomy. Thus it happened that, when I commenced work at the Naval Observatory in 1861, the question whether the moon exactly followed the course laid out for her by Hansen's tables was becoming of great importance.

The same question arose in the case of the planets. So from a survey of the whole field, I made observations of the sun, moon, and planets my specialty at the observatory. If the astronomical reader has before him the volume of observations for 1861, he will, by looking at pages 366–440, be able to infer with nearly astronomical precision the date when I reported for duty.

For a year or two our observations showed that the moon seemed to be falling a little behind her predicted motion. But this soon ceased, and she gradually forged ahead in a much more remarkable way. In five or six years it was evident that this was becoming permanent; she was a little farther ahead every year. What could it mean? To consider this question, I may add a word to what I have already said on the subject.

In comparing the observed and predicted motion

of the moon, mathematicians and astronomers, beginning with Laplace, have been perplexed by what are called " inequalities of long period." For a number of years, perhaps half a century, the moon would seem to be running ahead, and then she would gradually relax her speed and fall behind. Laplace suggested possible causes, but could not prove them. Hansen, it was supposed, had straightened out the tangle by showing that the action of Venus produced a swinging of this sort in the moon; for one hundred and thirty years she would be running ahead and then for one hundred and thirty years more falling back again, like a pendulum. Two motions of this sort were combined together. They were claimed to explain the whole difficulty. The moon, having followed Hansen's theory for one hundred years, would not be likely to deviate from it. Now, it was deviating. What could it mean?

Taking it for granted, on Hansen's authority, that his tables represented the motions of the moon perfectly since 1750, was there no possibility of learning anything from observations before that date? As I have already said, the published observations with the usual instruments were not of that refined character which would decide a question like this. But there is another class of observations which might possibly be available for the purpose.

Millions of stars, visible with large telescopes, are scattered over the heavens; tens of thousands are bright enough to be seen with small instruments, and several thousand are visible to any ordinary eye. The moon, in her monthly course around the heavens, often passes over a star, and of course hides it from view during the time required for the passage. The great majority of stars are so small that their light is obscured by the effulgence of the moon as the latter approaches them. But quite frequently the star passed over is so bright that the exact moment when the moon reaches it can be observed with the utmost precision. The star then disappears from view in an instant, as if its light were suddenly and absolutely extinguished. This is called an occultation. If the moment at which the disappearance takes place is observed, we know that at that instant the apparent angle between the centre of the moon and the star is equal to the moon's semi-diameter. By the aid of a number of such observations, the path of the moon in the heavens, and the time at which she arrives at each point of the path, can be determined. In order that the determination may be of sufficient scientific precision, the time of the occultation must be known within one or two seconds; otherwise, we shall be in doubt how much of the discrepancy may be due to the error of the observation, and how much to the error of the tables.

Occultations of some bright stars, such as Aldebaran and Antares, can be observed by the naked eye; and yet more easily can those of the planets be seen. It is therefore a curious historic fact that there is no certain record of an actual observation of this sort having been made until after the commencement of the seventeenth century. Even then the observations were of little or no use, because astronomers could not determine their time with sufficient precision. It was not till after the middle of the century, when the telescope had been made part of astronomical instruments for finding the altitude of a heavenly body, and after the pendulum clock had been invented by Huyghens, that the time of an occultation could be fixed with the required exactness. Thus it happens that from 1640 to 1670 somewhat coarse observations of the kind are available, and after the latter epoch those made by the French astronomers become almost equal to the modern ones in precision.

The question that occurred to me was: Is it not possible that such observations were made by astronomers long before 1750? Searching the published memoirs of the French Academy of Sciences and the Philosophical Transactions, I found that a few such observations were actually made between 1660 and 1700. I computed and reduced a few of them, finding with surprise that Hansen's tables were evidently much in error at that time. But neither the cause, amount, or nature of the error could be well determined without more observations than these. Was it not possible that these astronomers had made more than they published? The hope that material of this sort existed was encouraged by the discovery at the Pulkowa Observatory of an old manuscript by the French astronomer Delisle, containing some observations of this kind. I therefore planned a thorough search of the old records in Europe to see what could be learned.

The execution of this plan was facilitated by the occurrence, in December, 1870, of an eclipse of the sun in Spain and along the Mediterranean. A number of parties were going out from this country to observe it, two of which were fitted out at the Naval Observatory. I was placed in charge of one of these, consisting, practically, of myself. The results of my observation would be of importance in the question of the moon's motion, but, although the eclipse was ostensibly the main object, the proposed search of the records was what I really had most in view. In Paris was to be found the most promising mine; but the Franco-Prussian war was then going on, and I had to wait for its termination. Then I made a visit to Paris, which will be described in a later chapter.

At the observatory the old records I wished to consult were placed at my disposal, with full liberty not only to copy, but to publish anything of value I could find in them. The mine proved rich beyond the most sanguine expectation. After a little prospecting, I found that the very observations I wanted had been made in great numbers by the Paris astronomers, both at the observatory and at other points in the city.

And how, the reader may ask, did it happen that these observations were not published by the astronomers who made them? Why should they have lain unused and forgotten for two hundred years? The answer to these questions is made plain enough by an examination of the records. The astronomers had no idea of the possible usefulness and value of what they were recording. So far as we can infer from their work, they made the observations merely because an occultation was an interesting thing to see; and they were men of sufficient scientific experience and training to have acquired the excellent habit of noting the time at which a phenomenon was observed. But they were generally satisfied with simply putting down the clock time. How they could have expected their successors to make any use of such a record, or whether they had any expectations on the subject, we cannot say with confidence. It will be readily understood that no clocks of the present time (much less those of two hundred years ago) run with such precision that the moment read from the clock is exact within one or two seconds. The modern astronomer does not pretend to keep his clock correct within less than a minute; he determines by observation how far it is wrong, on each date of observation, and adds so much to the time given by the clock, or subtracts it, as the case may be, in order to get the correct moment of true time. In the case of the French astronomers, the clock would frequently be fifteen minutes or more in error, for the reason that they used apparent time, instead of mean time as we do. Thus when, as was often the case, the only record found was that, at a certain hour, minute, and second, by a certain clock, *une étoile se cache par la lune,* a number of very difficult problems were presented to the astronomer who was to make use of the observations two centuries afterward. First of all, he must find out what the error of the clock was at the designated hour, minute, and second; and for this purpose he must reduce the observations made by the observer in order to determine the error. But it was very clear that the observer did not expect any successor to take this trouble, and therefore did not supply him with any facilities for so doing. He did not even describe the particular instrument with which the observations were made, but only wrote down certain figures and symbols, of a more or less hieroglyphic character. It needed much comparison and examination to find out what sort of an instrument was used, how the observations were made, and how they should be utilized for the required purpose.

Generally the star which the moon hid was mentioned, but not in all cases. If it was not, the identification of the star was a puzzling problem. The only way to proceed was to calculate the apparent position of the centre of the moon as seen by an observer at the Paris Observatory, at the particular hour and minute of the observation. A star map was then taken; the points of a pair of dividers were separated by the length of the moon's radius, as it would appear on the scale of the map; one point of the dividers was put into the position of the moon's centre on the map, and with the other a circle was drawn. This circle represented the outline of the moon, as it appeared to the observer at the Paris Observatory, at the hour and minute in question, on a certain day in the seventeenth century. The star should be found very near the circumference of the circle, and in nearly all cases a star was there.

Of course all this could not be done on the spot. What had to be done was to find the observations, study their relations and the method of making them, and copy everything that seemed necessary for working them up. This took some six weeks, but the material I carried away proved the greatest find I ever made. Three or four years were spent in making all the calculations I have described. Then it was found that seventy-five years were added, at a single step, to the period during which the history of the moon's motion could be written. Previously this history was supposed to commence with the observations of Bradley, at Greenwich, about 1750; now it was extended back to 1675, and with a less degree of accuracy thirty years farther still. Hansen's tables were found to deviate from the truth, in 1675 and subsequent years, to a surprising extent; but the cause of the deviation is not entirely unfolded even now.

During the time I was doing this work, Paris was under the reign of the Commune and besieged by the national forces. The studies had to be made within hearing of the besieging guns; and I could sometimes go to a window and see flashes of artillery from one of the fortifications to the south. Nearly every day I took a walk through the town, occasionally as far as the Arc. As my observations during these walks have no scientific value, I shall postpone an account of what I saw to another chapter.

One curious result of this work is that the longitude of the moon may now be said to be known with greater accuracy through the last quarter of the seventeenth century than during the ninety years from 1750 to 1840. The reason is that, for this more modern period, no effective comparison has been made between observations and Hansen's tables.

Just as this work was approaching completion I was called upon to decide a question which would materially influence all my future activity. The lamented death of Professor Winlock in 1875 left vacant the directorship of the Harvard Observatory. A month or two later I was quite taken by surprise to receive a letter from President Eliot tendering me this position. I thus had to choose between two courses. One led immediately to a professorship in Harvard University, with all the distinction and worldly advantages associated with it, including complete freedom of action, an independent position, and the opportunity of doing such work as I deemed best with the limited resources at the disposal of the observatory. On the other hand was a position to which the official world attached no importance, and which brought with it no worldly advantages whatever.

I first consulted Mr. Secretary Robeson on the matter. The force with which he expressed himself took me quite by surprise. " By all means accept the place; don't remain in the government service a day longer than you have to. A scientific man here has no future before him, and the quicker he can get away the better." Then he began to descant on our miserable " politics " which brought about such a state of things.

Such words, coming from a sagacious head of a department who, one might suppose, would have been sorry to part with a coadjutor of sufficient importance to be needed by Harvard University, seemed to me very suggestive. And yet I finally declined the place, perhaps unwisely for myself, though no one who knows what the Cambridge Observatory has become under Professor Pickering can feel that Harvard has any cause to regret my decision. An apology for it on my own behalf will seem more appropriate.

On the Cambridge side it must be remembered that the Harvard Observatory was then almost nothing compared with what it is now. It was poor in means, meagre in instrumental outfit, and wanting in working assistants; I think the latter did not number more than three or four, with perhaps a few other temporary employees. There seemed little prospect of doing much.

On the Washington side was the fact that I was bound to Washington by family ties, and that, if Harvard needed my services, surely the government needed them much more. True, this argument was, for the time, annulled by the energetic assurance of Secretary Robeson, showing that the government felt no want of any one in its service able to command a university professorship. But I was still pervaded by the optimism of youth in everything that concerned the future of our government, and did not believe that, with the growth of intelligence in our country, an absence of touch between the scientific and literary classes on the one side, and " politics " on the other, could continue. In addition to this was the general feeling by which I have been actuated from youth — that one ought to choose that line of activity for which Nature had best fitted him, trusting that the operation of moral causes would, in the end, right every wrong, rather than look out for place and preferment. I felt that the conduct of government astronomy was that line of activity for which I was best fitted, and that, in the absence of strong reason to the contrary, it had better not be changed. In addition to these general considerations was the special point that, in the course of a couple of years, the directorship of the Nautical Almanac would become vacant, and here would be an unequaled opportunity for carrying on the work in mathematical astronomy I had most at heart. Yet, could I have foreseen that the want of touch which I have already referred to would not be cured, that I should be unable to complete the work I had mapped out before my retirement, or to secure active public interest in its continuance, my decision would perhaps have been different.

On September 15, 1877, I took charge of the Nautical Almanac Office. The change was one of the happiest of my life. I was now in a position of recognized responsibility, where my recommendations met with the respect due to that responsibility, where I could make plans with the assurance of being able to carry them out, and where the countless annoyances of being looked upon as an important factor in work where there was no chance of my being such would no longer exist. Practically I had complete control of the work of the office, and was thus, metaphorically speaking, able to work with untied hands. It may seem almost puerile to say this to men of business experience, but there is a current notion, spread among all classes, that because the Naval Observatory has able and learned professors, therefore they must be able to do good and satisfactory work, which may be worth correcting.

I found my new office in a rather dilapidated old dwelling-house, about half a mile or less from the observatory, in one of those doubtful regions on the border line between a slum and the lowest order of respectability. If I remember aright, the only occupants of the place were the superintendent, my old friend Mr. Loomis, senior assistant, who looked after current business, a proof-reader and a messenger. All the computers, including even one copyist, did their work at their homes.

A couple of changes had to be made in the interest of efficiency. The view taken of one of these may not only interest the reader, but give him an idea of what people used to think of government service before the era of civil service reform. The proof-reader was excellent in every respect except that of ability to perform his duty. He occupied a high position, I believe, in the Grand Army of the Republic, and thus wielded a good deal of influence. When his case was appealed to the Secretary of the Navy, apellant was referred to me. I stated the trouble to counsel, — he did not appear to see figures, or be able to distinguish whether they were right or wrong, and therefore was useless as a proof-reader.

"It is not his fault," was the reply; "he nearly lost his eyesight in the civil war, and it is hard for him to see at all." In the view of counsel that explanation ought to have settled the case in his favor. It did not, however, but "influence" had no difficulty in making itself more successful in another field.

Among my first steps was that of getting a new office in the top of the Corcoran Building, then just completed. It was large and roomy enough to allow quite a number of assistants around me.

Much of the work was then, as now, done by the piece, or annual job, the computers on it very generally working at their homes. This offers many advantages for such work; the government is not burdened with an officer who must be paid his regular monthly salary whether he supplies his work or not, and whom it is unpleasant and difficult to get rid of in case of sickness or breakdown of any sort. The work is paid for when furnished, and the main trouble of administration saved. It is only necessary to have a brief report from time to time, showing that the work is actually going on.

I began with a careful examination of the relation of prices to work, making an estimate of the time probably necessary to do each job. Among the performers of the annual work were several able and eminent professors at various universities and schools. I found that they were being paid at pretty high professional prices. I recall with great satisfaction that I was able to reduce the prices and, step by step, concentrate all the work in Washington, without detriment to the pleasant relations I sustained with these men, some of them old and intimate friends. These economies went on increasing year by year, and every dollar that was saved went into the work of making the tables necessary for the future use of the Ephemeris.

The programme of work which I mapped out, involved, as one branch of it, a discussion of all the observations of value on the positions of the sun, moon, and planets, and incidentally, on the bright fixed stars, made at the leading observatories of the world since 1750. One might almost say it involved repeating, in a space of ten or fifteen years, an important part of the world's work in astronomy for more than a century past. Of course, this was impossible to carry out in all its completeness. In most cases what I was obliged practically to confine myself to was a correction of the reductions already made and published. Still, the job was one with which I do not think any astronomical one ever before attempted by a single person could compare in extent. The number of meridian observations on the sun, Mercury, Venus, and Mars alone numbered 62,030. They were made at the observatories of Greenwich, Paris, Königsberg, Pulkowa, Cape of Good Hope, — but I need not go over the entire list, which numbers thirteen.

The other branches of the work were such as I have already described, — the computation of the formulæ for the perturbation of the various planets by each other. As I am writing for the general reader, I need not go into any further technical description of this work than I have already done. Something about my assistants may, however, be of interest. They were too numerous to be all recalled individually. In fact, when the work was at its height, the office was, in the number of its scientific employees, nearly on an equality with the three or four greatest observatories of the world.

One of my experiences has affected my judgment on the general morale of the educated young men of our country. In not a single case did I ever have an assistant who tried to shirk his duty to the government, nor do I think there was more than a single case in which one tried to contest my judgment of his own merits, or those of his work. I adopted the principle that promotion should be by merit rather than by seniority, and my decisions on that matter were always accepted without complaint. I recall two men who voluntarily resigned when they found that, through failure of health or strength, they were unable to properly go on with their work. In frankness I must admit that there was one case in which I had a very disagreeable contest in getting rid of a learned gentleman whose practical powers were so far inferior to his theoretical knowledge that he was almost useless in the office. He made the fiercest and most determined fight in which I was ever engaged, but I must, in justice to all concerned, say that his defect was not in will to do his work but in the requisite power. Officially I was not without fault, because, in the press of matters requiring my attention, I had entrusted too much to him, and did not discover his deficiencies until some mischief had been done.

Perhaps the most eminent and interesting man associated with me during this period was Mr. George W. Hill, who will easily rank as the greatest master of mathematical astronomy during the last quarter of the nineteenth century. The only defect of his make-up of which I have reason to complain is the lack of the teaching faculty. Had this been developed in him, I could have learned very much from him that would have been to my advantage. In saying this I have one especial point in mind. In beginning my studies in celestial mechanics, I lacked the guidance of some one conversant with the subject on its practical side. Two systems of computing planetary perturbations had been used, one by Leverrier, while the other was invented by Hansen. The former method was, in principle, of great simplicity, while the latter seemed to be very complex and even clumsy. I naturally supposed that the man who computed the direction of the planet Neptune before its existence was known, must be a master of the whole subject, and followed the lines he indicated. I gradually discovered the contrary, and introduced modified methods, but did not entirely break away from the old trammels. Hill had never been bound by them, and used Hansen's method from the beginning. Had he given me a few demonstrations of its advantages, I should have been saved a great deal of time and labor.

The part assigned to Hill was about the most difficult in the whole work, — the theory of Jupiter and Saturn. Owing to the great mass of these " giant planets," the inequalities of their motion, especially in the case of Saturn, affected by the attraction of Jupiter, is greater than in the case of the other planets. Leverrier failed to attain the necessary exactness in his investigation of their motion. Hill had done some work on the subject at his home in Nyack Turnpike before I took charge of the office. He now moved to Washington, and seriously began the complicated numerical calculations which his task involved. I urged that he should accept the assistance of less skilled computers; but he declined it from a desire to do the entire work himself. Computers to make the duplicate computations necessary to guard against accidental numerical errors on his part were all that he required. He labored almost incessantly for about ten years, when he handed in the manuscript of what now forms Volume IV. of the "Astronomical Papers."

A pleasant incident occurred in 1884, when the office was honored by a visit from Professor John C. Adams of England, the man who, independently of Leverrier, had computed the place of Neptune, but failed to receive the lion's share of the honor because it happened to be the computations of the Frenchman and not his which led immediately to the discovery of the planet. It was of the greatest interest to me to bring two such congenial spirits as Adams and Hill together.

It would be difficult to find a more impressive example than that afforded by Hill's career, of the difficulty of getting the public to form and act upon sane judgments in such cases as his. The world has the highest admiration for astronomical research, and in this sentiment our countrymen are foremost. They spend hundreds of thousands of dollars to promote it. They pay good salaries to professors who chance to get a certain official position where they may do good work. And here was perhaps the greatest living master in the highest and most difficult field of astronomy, winning worldwide recognition for his country in the science, and receiving the salary of a department clerk. I never wrestled harder with a superior than I did with Hon. R. W. Thompson, Secretary of the Navy, about 1880, to induce him to raise Mr. Hill's salary from $1200 to $1400. It goes without saying that Hill took even less interest in the matter than I did. He did not work for pay, but for the love of science. His little farm at Nyack Turnpike sufficed for his home, and supplied his necessities so long as he lived there, and all he asked in Washington was the means of going on with his work. The deplorable feature of the situation is, that this devotion to his science, instead of commanding due recognition on the public and official side, rather tended to create an inadequate impression of the importance of what he was doing. That I could not secure for him at least the highest official consideration is among the regretful memories of my official life.

Although, so far as the amount of labor is concerned, Mr. Hill's work upon Jupiter and Saturn is the most massive he ever undertook, his really great scientific merit consists in the development of a radically new method of computing the inequalities of the moon's motion, which is now being developed and applied by Professor E. W. Brown. His most marked intellectual characteristic is the eminently practical character of his researches. He does not aim so much at elegant mathematical formulæ, as to determine with the greatest precision the actual quantities of which mathematical astronomy stands in need. In this direction he has left every investigator of recent or present time far in the rear.

After the computations on Jupiter and Saturn were made, it was necessary to correct their orbits and make tables of their motions. This work I left entirely in Mr. Hill's hands, the only requirement being that the masses of the planets and other data which he adopted should be uniform with those I used in the rest of the work. His tables were practically completed in manuscript at the beginning of 1892. When they were through, doubtless feeling, as well he might, that he had done his whole duty to science and the government, Mr. Hill resigned his office and returned to his home. During the summer he paid a visit to Europe, and visiting the Cambridge University, was honored with the degree of Doctor of Laws, along with a distinguished company, headed by the Duke of Edinburgh. One of the pleasant things to recall was that, during the fifteen years of our connection, there was never the slightest dissension or friction between us.

I may add that the computations which he made on the theory of Jupiter and Saturn are all preserved complete and in perfect form at the Nautical Almanac Office, so that, in case any question should arise respecting them in future generations, the point can be cleared up by an inspection.

In 1874, three years before I left the observatory, I was informed by Dr. Henry Draper that he had a mechanical assistant who showed great fondness for and proficiency in some work in mathematical astronomy. I asked to see what he was doing, and received a collection of papers of a remarkable kind. They consisted mainly of some of the complicated developments of celestial mechanics. In returning them I wrote to Draper that, when I was ready to begin my work on the planetary theories, I must have his man, — could he possibly be spared? But he came to me before the time, while I was carrying on some investigations with aid afforded by the Smithsonian Institution. Of course, when I took charge of the Nautical Almanac Office, he was speedily given employment on its work. His name was John Meier, a Swiss by birth, evidently from the peasant class, but who had nevertheless been a pupil of Professor Rudolph Wolf at Zurich. Emigrating to this country, he was, during the civil war, an engineer's mate or something of that grade in the navy. He was the most perfect example of a mathematical machine that I ever had at command. Of original power, — the faculty of developing new methods and discovering new problems, he had not a particle. Happily for his peace of mind, he was totally devoid of worldly ambition. I had only to prepare the fundamental data for him, explain what was wanted, write down the matters he was to start with, and he ground out day after day the most complicated algebraic and trigonometrical computations with untiring diligence and almost unerring accuracy.

But a dark side of the picture showed itself very suddenly and unexpectedly in a few years. For the most selfish reasons, if for no others, I desired that his peace of mind should be undisturbed. The result was that I was from time to time appealed to as an arbitrator of family dissensions, in which it was impossible to say which side was right and which wrong. Then, as a prophylactic against malaria, his wife administered doses of whiskey. The rest of the history need not be told. It illustrates the maxim that "blood will tell," which I fear is as true in scientific work as in any other field of human activity.

A man of totally different blood, the best in fact, entered the office shortly before Meier broke down. This was Mr. Cleveland Keith, son of Professor Reuel Keith, who was one of the professors at the observatory when it was started. His patience and ability led to his gradually taking the place of a foreman in supervising the work pertaining to the reduction of the observations, and the construction of the tables of the planets. Without his help, I fear I should never have brought the tables to a conclusion. He died in 1896, just as the final results of the work were being put together.

High among the troublesome problems with which I had to deal while in charge of the Nautical Almanac, was that of universal time. All but the youngest of my readers will remember the period when every railway had its own meridian, by the time of which its trains were run, which had to be changed here and there in the case of the great trunk lines, and which seldom agreed with the local time of a place. In the Pennsylvania station at Pittsburg were three different times; one that of Philadelphia, one of some point farther west, and the third the local Pittsburg time. The traveler was constantly liable to miss a train, a connection, or an engagement by the doubt and confusion thus arising.

This was remedied in 1883 by the adoption of our present system of standard times of four different meridians, the introduction of which was one of the great reforms of our generation. When this change was made, I was in favor of using Washington time as the standard, instead of going across the ocean to Greenwich for a meridian. But those who were pressing the measure wanted to have a system for the whole world, and for this purpose the meridian of Greenwich was the natural one.

Practically our purpose was served as well by the Greenwich meridian as it would have been by that of Washington.

The year following this change an international meridian conference was held at Washington, on the invitation of our government, to agree upon a single prime meridian to be adopted by the whole world in measuring longitudes and indicating time.

Of course the meridian of Greenwich was the only one that would answer the purpose. This had already been adopted by several leading maritime nations, including ourselves as well as Great Britain. It was merely a question of getting the others to fall into line. No conference was really necessary for this purpose, because the dissentients caused much more inconvenience to themselves than to any one else by their divergent practice. The French held out against the adoption of the Greenwich meridian, and proposed one passing through Behring Strait. I was not a member of the conference, but was invited to submit my views, which I did orally. I ventured to point out to the Frenchmen that the meridian of Greenwich also belonged to France, passing near Havre and intersecting their country from north to south. It was therefore as much a French as an English meridian, and could be adopted without any sacrifice of national position. But they were not convinced, and will probably hold out until England adopts the metric system, on which occasion it is said that they will be prepared to adopt the Greenwich meridian.

One proceeding of the conference illustrates a general characteristic of reformers. Almost without debate, certainly without adequate consideration, the conference adopted a recommendation that astronomers and navigators should change their system of reckoning time. Both these classes have, from time immemorial, begun the day at noon, because this system was most natural and convenient, when the question was not that of a measure of time for daily life, but simply to indicate with mathematical precision the moment of an event. Navigators had begun the day at noon, because the observations of the sun, on which the latitude of a ship depends, are necessarily made at noon, and the run of the ship is worked up immediately afterward. The proposed change would have produced unending confusion in astronomical nomenclature, owing to the difficulty of knowing in all cases which system of time was used in any given treatise or record of observations. I therefore felt compelled, in the general interest of science and public convenience, to oppose the project with all my power, suggesting that, if the new system must be put into operation, we should wait until the beginning of a new century.

"I hope you will succeed in having its adoption postponed until 1900," wrote Airy to me, "and when 1900 comes, I hope you will further succeed in having it again postponed until the year 2000."

The German official astronomers, and indeed most of the official ones everywhere, opposed the change, but the efforts on the other side were vigorously continued. The British Admiralty was strongly urged to introduce the change into the Nautical Almanac, and the question of doing this was warmly discussed in various scientific journals.

One result of this movement was that, in 1886, Rear-Admiral George H. Belknap, superintendent of the Naval Observatory, and myself were directed to report on the question. I drew up a very elaborate report, discussing the subject especially in its relations to navigation, pointing out in the strongest terms I could the danger of placing in the hands of navigators an almanac in which the numbers were given in a form so different from that to which they were accustomed. If they chanced to forget the change, the results of their computations might be out to any extent, to the great danger and confusion of their reckoning, while not a solitary advantage would be gained by it.

There is some reason to suppose that this document found its way to the British Admiralty, but I never heard a word further on the subject except that it ceased to be discussed in London. A few years later some unavailing efforts were made to revive the discussion, but the twentieth century is started without this confusing change being introduced into the astronomical ephemerides and nautical almanacs of the world, and navigators are still at liberty to practice the system they find most convenient.

In 1894 I had succeeded in bringing so much of the work as pertained to the reduction of the observations and the determination of the elements of the planets to a conclusion. So far as the larger planets were concerned, it only remained to construct the necessary tables, which, however, would be a work of several years.

With the year 1896 came what was perhaps the most important event in my whole plan. I have already remarked upon the confusion which pervaded the whole system of exact astronomy, arising from the diversity of the fundamental data made use of by the astronomers of foreign countries and various institutions in their work. It was, I think, rather exceptional that any astronomical result was based on entirely homogeneous and consistent data. To remedy this state of things and start the exact astronomy of the twentieth century on one basis for the whole world, was one of the objects which I had mapped out from the beginning. Dr. A. M. W. Downing, superintendent of the British Nautical Almanac, was struck by the same consideration and animated by the same motive. He had especially in view to avoid the duplication of work which arose from the same computations being made in different countries for the same result, whereby much unnecessary labor was expended. The field of astronomy is so vast, and the quantity of work urgently required to be done so far beyond the power of any one nation, that a combination to avoid all such waste was extremely desirable.

When, in 1895, my preliminary results were published, he took the initiative in a project for putting the idea into effect, by proposing an international conference of the directors of the four leading ephemerides, to agree upon a uniform system of data for all computations pertaining to the fixed stars. This conference was held in Paris in May, 1896. After several days of discussion, it resolved that, beginning with 1901, a certain set of constants should be used in all the ephemerides, substantially the same as those I had worked out, but without certain ulterior, though practically unimportant, modifications which I had applied for the sake of symmetry. My determination of the positions and motions of the bright fixed stars, which I had not yet completed, was adopted in advance for the same purpose, I agreeing to complete it if possible in time for use in 1901. I also agreed to make a new determination of the constant of precession, that which I had used in my previous work not being quite satisfactory. All this by no means filled the field of exact astronomy, yet what was left outside of it was of comparatively little importance for the special object in view.

More than a year after the conference I was taken quite by surprise by a vigorous attack on its work and conclusions on the part of Professor Lewis Boss, director of the Dudley Observatory, warmly seconded by Mr. S. C. Chandler of Cambridge, the editor of the "Astronomical Journal." The main grounds of attack were two in number. The time was not ripe for concluding upon a system of permanent astronomical standards. Besides this, the astronomers of the country should have been consulted before a decision was reached.

Ultimately the attack led to a result which may appear curious to the future astronomer. He will find the foreign ephemerides using uniform data worked out in the office of the "American Ephemeris and Nautical Almanac" at Washington for the years beginning with 1901. He will find that these same data, after being partially adopted in the ephemeris for 1900, were thrown out in 1901, and the antiquated ones reintroduced in the main body of the ephemeris. The new ones appear simply in an appendix.

As, under the operation of law, I should be retired from active service in the March following the conference, it became a serious question whether I should be able to finish the work that had been mapped out, as well as the planetary tables. Mr. Secretary Herbert, on his own motion so far as I know, sent for me to inquire into the subject. The result of the conference was a movement on his part to secure an appropriation somewhat less than the highest salary of a professor, to compensate me for the completion of the work after my retirement. The House Committee on Appropriations, ever mindful of economy in any new item, reduced the amount to a clerical salary. The committee of conference compromised on a mean between the two. It happened that the work on the stars was not specified in the law, — only the tables of the planets. In consequence I had no legal right to go on with the former, although the ephemerides of Europe were waiting for the results. After much trouble an arrangement was effected under which the computers on the work were not to be prohibited from consulting me in its prosecution.

Astronomical work is never really done and finished. The questions growing out of the agreement or non-agreement of the tables with observations still remain to be studied, and require an immense amount of computation. In what country and by whom these computations will be made no one can now tell. The work which I most regretted to leave unfinished was that on the motion of the moon. As I have already said, this work is complete to 1750. The computations for carrying it on from 1750 to the present time were perhaps three fourths done when I had to lay them aside. In 1902, when the Carnegie Institution was organized, it made a grant for supplying me with the computing assistance and other facilities necessary for the work, and the Secretary of the Navy allowed me the use of the old computations. Under such auspices the work was recommenced in March, 1903.

So far as I can recall, I never asked anything from the government which would in any way promote my personal interests. The only exception, if such it is, is that during the civil war I joined with other professors in asking that we be put on the same footing with other staff corps of the navy as regarded pay and rank. So far as my views were concerned, the rank was merely a *pro forma* matter, as I never could see any sound reason for a man pursuing astronomical duties caring to have military rank.

In conducting my office also, the utmost economy was always studied. The increase in the annual appropriations for which I asked was so small that, when I left the office in 1877, they were just about the same as they were back in the fifties, when it was first established. The necessary funds were saved by economical administration. All this was done with a feeling that, after my retirement, the satisfaction with which one could look back on such a policy would be enhanced by a feeling on the part of the representatives of the public that the work I had done must be worthy of having some pains taken to secure its continuance in the same spirit.

I do not believe that the men who conduct our own government are a whit behind the foremost of other countries in the desire to promote science. If after my retirement no special measures were deemed necessary to secure the continuance of the work in which I had been engaged, I prefer to attribute it to adventitious circumstances rather than to any undervaluation of scientific research by our authorities.

IX

SCIENTIFIC WASHINGTON

It is sometimes said that no man, in passing away, leaves a place which cannot be equally well filled by another. This is doubtless true in all ordinary cases. But scientific research, and scientific affairs generally at the national capital, form an exception to many of the rules drawn from experience in other fields.

Professor Joseph Henry, first secretary of the Smithsonian Institution, was a man of whom it may be said, without any reflection on men of our generation, that he held a place which has never been filled. I do not mean his official place, but his position as the recognized leader and exponent of scientific interests at the national capital. A worldwide reputation as a scientific investigator, exalted character and inspiring presence, broad views of men and things, the love and esteem of all, combined to make him the man to whom all who knew him looked for counsel and guidance in matters affecting the interests of science. Whether any one could since have assumed this position, I will not venture to say; but the fact seems to be that no one has been at the same time able and willing to assume it.

On coming to Washington I soon became very intimate with Professor Henry, and I do not think there was any one here to whom he set forth his personal wishes and convictions respecting the policy of the Smithsonian Institution and its relations to the government more freely than he did to me. As every point connected with the history and policy of this establishment is of world-wide interest, and as Professor Henry used to put some things in a different light from that shed upon the subject by current publications, I shall mention a few points that might otherwise be overlooked.

It has always seemed to me that a deep mystery enshrouded the act of Smithson in devising his fortune as he did. That an Englishman, whose connections and associations were entirely with the intellectual classes, — who had never, so far as is known, a single American connection, or the slightest inclination toward democracy, — should, in the intellectual condition of our country during the early years of the century, have chosen its government as his trustee for the foundation of a scientific institution, does of itself seem singular enough. What seems yet more singular is that no instructions whatever were given in his will or found in his papers beyond the comprehensive one " to found an institution at Washington to be called the Smithsonian Institution for the increase and diffusion of knowledge among men." No plan of the institution, no scrap of paper which might assist in the interpretation of the mandate, was ever discovered.

Not a word respecting his intention was ever known to have been uttered. Only a single remark was ever recorded which indicated that he had anything unusual in view. He did at one time say, "My name shall live in the memory of men when the titles of the Northumberlands and the Percys are extinct and forgotten."

One result of this failure to indicate a plan for the institution was that, when the government received the money, Congress was at a loss what to do with it. Some ten years were spent in discussing schemes of various kinds, among them that of declining the gift altogether. Then it was decided that the institution should be governed by a Board of Regents, who should elect a secretary as their executive officer and the administrator of the institution. The latter was to include a library, a museum, and a gallery of art. The plans for the fine structure, so well known to every visitor to the capital, were prepared, the building was started, the regents organized, and Professor Henry made secretary.

We might almost say that Henry was opposed to every special function assigned to the institution by the organic law. He did not agree with me as to any mystery surrounding the intentions of the founder. To him they were perfectly clear. Smithson was a scientific investigator; and the increase and diffusion of knowledge among men could be best promoted on the lines that he desired, by scientific investigation and the publication of scientific researches. For this purpose a great building was not necessary, and he regretted all the money spent on it. The library, museum, and gallery of art would be of only local advantage, whereas " diffusion among men " implied all men, whether they could visit Washington or not. It was clearly the business of the government to supply purely local facilities for study and research, and the endowment of Smithson should not be used for such a purpose.

His opposition to the building tinged the whole course of his thought. I doubt whether he was ever called upon by founders of institutions of any sort for counsel without his warning them to beware of spending their money in bricks and mortar. The building being already started before he took charge, and the three other objects being sanctioned by law, he was, of course, hampered in carrying out his views. But he did his utmost to reduce to a minimum the amount of the fund that should be devoted to the objects specified.

This policy brought on the most animated contest in the history of the institution. It was essential that his most influential assistants should share his views or at least not thwart them. This, he found, was not the case. The librarian, Mr. C. C. Jewett, an able and accomplished man in the line of his profession, was desirous of collecting one of the finest scientific libraries. A contest arose, to which Professor Henry put an end by the bold course of removing the librarian from office. Mr.

Jewett denied his power to do this, and the question came before the board of regents. The majority of these voted that the secretary had the power to remove his assistants. Among the minority was Rufus Choate, who was so strongly opposed to the action that he emphasized his protest against it by resigning from the board.

A question of legal interpretation came in to make the situation yet more difficult. The regents had resolved that, after the completion of the building, one half the income should be devoted to those objects which Professor Henry considered most appropriate. Meanwhile there was no limit to the amount that might be appropriated to these objects, but Mr. Jewett and other heads of departments wished to apply the rule from the beginning. Henry refused to do so, and looked with entire satisfaction on the slowness of completion of what was, in his eyes, an undesirable building.

It must be admitted that there was one point which Professor Henry either failed to appreciate, or perhaps thought unworthy of consideration. This is, the strong hold on the minds of men which an institution is able to secure through the agency of an imposing building. Saying nothing of the artistic and educational value of a beautiful piece of architecture, it would seem that such a structure has a peculiar power of impressing the minds of men with the importance of the object to which it is devoted, or of the work going on within it. Had Professor Henry been allowed to perform all the functions of the Smithsonian Institution in a moderate-sized hired house, as he felt himself abundantly able to do, I have very serious doubts whether it would have acquired its present celebrity and gained its present high place in the estimation of the public.

In the winter of 1865 the institution suffered an irreparable loss by a conflagration which destroyed the central portion of the building. At that time the gallery of art had been confined to a collection of portraits of Indians by Stanley. This collection was entirely destroyed. The library, being at one end, remained intact. The lecture room, where courses of scientific lectures had been delivered by eminent men of science, was also destroyed. This event gave Professor Henry an opportunity of taking a long step in the direction he desired. He induced Congress to take the Smithsonian library on deposit as a part of its own, and thus relieve the institution of the cost of supporting this branch. The Corcoran Art Gallery had been founded in the mean time, and relieved the institution of all necessity for supporting a gallery of art. He would gladly have seen the National Museum made a separate institution, and the Smithsonian building purchased by the government for its use, but he found no chance of carrying this out.

After the death of Professor Henry the Institution grew rapidly into a position in which it might almost claim to be a scientific department of the government. The National Museum, remaining under its administration, was greatly enlarged, and one of its ramifications was extended into the National Zoölogical Park. The studies of Indian ethnology, begun by Major J. W. Powell, grew into the Bureau of Ethnology. The Astrophysical Observatory was established, in which Professor Langley has continued his epoch-making work on the sun's radiant heat with his wonderful bolometer, an instrument of his own invention.

Before he was appointed to succeed Professor Henry, Professor Baird was serving as United States Fish Commissioner, and continued to fill this office, without other salary than that paid by the Smithsonian Institution. The economic importance of the work done and still carried on by this commission is too well known to need a statement. About the time of Baird's death, the work of the commission was separated from that of the Institution by providing a salary for the commissioner.

We have here a great extension of the idea of an institution for scientific publications and research. I recall once suggesting to Professor Baird the question whether the utilization of the institution founded by Smithson for carrying on and promoting such government work as that of the National Museum was really the right thing to do. He replied, "It is not a case of using the Smithsonian fund for government work, but of the government making appropriations for the work of the Smithsonian Institution." Between the two sides of the question thus presented, — one emphasizing the honor done to Smithson by expanding the institution which bears his name, and the other aiming solely at the best administration of the fund which we hold in trust for him, — I do not pretend to decide.

On the academic side of social life in Washington, the numerous associations of alumni of colleges and universities hold a prominent place. One of the earliest of these was that of Yale, which has held an annual banquet every year, at least since 1877, when I first became a member. Its membership at this time included Mr. W. M. Evarts, then Secretary of State, Chief Justice Waite, Senator Dawes, and a number of other men prominent in political life. The most attractive speaker was Mr. Evarts, and the fact that his views of education were somewhat conservative added much to the interest of his speeches. He generally had something to say in favor of the system of a prescribed curriculum in liberal education, which was then considered as quite antiquated. When President Dwight, shortly after his accession to office, visited the capital to explain the modernizing of the Yale educational system, he told the alumni that the college now offered ninety-five courses to undergraduates. Evarts congratulated the coming students on sitting at a banquet table where they had their choice of ninety-five courses of intellectual aliment.

Perhaps the strongest testimonial of the interest

attached to these reunions was unconsciously given by President Hayes. He had received an honorary degree from Yale, and I chanced to be on the committee which called to invite him to the next banquet. He pleaded, as I suppose Presidents always do, the multiplicity of his engagements, but finally said, —

"Well, gentlemen, I will come, but it must be on two well-understood conditions. In the first place, I must not be called to my feet. You must not expect a speech of me. The second condition is, I must be allowed to leave punctually at ten o'clock."

"We regret your conditions, Mr. President," was the reply, "but must, of course, accede to them, if you insist."

He came to the banquet, he made a speech, — a very good, and not a very short one, — and he remained, an interested hearer, until nearly two o'clock in the morning.

In recent years I cannot avoid a feeling that a change has come over the spirit of such associations. One might gather the impression that the apothegm of Sir William Hamilton needed a slight amendment.

> On earth is nothing great but Man,
> In Man is nothing great but Mind.

Strike out the last word, and insert "Muscle." The reader will please not misinterpret this remark. I admire the physically perfect man, loving everything out of doors, and animated by the spirit that takes him through polar snows and over mountain tops. But I do not feel that mere muscular practice during a few years of college life really fosters this spirit.

Among the former institutions of Washington of which the memory is worth preserving, was the Scientific Club. This was one of those small groups, more common in other cities than in Washington, of men interested in some field of thought, who meet at brief intervals at one another's houses, perhaps listen to a paper, and wind up with a supper. When or how the Washington Club originated, I do not know, but it was probably sometime during the fifties. Its membership seems to have been rather ill defined, for, although I have always been regarded as a member, and am mentioned in McCulloch's book as such,[1] I do not think I ever received any formal notice of election. The club was not exclusively scientific, but included in its list the leading men who were supposed to be interested in scientific matters, and whose company was pleasant to the others. Mr. McCulloch himself, General Sherman, and Chief Justice Chase are examples of the members of the club who were of this class.

It was at the club meetings that I made the acquaintance of General Sherman. His strong characteristics were as clearly seen at these evening

[1] *Men and Measures of Half a Century*, by Hugh McCulloch. New York: Chas. Scribner's Sons, 1889.

gatherings as in a military campaign. His restlessness was such that he found it hard to sit still, especially in his own house, two minutes at a time. His terse sentences, leaving no doubt in the mind of the hearer as to what he meant, always had the same snap. One of his military letters is worth reviving. When he was carrying on his campaign in Georgia against Hood, the latter was anxious that the war should damage general commercial interests as little as possible; so he sent General Sherman a letter setting forth the terms and conditions on which he, Hood, would refrain from burning the cotton in his line of march, but leave it behind, — at as great length and with as much detail as if it were a treaty of peace between two nations. Sherman's reply was couched in a single sentence: "I hope you will burn all the cotton you can, for all you don't burn I will." When he introduced two people, he did not simply mention their names, but told who each one was. In introducing the adjutant-general to another officer who had just come into Washington, he added, " You know his signature."

Mr. McCulloch, who succeeded Mr. Chase as Secretary of the Treasury, was my beau idéal of an administrator. In his personal make-up, he was as completely the opposite of General Sherman as a man well could be. Deliberate, impassive, heavy of build, slow in physical movement, he would have been supposed, at first sight, a man who would take life easy, and concern himself as little as possible about public affairs. But, after all, there is a quality in the head of a great department which is quite distinct from sprightliness, and that is wisdom. This he possessed in the highest degree. The impress which he made on our fiscal system was not the product of what looked like energetic personal action, but of a careful study of the prevailing conditions of public opinion, and of the means at his disposal for keeping the movement of things in the right direction. His policy was what is sometimes claimed, and correctly, I believe, to embody the highest administrative wisdom : that of doing nothing himself that he could get others to do for him. In this way all his energies could be devoted to his proper work, that of getting the best men in office, and of devising measures from time to time calculated to carry the government along the lines which he judged to be best for the public interests.

The name of another attendant at the meetings of the club has from time to time excited interest because of its connection with a fundamental principle of evolutionary astronomy. This principle, which looks paradoxical enough, is that up to a certain stage, as a star loses heat by radiation into space, its temperature becomes higher. It is now known as Lane's Law. Some curiosity as to its origin, as well as the personality of its author, has sometimes been expressed. As the story has never been printed, I ask leave to tell it.

Among the attendants at the meetings of the Scientific Club was an odd-looking and odd-mannered little man, rather intellectual in appearance, who listened attentively to what others said, but who, so far as I noticed, never said a word himself. Up to the time of which I am speaking, I did not even know his name, as there was nothing but his oddity to excite any interest in him.

One evening about the year 1867, the club met, as it not infrequently did, at the home of Mr. McCulloch. After the meeting Mr. W. B. Taylor, afterward connected with the Smithsonian Institution in an editorial capacity, accompanied by the little man, set out to walk to his home, which I believe was somewhere near the Smithsonian grounds. At any rate, I joined them in their walk, which led through these grounds. A few days previous there had appeared in the "Reader," an English weekly periodical having a scientific character, an article describing a new theory of the sun. The view maintained was that the sun was not a molten liquid, as had generally been supposed up to that time, but a mass of incandescent gas, perhaps condensed at its outer surface, so as to form a sort of immense bubble. I had never before heard of the theory, but it was so plausible that there could be no difficulty in accepting it. So, as we wended our way through the Smithsonian grounds, I explained the theory to my companions in that *ex cathedra* style which one is apt to assume in setting forth a new idea to people who know little or nothing of the subject. My talk was mainly designed for Mr. Taylor, because I did not suppose the little man would take any interest in it. I was, therefore, much astonished when, at a certain point, he challenged, in quite a decisive tone, the correctness of one of my propositions. In a rather more modest way, I tried to maintain my ground, but was quite silenced by the little man informing us that he had investigated the whole subject, and found so and so — different from what I had been laying down.

I immediately stepped down from the pontifical chair, and asked the little man to occupy it and tell us more about the matter, which he did. Whether the theorem to which I have alluded was included in his statement, I do not recall. If it was not, he told me about it subsequently, and spoke of a paper he had published, or was about to publish, in the "American Journal of Science." I find that this paper appeared in Volume L. in 1870.

Naturally I cultivated the acquaintance of such a man. His name was J. Homer Lane. He was quite alone in the world, having neither family nor near relative, so far as any one knew. He had formerly been an examiner or something similar in the Patent Office, but under the system which prevailed in those days, a man with no more political influence than he had was very liable to lose his position, as he actually did. He lived in a good deal such a habitation and surroundings as men like Johnson and Goldsmith lived in in their time. If his home was not exactly a garret, it came as near it as a lodging of the present day ever does.

After the paper in question appeared, I called Mr. Lane's attention to the fact that I did not find any statement of the theorem which he had mentioned to me to be contained in it. He admitted that it was contained in it only impliedly, and proceded to give me a very brief and simple demonstration.

So the matter stood, until the centennial year, 1876, when Sir William Thomson paid a visit to this country. I passed a very pleasant evening with him at the Smithsonian Institution, engaged in a discussion, some points of which he afterwards mentioned in an address to the British Association. Among other matters, I mentioned this law, originating with Mr. J. Homer Lane. He did not think it could be well founded, and when I attempted to reproduce Mr. Lane's verbal demonstration, I found myself unable to do so. I told him I felt quite sure about the matter, and would write to him on the subject. When I again met Mr. Lane, I told him of my difficulty, and asked him to repeat the demonstration. He did so at once, and I sent it off to Sir William. The latter immediately accepted the result, and published a paper on the subject, in which the theorem was made public for the first time.

It is very singular that a man of such acuteness never achieved anything else of significance. He was at my station on one occasion when a total eclipse of the sun was to be observed, and made a report on what he saw. At the same time he called my attention to a slight source of error with which photographs of the transit of Venus might be affected. The idea was a very ingenious one, and was published in due course.

Altogether, the picture of his life and death remains in my memory as a sad one, the brightest gleam being the fact that he was elected a member of the National Academy of Sciences, which must have been to him a very grateful recognition of his work on the part of his scientific associates. When he died, his funeral was attended only by a few of his fellow members of the academy. Altogether, I feel it eminently appropriate that his name should be perpetuated by the theorem of which I have spoken.

If the National Academy of Sciences has not proved as influential a body as such an academy should, it has still taken such a place in science, and rendered services of such importance to the government, that the circumstances connected with its origin are of permanent historic interest. As the writer was not a charter member, he cannot claim to have been "in at the birth," though he became, from time to time, a repository of desultory information on the subject. There is abundant internal and circumstantial evidence that Dr. B. A. Gould, although his name has, so far as I am aware, never been mentioned in this connection, was a leading spirit in the first organization. On the other hand, curiously enough, Professor Henry was not. I was quite satisfied that Bache took an active part, but Henry assured me that he could not believe this, because he was so intimate with Bache that, had the latter known anything of the matter, he would surely have consulted him. Some recent light is thrown on the subject by letters of Rear-Admiral Charles H. Davis, found in his "Life," as published by his son. Everything was carried on in the greatest secrecy, until the bill chartering the body was introduced by Senator Henry Wilson of Massachusetts. Fifty charter members were named, and this number was fixed as the permanent limit to the membership. The list did not include either George P. Bond, director of the Harvard Observatory, perhaps the foremost American astronomer of the time in charge of an observatory, nor Dr. John W. Draper. Yet the total membership in the section of astronomy and kindred sciences was very large. A story to which I give credence was that the original list, as handed to Senator Wilson, did not include the name of William B. Rogers, who was then founding the Institute of Technology. The senator made it a condition that room for Rogers should be found, and his wish was acceded to. It is of interest that the man thus added to the academy by a senator afterward became its president, and proved as able and popular a presiding officer as it ever had.

The governmental importance of the academy arose from the fact that its charter made it the scientific adviser of the government, by providing that it should "investigate, examine, experiment, and report upon any subject of science or art" whenever called upon by any department of the government. In this respect it was intended to perform the same valuable functions for the government that are expected of the national scientific academies or societies of foreign countries.

The academy was empowered to make its own constitution. That first adopted was sufficiently rigid and complex. Following the example of European bodies of the same sort, it was divided into two classes, one of mathematical and physical, the other of natural science. Each of these classes was divided into sections. A very elaborate system of procedure for the choice of new members was provided. Any member absent from four consecutive stated meetings of the academy had his name stricken from the roll unless he communicated a valid reason for his absence. Notwithstanding this requirement, the academy had no funds to defray the traveling expenses of members, nor did the government ever appropriate money for this purpose.

For seven years it became increasingly doubtful whether the organization would not be abandoned. Several of the most eminent members took no interest whatever in the academy, — did not attend the meetings, but did tender their resignations, which, however, were not accepted. This went on at such a rate that, in 1870, to avoid a threatened dissolution, a radical change was made in the constitution. Congress was asked to remove the restriction upon the number of members, which it promptly did. Classes and sections were entirely abandoned. The members formed but a single body. The method of election was simplified, — too much simplified, in fact.

The election of new members is, perhaps, the most difficult and delicate function of such an organization. It is one which cannot be performed to public satisfaction, nor without making many mistakes; and the avoidance of the latter is vastly more difficult when the members are so widely separated and have little opportunity to discuss in advance the merits of the men from whom a selection is to be made. An ideal selection cannot be made until after a man is dead, so that his work can be summed up ; but I think it may fairly be said that, on the whole, the selections have been as good as could be expected under the conditions.

Notwithstanding the indifference of the government to the possible benefits that the academy might render it, it has — in addition to numerous reports on minor subjects — made two of capital importance to the public welfare. One of these was the planning of the United States Geological Survey, the other the organization of a forestry system for the United States.

During the years 1870–77, besides several temporary surveys or expeditions which had from time to time been conducted under the auspices of the government, there were growing up two permanent surveys of the territories. One of these was the Geographical Survey of territories west of the 100th meridian, under the Chief of Engineers of the Army; the other was the Geological Survey of the territories under the Interior Department, of which the chief was Professor F. V. Hayden.

The methods adopted by the two chiefs to gain the approval of the public and the favoring smiles of Congress were certainly very different. Wheeler's efforts were made altogether by official methods and through official channels. Hayden considered it his duty to give the public every possible opportunity to see what he was doing and to judge his work. His efforts were chronicled at length in the public prints. His summers were spent in the field, and his winters were devoted to working up results and making every effort to secure influence. An attractive personality and extreme readiness to show every visitor all that there was to be seen in his collections, facilitated his success. One day a friend introduced a number of children with an expression of doubt as to the little visitors being welcome. "Oh, I always like to have the children come here," he replied, "they influence their parents." He was so successful in his efforts that his organization grew apace, and soon developed into the Geological Survey of the Territories.

Ostensibly the objects of the two organizations were different. One had military requirements mainly in view, especially the mapping of routes. Hayden's survey was mainly in the interests of geology. Practically, however, the two covered the same field in all points. The military survey extended its scope by including everything necessary for a complete geographical and geological atlas. The geological survey was necessarily a complete topographical and geological survey from the beginning. Between 1870 and 1877, both were engaged in making an atlas of Colorado, on the maps of which were given the same topographical features and the same lines of communication. Parties of the two surveys mounted their theodolites on the same mountains, and triangulated the same regions. The Hayden survey published a complete atlas of Colorado, probably more finely gotten up than any atlas of a State in the Union, while the Wheeler survey was vigorously engaged in issuing maps of the same territory. No effort to prevent this duplication of work by making an arrangement between the two organizations led to any result. Neither had any official knowledge of the work of the other. Unofficially, the one was dissatisfied with the political methods of the other, and claimed that the maps which it produced were not fit for military purposes. Hayden retorted with unofficial reflections on the geological expertness of the engineers, and maintained that their work was not of the best. He got up by far the best maps; Wheeler, in the interests of economy, was willing to sacrifice artistic appearance to economy of production. We thus had the curious spectacle of the government supporting two independent surveys of the same region. Various compromises were attempted, but they all came to nothing. The state of things was clear enough to Congress, but the repugnance of our national legislature to the adoption of decisive measures of any sort for the settlement of a disputed administrative question prevented any effective action. Infant bureaus may quarrel with each other and eat up the paternal substance, but the parent cannot make up his mind to starve them outright, or even to chastise them into a spirit of conciliation. Unable to decide between them, Congress for some years pursued the policy of supporting both surveys.

The credit for introducing a measure which would certainly lead to unification is due to Mr. A. S. Hewitt, of New York, then a member of the Committee on Appropriations. He proposed to refer the whole subject to the National Academy of Sciences. His committee accepted his view, and a clause was inserted in the Sundry Civil Bill of June 30, 1878, requiring the academy at its next meeting to take the matter into consideration and report to Congress " as soon thereafter as may be practicable, a plan for surveying and mapping the territory of the United States on such general system as will, in their judgment, secure the best results at the least possible cost."

Several of the older and more conservative members of the academy objected that this question was not one of science or art, with which alone the academy was competent to deal, but was a purely administrative question which Congress should settle for itself. They feared that the academy would be drawn into the arena of political discussion to an extent detrimental to its future and welfare and usefulness. Whether the exception was or was not well taken, it was felt that the academy, the creature of Congress, could not join issue with the latter as to its functions, nor should an opportunity of rendering a great service to the government be lost for such a reason as this.

The plan reported by the academy was radical and comprehensive. It proposed to abolish all the existing surveys of the territories except those which, being temporary, were completing their work, and to substitute for them a single organization which would include the surveys of the public lands in its scope. The interior work of the Coast and Geodetic Survey was included in the plan, it being proposed to transfer this bureau to the Interior Department, with its functions so extended as to include the entire work of triangulation.

When the proposition came up in Congress at the following session, it was vigorously fought by the Chief of Engineers of the army, and by the General Land Office, of which the surveying functions were practically abolished. The Land Office carried its point, and was eliminated from the scheme. General Humphreys, the Chief of Engineers, was a member of the academy, but resigned on the ground that he could not properly remain a member while contesting the recommendations of the body. But the academy refused to accept the resignation, on the very proper ground that no obligation was imposed on the members to support the views of the academy, besides which, the work of the latter in the whole matter was terminated when its report was presented to Congress.

Although this was true of the academy, it was not true of the individual members who had taken part in constructing the scheme. They were naturally desirous of seeing the plan made a success, and, in the face of such vigorous opposition, this required constant attention. A dexterous movement was that of getting the measure transferred from one appropriation bill to another when it passed over to the Senate. The measure at length became a law, and thus was established the Geological Survey of the United States, which was to be governed by a Director, appointed by the President, by and with the advice and consent of the Senate.

Then, on March 4, 1879, an important question arose. The right man must be placed at the head of the new bureau. Who is he? At first there seemed to be but one voice on the subject. Professor Hayden had taken the greatest pains to make known the work of his survey, not only to Congress, but to every scientific society, small and great, the world over. Many of these had bestowed their approbation upon it by electing its director to honorary membership. It has been said, I do not know how truly, that the number of these testimonials exceeded that received by any other scientific man in America. If this were so, they would have to be counted, not weighed. It was, therefore, not surprising that two thirds of the members of Congress were said to have sent a recommendation to the President for the appointment of so able and successful a man to the new position. The powerful backing of so respectable a citizen as Hon. J. D. Cox, formerly Secretary of the Interior, was also heartily proffered. To these forces were added that of a certain number of geologists, though few or none of them were leaders in the science. Had it not been for a private intimation conveyed to Secretary Schurz that the scientific men interested might have something to say on the subject, Hayden might have been appointed at the very moment the bill was signed by the President.

Notwithstanding all of Hayden's merits as the energetic head of a survey, the leaders in the movement considered that Mr. Clarence King was the better qualified for the duties of the new position. It is not unlikely that a preference for a different method of influencing Congress than that which I have described, was one of the reasons in favor of Mr. King. He was a man of charming personality and great literary ability. Some one said of him that he could make a more interesting story out of what he saw during a ride in a street car than most men could with the best material at their disposal. His "Mountaineering in the Sierra Nevadas" was as interesting an account of Western exploration as has ever been published. I understand it was suppressed by the author because some of the characters described in it were much hurt by finding themselves painted in the book.

Hopeless though the contest might have seemed, an effort was made by three or four of the men most interested to secure Mr. King's appointment. If I wanted to show the fallacy of the common impression that scientific men are not fitted for practical politics, I could not do it better than by giving the internal history of the movement. This I shall attempt only in the briefest way. The movers in the matter divided up the work, did what they could in the daytime, and met at night at Wormley's Hotel to compare notes, ascertain the effect of every shot, and decide where the next one should be fired. As all the parties concerned in the matter have now passed off the stage, I shall venture to mention one of these shots. One eminent geologist, whose support was known to be available, had not been called in, because an impression had been formed that President Hayes would not be willing to consider favorably what he might say. After the matter had been discussed at one or two meetings, one of the party proposed to sound the President on the subject at his next interview. So, when the occasion arose, he gently introduced the name of the gentleman.

"What view does he take?" inquired the President.

"I think he will be favorable to Mr. King," was the reply; "but would you give great weight to his opinion?"

"I would give great weight to it, very great weight, indeed," was the reply.

This expression was too decided in its tone to leave any doubt, and the geologist in question was on his way to Washington as soon as electricity could tell him that he was wanted. When the time finally came for a decision, the President asked Secretary Schurz for his opinion. Both agreed that King was the man, and he was duly appointed.

The new administration was eminently successful. But King was not fond of administrative work, and resigned the position at the end of a year or so. He was succeeded by John W. Powell, under whom the survey grew with a rapidity which no one had anticipated. As originally organized, the survey was one of the territories only, but the question whether it should not be extended to the States as well, and prepare a topographical atlas of the whole country, was soon mooted, and decided by Congress in the affirmative. For this extension, however, the original organizers of the survey were in no way responsible. It was the act of Congress, pure and simple.

If the success of an organization is to be measured by the public support which it has commanded, by the extension of its work and influence, and by the gradual dying out of all opposition, it must be admitted that the plan of the academy was a brilliant success. It is true that a serious crisis had once to be met. While Mr. Cleveland was governor of New York, his experience with the survey of that State had led him to distrust the methods on which the surveys of the United States were being conducted. This distrust seems to have pervaded the various heads of the departments under his administration, and led to serious charges against the conduct of both the Coast and Geological surveys. An unfavorable report upon the administration of the former was made by a committee especially appointed by the Secretary of the Treasury, and led to the resignation of its superintendent. But, in the case of the Geological Survey, the attacks were mostly conducted by the newspapers. At length, Director Powell asked permission of Secretary Lamar to write him a letter in reply. His answers were so sweeping, and so conclusive on every point, that nothing more was heard of the criticisms.

The second great work of the academy for the government was that of devising a forestry system for the United States. The immediate occasion for action in this direction was stated by Secretary Hoke Smith to be the " inadequacy and confusion of existing laws relating to the public timber lands and consequent absence of an intelligent policy in their administration, resulting in such conditions as may, if not speedily stopped, prevent the proper development of a large part of our country."

Even more than in the case of the Geological

Survey might this work seem to be one of administration rather than of science. But granting that such was the case, the academy commanded great advantages in taking up the subject. The commission which it formed devoted more than a year to the study, not only of the conditions in our own country, but of the various policies adopted by foreign countries, especially Germany, and their results. As in the case of the Geological Survey, a radically new and very complete system of forestry administration was proposed. Interests having other objects than the public good were as completely ignored as they had been before.

The soundness of the conclusions reached by the Academy Commission were challenged by men wielding great political power in their respective States. For a time it was feared that the academy would suffer rather than gain in public opinion by the report it had made. But the moral force behind it was such that, in the long run, some of the severest critics saw their error, and a plan was adopted which, though differing in many details from that proposed, was, in the main, based on the conclusion of the commission. The Interior Department, the Geological Survey, and the Department of Agriculture all have their part in the work.

Notwithstanding these signal demonstrations of the valuable service which the academy may render to the government, the latter has done nothing for it. The immediate influence of the leading scientific men in public affairs has perhaps been diminished as much in one direction as it has been increased in another by the official character of the organization. The very fact that the members of the academy belong to a body which is, officially, the scientific adviser of the government, prevents them from coming forward to exercise that individual influence which they might exercise were no such body in existence.

The academy has not even a place of meeting, nor is a repository for its property and records provided for it. Although it holds in trust large sums which have been bequeathed from time to time by its members for promoting scientific investigation, and is, in this way, rendering an important service to the progress of knowledge, it has practically no income of its own except the contributions of its own members, nearly all of whom are in the position described by the elder Agassiz, of having "no time to make money."

Among the men who have filled the office of president of the academy, Professor O. C. Marsh was perhaps the one whose activity covered the widest field. Though long well known in scientific circles, he first came into public prominence by his exposure of the frauds practiced by contractors in furnishing supplies for the Indians. This business had fallen into the hands of a small ring of contractors known as the "Indian ring," who knew the ropes so well that they could bid below any competitor and yet manage things so as to gain a handsome profit out of the contracts. In the course of his explorations Marsh took pains to investigate the whole matter, and published his conclusions first in the New York "Tribune," and then more fully in pamphlet form, taking care to have public attention called to the subject so widely that the authorities would have to notice it. In doing so, Mr. Delano, Secretary of the Interior, spoke of them as charges made by "a Mr. Marsh." This method of designating such a man was made effective use of by Mr. Delano's opponents in the case.

Although the investigation which followed did not elicit all the facts, it had the result of calling the attention of succeeding Secretaries of the Interior to the necessity of keeping the best outlook on the administration of Indian affairs. What I believe to have been the final downfall of the ring was not brought about until Cleveland's first administration. Then it happened in this way. Mr. Lamar, the Secretary of the Interior, was sharply on the lookout for frauds of every kind. As usual, the lowest bid for a certain kind of blanket had been accepted, and the Secretary was determined to see whether the articles furnished actually corresponded with the requirements of the contract. It chanced that he had as his appointment clerk Mr. J. J. S. Hassler, a former manufacturer of woolen goods. Mr. Hassler was put on the board to inspect the supplies, and found that the blankets, although to all ordinary appearance of the kind and quality required, were really of a much inferior and cheaper material. The result was the enforced failure of the contractor, and, I believe, the end of the Indian ring.

Marsh's explorations in search of fossil remains of the animals which once roamed over the western parts of our continent were attended by adventures of great interest, which he long had the intention of collecting and publishing in book form. Unfortunately, he never did it, nor, so far as I am aware, has any connected narrative of his adventures ever appeared in print. This is more to be regretted, because they belong to a state of things which is rapidly passing away, leaving few records of that lifelike sort which make the most impressive picture.

His guide during his early explorations was a character who has since become celebrated in America and Europe by the vivid representations of the "Wild West" with which he has amused and instructed the dwellers on two continents. Marsh was on his way to explore the region in the Rocky Mountains where he was to find the fossils which have since made his work most celebrated. The guide was burning with curiosity as to the object of the expedition. One night over the campfire he drew his chief into a conversation on the subject. The latter told him that there was once a time when the Rocky Mountains did not exist, and that part of the continent was a level plain. In the course of long ages mountains rose, and animals ran over them. Then the mountains split open;

the animals died and left their bones in the clefts. The object of his expedition was now to search for some of these bones.

The bones were duly discovered, and it was not many years thereafter before the Wild West Exhibition was seen in the principal Eastern cities. When it visited New Haven, its conductor naturally renewed the acquaintance of his former patron and supporter.

"Do you remember, professor," said he, "our talk as we were going on your expedition to the Rockies, — how you told me about the mountains rising up and being split open and the bones of animals being lost in there, and how you were going to get them?"

"Oh, yes," said the other, "I remember it very well."

"Well, professor, do you know, when you told me all that I r'ally thought you was puttin' up a job on me."

The result was a friendship between the two men, which continued during Marsh's whole life. When the one felt that he ought no longer to spend all the money he earned, he consulted Marsh on the subject of "salting it down," and doubtless got good advice.

As an exposer of humbugs Marsh took a prominent place. One of these related to the so-called "Cardiff Giant." Sometime in 1869 the newspapers announced the discovery in northern New York, near the Canadian border, of an extraordinary fossil man, or colossal statue, people were not sure which, eight or ten feet high. It was found several feet below the ground while digging a well. Men of some scientific repute, including even one so eminent as Professor James Hall, had endorsed the genuineness of the find, and, on the strength of this, it was taken around to show the public. In the course of a journey through New York State, Marsh happened to pass through the town where the object was on exhibition. His train stopped forty minutes for dinner, which would give him time to drive to the place and back, and leave a margin of about fifteen minutes for an examination of the statue. Hardly more than a glance was necessary to show its fraudulent character. Inside the ears the marks of a chisel were still plainly visible, showing that the statue had been newly cut. One of the most curious features was that the stone had not been large enough to make the complete statue, so that the surface was, in one place, still in the rough. The object had been found in wet ground. Its material was sulphate of lime, the slight solubility of which would have been sufficient to make it dissolve entirely away in the course of centuries. The absence of any degradation showed that the thing was comparatively new. On the strength of this, Marsh promptly denounced the affair as a humbug. Only a feeble defense was made for it, and, a year or two later, the whole story came out. It had been designed and executed somewhere in the Northwest, transported to the place where discovered, and buried, to be afterward dug up and reported as a prehistoric wonder.

Only a few years ago the writer had an opportunity of seeing with what wonderful ease intelligent men can be imposed upon by these artificial antiquities. The would-be exhibitor of a fossil woman, found I know not where, appeared in Washington. He had not discovered the fossil himself, but had purchased it for some such sum as $100, on the assurance of its genuine character. He seems, however, to have had some misgivings on the subject, and, being an honest fellow, invited some Washington scientific men to examine it in advance of a public exhibition. The first feature to strike the critical observer was that the arms of the fossil were crossed over the breast in the most approved undertaker's fashion, showing that if the woman had ever existed, she had devoted her dying moments to arranging a pose for the approval of posterity. Little more than a glance was necessary to show that the fossil was simply baked clay. Yet the limbs were hard and stiff. One of the spectators therefore asked permission of the owner to bore with an auger into the leg and see what was inside. A few moments' work showed that the bone of the leg was a bar of iron, around which clay had been moulded and baked. I must do the crestfallen owner the justice to say that his anxiety to convince the spectators of his own good faith in the matter far exceeded his regret at the pecuniary loss which he had suffered.

Another amusing experience that Marsh had with a would-be fossil arose out of the discovery here and there in Connecticut of the fossil footprints of birds. Shortly after a find of this kind had been announced, a farmer drove his wagon up in front of the Peabody Museum, called on the professor, and told him he had dug up something curious on his farm, and he wished the professor would tell him what it was. He thought it looked like the footprints of a bird in a stone, but he was not quite sure.

Marsh went out and looked at the stone. A single glance was enough.

"Oh, I see what they are. They are the footprints of the domestic turkey. And the oddest part of it is, they are all made with the right foot."

The simple-minded countryman, in making the prints with the turkey's foot, had overlooked the difference between the right and left foot, and the consequent necessity of having the tracks which pertained to the two feet alternate.

Washington is naturally a centre of information on all subjects relating to the aboriginal tribes of America and to life on the plains generally. Besides the Geological Survey, the Bureau of Ethnology has been an active factor in this line. An official report cannot properly illustrate life in all its aspects, and therefore should be supplemented by the experiences of leading explorers. This is all the more necessary if, as seems to be the case, the peculiar characteristics of the life in question are being replaced by those more appropriate to civilization. Yet the researches of the bureau in question are not carried on in any narrow spirit, and will supply the future student of humanity with valuable pictures of the most heroic of all races, and yet doomed, apparently, to ultimate extinction. I do not think I ever saw a more impressive human figure and face than those of Chief Joseph as he stood tall, erect, and impassive, at a President's reception in the winter of 1903. He was attired in all the brilliancy of his official costume; but not a muscle of his strongly marked face betrayed the sentiments with which he must have gazed on the shining uniforms passing before him.

X

SCIENTIFIC ENGLAND

My first trip to Europe, mentioned in the last chapter, was made with my wife, when the oldest transatlantic line was still the fashionable one. The passenger on a Cunarder felt himself amply compensated for poor attendance, coarse food, and bad coffee by learning from the officers on the promenade deck how far the ships of their line were superior to all others in strength of hull, ability of captain, and discipline of crew. Things have changed on both sides since then. Although the Cunard line has completed its half century without having lost a passenger, other lines are also carefully navigated, and the Cunard passenger, so far as I know, fares as well as any other. Captain McMickan was as perfect a type of the old-fashioned captain of the best class as I ever saw. His face looked as if the gentlest zephyr that had ever fanned it was an Atlantic hurricane, and yet beamed with Hibernian good humor and friendliness. He read prayers so well on Sunday that a passenger assured him he was born to be a bishop. One day a ship of the North German Lloyd line was seen in the offing slowly gaining on us. A passenger called the captain's attention to the fact that we were being left behind. "Oh, they're very lightly built, them German ships; built to carry German dolls and such like cargo."

In London one of the first men we met was Thomas Hughes, of Rugby fame, who made us feel how worthy he was of the love and esteem bestowed upon him by Americans. He was able to make our visit pleasant in more ways than one. Among the men I wanted to see was Mr. John Stuart Mill, to whom I was attracted not only by his fame as a philosopher and the interest with which I had read his books, but also because he was the author of an excellent pamphlet on the Union side during our civil war.

On my expressing a desire to make Mr. Mill's acquaintance, Mr. Hughes immediately offered to give me a note of introduction. Mill lived at Blackheath, which, though in an easterly direction down the Thames, is one of the prettiest suburbs of the great metropolis. His dwelling was a very modest one, entered through a passage of trellis-work in a little garden. He was by no means the grave and distinguished-looking man I had expected to see. He was small in stature and rather spare, and did not seem to have markedly intellectual features. The cordiality of his greeting was more than I could have expected; and he was much pleased to know that his work in moulding English sentiment in our favor at the commencement of the civil war was so well remembered and so highly appreciated across the Atlantic.

As a philosopher, it must be conceded that Mr. Mill lived at an unfortunate time. While his vigor and independence of thought led him to break loose from the trammels of the traditional philosophy, modern scientific generalization had not yet reached a stage favorable to his becoming a leader in developing the new philosophy. Still, whatever may be the merits of his philosophic theories, I believe that up to a quite recent time no work on scientific method appeared worthy to displace his " System of Logic."

A feature of London life that must strongly impress the scientific student from our country is the closeness of touch, socially as well as officially, between the literary and scientific classes on the one side and the governing classes on the other. Mr. Hughes invited us to make an evening call with him at the house of a cabinet minister, — I think it was Mr. Goschen, — where we should find a number of persons worth seeing. Among those gathered in this casual way were Mr. Gladstone, Dean Stanley, and our General Burnside, then grown quite gray. I had never before met General Burnside, but his published portraits were so characteristic that the man could scarcely have been mistaken. The only change was in the color of his beard. Then and later I found that a pleasant feature of these informal " at homes," so universal in London, is that one meets so many people he wants to see, and so few he does not want to see.

Congress had made a very liberal appropriation for observations of the solar eclipse, — the making of which was one object of my visit, — to be expended under the direction of Professor Peirce, superintendent of the Coast Survey. Peirce went over in person to take charge of the arrangements. He arrived in London with several members of his party a few days before we did, and about the same time came an independent party of my fellow astronomers from the Naval Observatory, consisting of Professors Hall, Harkness, and Eastman. The invasion of their country by such an army of American astronomers quite stirred up our English colleagues, who sorrowfully contrasted the liberality of our government with the parsimony of their own, which had, they said, declined to make any provision for the observations of the eclipse. Considering that it was visible on their own side of the Atlantic, they thought their government might take a lesson from ours. Of course we could not help them directly; and yet I suspect that our coming, or at least the coming of Peirce, really did help them a great deal. At any rate, it was a curious coincidence that no sooner did the American invasion occur than it was semi-officially discovered that no application of which her Majesty's government could take cognizance had been made by the scientific authorities for a grant of money with which to make preparations for observing the eclipse. That the scientific authorities were not long in catching so broad a hint as this goes without saying. A little more of the story came out a few days later in a very unexpected way.

In scientific England, the great social event of the year is the annual banquet of the Royal Society, held on St. Andrew's day, the date of the annual meeting of the society, and of the award of its medals for distinguished work in science. At the banquet the scientific outlook is discussed not only by members of the society, but by men high in political and social life. The medalists are toasted, if they are present; and their praises are sung, if, as is apt to be the case with foreigners, they are absent. First in rank is the Copley medal, founded by Sir Godfrey Copley, a contemporary of Newton. This medal has been awarded annually since 1731, and is now considered the highest honor that scientific England has to bestow. The recipient is selected with entire impartiality as to country, not for any special work published during the year, but in view of the general merit of all that he has done. Five times in its history the medal has crossed the Atlantic. It was awarded to Franklin in 1753, Agassiz in 1861, Dana in 1877, and J. Willard Gibbs in 1902. The long time that elapsed between the first and the second of these awards affords an illustration of the backwardness of scientific research in America during the greater part of the first century of our independence. The year of my visit the medal was awarded to Mr. Joule, the English physicist, for his work on the relation of heat and energy.

I was a guest at the banquet, which was the most brilliant function I had witnessed up to that time. The leaders in English science and learning sat around the table. Her Majesty's government was represented by Mr. Gladstone, the Premier, and Mr. Lowe, afterward Viscount Sherbrooke, Chancellor of the Exchequer. Both replied to toasts. Mr. Lowe as a speaker was perhaps a little dull, but not so Mr. Gladstone. There was a charm about the way in which his talk seemed to display the inner man. It could not be said that he had either the dry humor of Mr. Evarts or the wit of Mr. Depew; but these qualities were well replaced by the vivacity of his manner and the intellectuality of his face. He looked as if he had something interesting he wanted to tell you; and he proceeded to tell it in a very felicitous way as regarded both manner and language, but without anything that savored of eloquence. He was like Carl Schurz in talking as if he wanted to inform you, and not because he wanted you to see what a fine speaker he was. With this he impressed one as having a perfect command of his subject in all its bearings.

I did not for a moment suppose that the Premier of England could have taken any personal interest in the matter of the eclipse. Great, therefore, was my surprise when, in speaking of the relations of the government to science, he began to talk about the coming event. I quote a passage from memory, after twenty-seven years: " I had the pleasure of a visit, a few days since, from a very distinguished American professor, Professor Peirce of Harvard. In the course of the interview, the learned gentleman expressed his regret that her Majesty's government had declined to take any measures to promote observations of the coming eclipse of the sun by British astronomers. I replied that I was not aware that the government had declined to take such measures. Indeed, I went further, and assured him that any application from our astronomers for aid in making these observations would receive respectful consideration." I felt that there might be room for some suspicion that this visit of Professor Peirce was a not unimportant factor in the changed position of affairs as regarded British observations of the eclipse.

Not only the scene I have described, but subsequent experience, has impressed me with the high appreciation in which the best scientific work is held by the leading countries of Europe, especially England and France, as if the prosecution were something of national importance which men of the highest rank thought it an honor to take part in. The Marquis of Salisbury, in an interval between two terms of service as Premier of England, presided over the British Association for the Advancement of Science, and delivered an address showing a wide and careful study of the generalizations of modern science.

In France, also, one great glory of the nation is felt to be the works of its scientific and learned men of the past and present. Membership of one of the five academies of the Institute of France is counted among the highest honors to which a Frenchman can aspire. Most remarkable, too, is the extent to which other considerations than that of merit are set aside in selecting candidates for this honor. Quite recently a man was elected a member of the Academy of Sciences who was without either university or official position, and earned a modest subsistence as a collaborator of the " Revue des Deux Mondes." But he had found time to make investigations in mathematical astronomy of such merit that he was considered to have fairly earned this distinction, and the modesty of his social position did not lie in his way.

At the time of this visit Lister was an eminent member of the medical profession, but had not, so far as I am aware, been recognized as one who was to render incalculable service to suffering humanity. From a professional point of view there are no two walks in life having fewer points of contact than those of the surgeon and the astronomer. It is therefore a remarkable example of the closeness of touch among eminent Englishmen in every walk of life, that, in subsequent visits, I was repeatedly thrown into contact with one who may fairly be recommended as among the greatest benefactors of the human race that the nineteenth century has given us. This was partly, but not wholly, due to his being, for several years, the president of the Royal Society. I would willingly say much more, but I am unable to write authoritatively upon the life and work of such a man, and must leave gossip to the daily press.

For the visiting astronomer at London scarcely a place in London has more attractions than the modest little observatory and dwelling house on Upper Tulse Hill, in which Sir William Huggins has done so much to develop the spectroscopy of the fixed stars. The owner of this charming place was a pioneer in the application of the spectroscope to the analysis of the light of the heavenly bodies, and after nearly forty years of work in this field, is still pursuing his researches. The charm of sentiment is added to the cold atmosphere of science by the collaboration of Lady Huggins. Almost at the beginning of his work Mr. Huggins, analyzing the light of the great nebula of Orion, showed that it must proceed from a mass of gas, and not from solid matter, thus making the greatest step possible in our knowledge of these objects. He was also the first to make actual measures of the motions of bright stars to or from our system by observing the wave length of the rays of light which they absorbed. Quite recently an illustrated account of his observatory and its work has appeared in a splendid folio volume, in which the rigor of science is tempered with a gentle infusion of art which tempts even the non-scientific reader to linger over its pages.

In England, the career of Professor Cayley affords an example of the spirit that impels a scientific worker of the highest class, and of the extent to which an enlightened community may honor him for what he is doing. One of the creators of modern mathematics, he never had any ambition beyond the prosecution of his favorite science. I first met him at a dinner of the Astronomical Society Club. As the guests were taking off their wraps and assembling in the anteroom, I noticed, with some surprise, that one whom I supposed to be an attendant was talking with them on easy terms. A moment later the supposed attendant was introduced as Professor Cayley. His garb set off the seeming haggardness of his keen features so effectively that I thought him either broken down in health or just recovering from some protracted illness. The unspoken words on my lips were, "Why, Professor Cayley, what has happened to you?" Being now in the confessional, I must own that I did not, at the moment, recognize the marked intellectuality of a very striking face. As a representation of a mathematician in the throes of thought, I know nothing to equal his portrait by Dickenson, which now hangs in the hall of Trinity College, Cambridge, and is reproduced in the sixth volume of Cayley's collected works. His life was that of a man moved to investigation by an uncontrollable impulse; the only sort of man whose work is destined to be imperishable. Until forty years of age he was by profession a conveyancer. His ability was such that he might have gained a fortune by practicing the highest branch of English law, if his energies had not been diverted in another direction. The spirit in which he pursued his work may be judged from an anecdote related by his friend and co-worker, Sylvester, who, in speaking of Cayley's even and placid temper, told me that he had never seen him ruffled but once. Entering his office one morning, intent on some new mathematical thought which he was discussing with Sylvester, he opened the letter-box in his door and found a bundle of papers relating to a law case which he was asked to take up. The interruption was too much. He flung the papers on the table with remarks more forcible than complimentary concerning the person who had distracted his attention at such an inopportune moment. In 1863 he was made a professor at Cambridge, where, no longer troubled with the intricacies of land tenure, he published one investigation after another with ceaseless activity, to the end of his life.

Among my most interesting callers was Professor John C. Adams, of whom I have spoken as sharing with Leverrier the honor of having computed the position of the planet Neptune before its existence was otherwise known. The work of the two men was prosecuted at almost the same time, but adopting the principle that priority of publication should be the sole basis of credit, Arago had declared that no other name than that of Leverrier should even be mentioned in connection with the work. If repute was correct, Leverrier was not distinguished for those amiable qualities that commonly mark the man of science and learning. His attitude toward Adams had always been hostile. Under these conditions chance afforded the latter a splendid opportunity of showing his superiority to all personal feeling. He was president of the Royal Astronomical Society when its annual medal was awarded to his French rival for his work in constructing new tables of the sun and planets. It thus became his duty to deliver the address setting forth the reasons for the award. He did this with a warmth of praise for Leverrier's works which could not have been exceeded had the two men been bosom friends.

Adams's intellect was one of the keenest I ever knew. The most difficult problems of mathematical astronomy and the most recondite principles that underlie the theory of the celestial motions were to him but child's play. His works place him among the first mathematical astronomers of the age, and yet they do not seem to do his ability entire justice. Indeed, for fifteen years previous to the time of my visit his published writings had been rather meagre. But I believe he was justly credited with an elaborate witticism to the following effect: " In view of the fact that the only human being ever known to have been killed by a meteorite was a monk, we may concede that after four hundred years the Pope's bull against the comet has been justified by the discovery that comets are made up of meteorites."

Those readers who know on what imperfect data men's impressions are sometimes founded will not be surprised to learn of my impression that an Englishman's politics could be inferred from his mental and social make-up. If all men are born either Aristotelians or Platonists, then it may be supposed that all Englishmen are born Conservatives or Liberals.

The utterances of English journalists of the Conservative party about American affairs during and after our civil war had not impressed me with the idea that one so unfortunate as to be born in that party would either take much interest in meeting an American or be capable of taking an appreciative view of scientific progress. So confident was I of my theory that I remarked to a friend with whom I had become somewhat intimate, that no one who knew Mr. Adams could have much doubt that he was a Liberal in politics.

An embarrassed smile spread over the friend's features. "You would not make that conclusion known to Mr. Adams, I hope," said he.

"But is he not a Liberal?"

"He is not only a Conservative, but declares himself 'a Tory of the Tories.'"

I afterward found that he fully justified his own description. At the university, he was one of the leading opponents of those measures which freed the academic degrees from religious tests. He was said to have been among those who objected to Sylvester, a Jew, receiving a degree.

I had decided to observe the eclipse at Gibraltar. In order that my results, if I obtained any, might be utilized in the best way, it was necessary that the longitude of the station should be determined by telegraph. This had never been done for Gibraltar. How great the error of the supposed longitude might have been may be inferred from the fact that a few years later, Captain F. Green of the United States Navy found the longitude of Lisbon on the Admiralty charts to be two miles in error. The first arrangements I had to make in England were directed to this end. Considering the relation of the world's great fortress to British maritime supremacy, it does seem as if there were something presumptuous in the coolness with which I went among the authorities to make arrangements for the enterprise. Nevertheless, the authorities permitted the work, with a cordiality which was of itself quite sufficient to remove any such impression, had it been entertained. The astronomers did, indeed, profess to feel it humiliating that the longitude of such a place as Gibraltar should have to be determined from Greenwich by an American. They did not say " by a foreigner," because they always protested against Americans looking upon themselves as such. Still, it would not be an English enterprise if an American carried it out. I suspect, however, that my proceedings were not looked upon with entire dissatisfaction even by the astronomers. They might prove as good a stimulant to their government in showing a little more enterprise in that direction as the arrival of our eclipse party did.

The longitude work naturally took me to the Royal Observatory which has made the little town of Greenwich so famous. It is situated some eight miles east from Charing Cross, on a hill in Greenwich Park, with a pleasant outlook toward the Thames. From my youth up I had been working with its observations, and there was no institution in the world which I had approached, or could approach, with the interest I felt in ascending the little hill on which it is situated. When the Calabria was once free from her wharf in New York harbor, and on her way down the Narrows, the foremost thought was, "Off for Europe; we shall see Greenwich!" The day of my arrival in London I had written to Professor Airy, and received an answer the same evening, inviting us to visit the observatory and spend an afternoon with him a day or two later.

I was shown around the observatory by an assistant, while my wife was entertained by Mrs. Airy and the daughters inside the dwelling. The family dined as soon as the day's work was over, about the middle of the afternoon. After the meal, we sat over a blazing fire and discussed our impressions of London.

"What place in London interested you most?" said Airy to my wife.

"The first place I went to see was Cavendish Square."

"What was there in Cavendish Square to interest you?"

"When I was a little girl, my mother once gave me, as a birthday present, a small volume of poems. The first verse in the book was:—

"'Little Ann and her mother were walking one day
  Through London's wide city so fair;
  And business obliged them to go by the way
  That led them through Cavendish Square.'"

To our astonishment the Astronomer Royal at once took up the thread:—

"'And as they passed by the great house of a lord,
  A beautiful chariot there came,
  To take some most elegant ladies abroad,
  Who straightway got into the same,'"

and went on to the end. I do not know which of the two was more surprised: Airy, to find an American woman who was interested in his favorite ballad, or she to find that he could repeat it by heart. The incident was the commencement of a family friendship which has outlived both the heads of the Airy family.

We may look back on Airy as the most commanding figure in the astronomy of our time. He owes this position not only to his early works in mathematical astronomy, but also to his ability as an organizer. Before his time the working force of an observatory generally consisted of individual observers, each of whom worked to a greater or less extent in his own way. It is true that organization was not unknown in such institutions. Nominally, at least, the assistants in a national observatory were supposed to follow the instructions of a directing head. This was especially the case at Greenwich. Still, great dependence was placed upon the judgment and ability of the observer himself, who was generally expected to be a man well trained in his specialty, and able to carry on good work without much help. From Airy's point of view, it was seen that a large part of the work necessary to the attainment of the traditional end of the Royal Observatory was of a kind that almost any bright schoolboy could learn to do in a few weeks, and that in most of the remaining part plodding industry, properly directed, was more important than scientific training. He could himself work out all the mathematical formulæ and write all the instructions required to keep a small army of observers and computers employed, and could then train in his methods a few able lieutenants, who would see that all the details were properly executed. Under these lieutenants was a grade comprising men of sufficient technical education to enable them to learn how to point the telescope, record a transit, and perform the other technical operations necessary in an astronomical observation. A third grade was that of computers: ingenious youth, quick at figures, ready to work for a compensation which an American laborer would despise, yet well enough schooled to make simple calculations. Under the new system they needed to understand only the four rules of arithmetic; indeed, so far as possible Airy arranged his calculations in such a way that subtraction and division were rarely required. His boys had little more to do than add and multiply. Thus, so far as the doing of work was concerned, he introduced the same sort of improvement that our times have witnessed in great manufacturing establishments, where labor is so organized that unskilled men bring about results that formerly demanded a high grade of technical ability. He introduced production on a large scale into astronomy.

At the time of my visit, it was much the fashion among astronomers elsewhere to speak slightingly of the Greenwich system. The objections to it were, in substance, the same that have been made to the minute subdivision of labor. The intellect of the individual was stunted for the benefit of the work. The astronomer became a mere operative. Yet it must be admitted that the astronomical work done at Greenwich during the sixty years since Airy introduced his system has a value and an importance in its specialty that none done elsewhere can exceed. All future conclusions as to the laws of motion of the heavenly bodies must depend largely upon it.

The organization of his little army necessarily involved a corresponding change in the instruments they were to use. Before his time the trained astronomer worked with instruments of very delicate construction, so that skill in handling them was one of the requisites of an observer. Airy made them in the likeness of heavy machinery, which could suffer no injury from a blow of the head of a careless observer. Strong and simple, they rarely got out of order. It is said that an assistant who showed a visiting astronomer the transit circle sometimes hit it a good slap to show how solid it was; but this was not done on the present occasion. The little army had its weekly marching orders and made daily reports of progress to its commander, who was thus enabled to control the minutest detail of every movement.

In the course of the evening Airy gave me a lesson in method, which was equally instructive and entertaining. In order to determine the longitude of Gibraltar, it was necessary that time signals should be sent by telegraph from the Royal Observatory. Our conversation naturally led us into a discussion of the general subject of such operations. I told him of the difficulties we had experienced in determining a telegraphic longitude, — that of the Harvard Observatory from Washington, for example, — because it was only after a great deal of talking and arranging on the evening of the observation that the various telegraph stations between the two points could have their connections successfully made at the same moment. At the appointed hour the Washington operator would be talking with the others, to know if they were ready, and so a general discussion about the arrangements might go on for half an hour before the connections were all reported good. If we had such trouble in a land line, how should we get a connection from London to the Gibraltar cable through lines in constant use?

"But," said Airy, "I never allow an operator who can speak with the instruments to take part in determining a telegraphic longitude."

"Then how can you get the connections all made from one end of the line to the other, at the same moment, if your operators cannot talk to one another?"

"Nothing is simpler. I fix in advance a moment, say eight o'clock Greenwich mean time, at which signals are to commence. Every intermediate office through which the signals are to pass is instructed to have its wires connected in both directions exactly at the given hour, and to leave them so connected for ten minutes, without asking any further instructions. At the end of the line the instruments must be prepared at the appointed hour to receive the signals. All I have to do here is to place my clock in the circuit and send on the signals for ten minutes, commencing at eight o'clock. They are recorded at the other end of the line without further trouble."

"But have you never met with a failure to understand the instructions?"

"No; they are too simple to be mistaken, once it is understood that no one has anything to do but make his connections at the designated moment, without asking whether any one else is ready."

Airy was noted not less for his ability as an organizer than for his methodical habits. The care with which he preserved every record led Sir William Rowan Hamilton to say that when Airy wiped his pen on a blotter, he fancied him as always taking a press copy of the mark. His machinery seemed to work perfectly, whether it was constructed of flesh or of brass. He could prepare instructions for the most complicated piece of work with such effective provision against every accident and such completeness in every detail that the work would go on for years without further serious attention from him. The instruments which he designed half a century ago are mostly in use to this day, with scarcely an alteration.

Yet there is some reason to fear that Airy carried method a little too far to get the best results. Of late years his system has been greatly changed, even at Greenwich. It was always questionable whether so rigid a military routine could accomplish the best that was possible in astronomy; and Airy himself, during his later years, modified his plan by trying to secure trained scientific men as his assistants, giving them liberty to combine independent research, on their own account, with the work of the establishment. His successor has gone farther in the same direction, and is now gathering around him a corps of young university men, from whose ability much may be expected.

Observations with the spectroscope have been pursued, and the observatory has taken a prominent part in the international work of making a photographic map of the heavens. Of special importance are the regular discussions of photographs of the sun, taken in order to determine the law of the variation of the spots. The advantage of the regular system which has been followed for more than fifty years is seen in the meteorological observations; these disprove some theories of the relation between the sun and the weather, in a way that no other set of meteorological records has done. While delicate determinations of the highest precision, such as those made at Pulkova, are not yet undertaken to any great extent, a regular even if slow improvement is going on in the general character of the observations and researches, which must bear fruit in due time.

One of the curious facts we learned at Greenwich was that astronomy was still supposed to be astrology by many in England. That a belief in astrology should survive was perhaps not remarkable, though I do not remember to have seen any evidence of it in this country. But applications received at the Royal Observatory, from time to time, showed a widespread belief among the masses that one of the functions of the astronomer royal was the casting of horoscopes.

We went to Edinburgh. Our first visit was to the observatory, then under the direction of Professor C. Piazzi Smyth, who was also an Egyptologist of repute, having made careful measurements of the Pyramids, and brought out some new facts regarding their construction. He was thus led to the conclusion that they bore marks of having been built by a people of more advanced civilization than was generally supposed, — so advanced, indeed, that we had not yet caught up to them in scientific investigation. These views were set forth with great fullness in his work on "The Antiquity of Intellectual Man," as well as in other volumes describing his researches. He maintained that the builders of the Pyramids knew the distance of the sun rather better than we did, and that the height of the Great Pyramid had been so arranged that if it was multiplied by a thousand millions we should get this distance more exactly than we could measure it in these degenerate days. With him, to believe in the Pyramid was to believe this, and a great deal more about the civilization which it proved. So, when he asked me whether I believed in the Pyramid, I told him that I did not think I would depend wholly upon the Pyramid for the distance of the sun to be used in astronomy, but should want its indications at least confirmed by modern researches. The hint was sufficient, and I was not further pressed for views on this subject.

He introduced us to Lady Hamilton, widow of the celebrated philosopher, who still held court at Edinburgh. The daughter of the family was in repute as a metaphysician. This was interesting, because I had never before heard of a female metaphysician, although there were several cases of female mathematicians recorded in history. First among them was Donna Maria Agnesi, who wrote one of the best eighteenth-century books on the calculus, and had a special dispensation from the Pope to teach mathematics at Bologna. We were therefore very glad to accept an invitation from Lady Hamilton to spend an evening with a few of her friends. Her rooms were fairly filled with books, the legacy of one of whom it was said that " scarcely a thought has come down to us through the ages which he has not mastered and made his own."

The few guests were mostly university people and philosophers. The most interesting of them was Professor Blackie, the Grecian scholar, who was the liveliest little man of sixty I ever saw; amusing us by singing German songs, and dancing about the room like a sprightly child among its playmates. I talked with Miss Hamilton about Mill, whose " Examination of Sir William Hamilton's Philosophy " was still fresh in men's minds. Of course she did not believe in this book, and said that Mill could not understand her father's philosophy. With all her intellect, she was a fine healthy-looking young lady, and it was a sad surprise, a few years later, to hear of her death. Madame Sophie Kovalevsky afterward appeared on the stage as the first female mathematician of our time, but it may be feared that the woman philosopher died with Miss Hamilton.

A large party of English astronomers were going to Algeria to observe the eclipse. The government had fitted up a naval transport for their use, and as I was arranging for a passage on a ship of the Peninsular and Oriental Line we received an invitation to become the guests of the English party. Among those on board were Professor Tyndall; Mr. Huggins, the spectroscopist; Sir Erastus Ommaney, a retired English admiral, and a fellow of the Royal Society; Father Perry, S. J., a well-known astronomer; and Lieutenant Wharton, who afterward became hydrographer to the Admiralty.

The sprightliest man on board was Professor Tyndall. He made up for the absence of mountains by climbing to every part of the ship he could reach. One day he climbed the shrouds to the maintop, and stood surveying the scene as if looking out from the top of the Matterhorn. A sailor followed him, and drew a chalk-line around his feet. I assume the reader knows what this means; if he does not, he can learn by straying into the sailors' quarters the first time he is on board an ocean steamer. But the professor absolutely refused to take the hint.

We had a rather rough passage, from which Father Perry was the greatest sufferer. One day he heard a laugh from the only lady on board, who was in the adjoining stateroom. "Who can laugh at such a time as this!" he exclaimed. He made a vow that he would never go on the ocean again, even if the sun and moon fought for a month. But the vows of a seasick passenger are forgotten sooner than any others I know of; and it was only four years later that Father Perry made a voyage to Kerguelen Island, in the stormiest ocean on the globe, to observe a transit of Venus.

Off the coast of Spain, the leading chains of the rudder got loose, during a gale in the middle of the night, and the steering apparatus had to be disconnected in order to tighten them. The ship veered round into the trough of the sea, and rolled so heavily that a table, twenty or thirty feet long, in the saloon, broke from its fastenings, and began to dance around the cabin with such a racket that some of the passengers feared for the safety of the ship.

Just how much of a storm there was I cannot say, believing that it is never worth while for a passenger to leave his berth, if there is any danger of a ship foundering in a gale. But in Professor Tyndall's opinion we had a narrow escape. On arriving at Gibraltar, he wrote a glowing account of the storm to the London Times, in which he described the feelings of a philosopher while standing on the stern of a rolling ship in an ocean storm, without quite knowing whether she was going to sink or swim. The letter was anonymous, which gave Admiral Ommaney an excellent opportunity to write as caustic a reply as he chose, under the signature of "A Naval Officer." He said that sailor was fortunate who could arrange with the clerk of the weather never to have a worse storm in crossing the Bay of Biscay than the one we had experienced.

We touched at Cadiz, and anchored for a few hours, but did not go ashore. The Brooklyn, an American man-of-war, was in the harbor, but there was no opportunity to communicate with her, though I knew a friend of mine was on board.

Gibraltar is the greatest babel in the world, or, at least, the greatest I know. I wrote home: "The principal languages spoken at this hotel are English, Spanish, Moorish, French, Italian, German, and Danish. I do not know what languages they speak at the other hotels." Moorish and Spanish are the local tongues, and of course English is the official one; but the traders and commercial travelers speak nearly every language one ever heard.

I hired a Moor — who bore some title which indicated that he was a descendant of the Caliphs, and by which he had to be addressed — to do chores and act as general assistant. One of the first things I did, the morning after my arrival, was to choose a convenient point on one of the stone parapets for "taking the sun," in order to test the running of my chronometer. I had some suspicion as to the result, but was willing to be amused. A sentinel speedily informed me that no sights were allowed to be taken on the fortification. I told him I was taking sights on the sun, not on the fortification. But he was inexorable; the rule was that no sights of any sort could be taken without a permit. I soon learned from Mr. Sprague, the

American consul, who the proper officer was to issue the permit, which I was assured would be granted without the slightest difficulty.

The consul presented me to the military governor of the place, General Sir Fenwick Williams of Kars. I did not know till long afterward that he was born very near where I was. He was a man whom it was very interesting to meet. His heroic defense of the town whose name was added to his own as a part of his title was still fresh in men's minds. It had won him the order of the Bath in England, the Grand Cross of the Legion of Honor and a sword from Napoleon III., and the usual number of lesser distinctions. The military governor, the sole authority and viceroy of the Queen in the fortress, is treated with the deference due to an exalted personage; but this deference so strengthens the dignity of the position that the holder may be frank and hearty at his own pleasure, without danger of impairing it. Certainly, we found Sir Fenwick a most genial and charming gentleman. The Alabama claims were then in their acute stage, and he expressed the earnest hope that the two nations would not proceed to cutting each other's throats over them.

There was no need of troubling the governor with such a detail as that of a permit to take sights; but the consul ventured to relate my experience of the morning. He took the information in a way which showed that England, in making him a general, had lost a good diplomatist. Instead of treating the matter seriously, which would have implied that we did not fully understand the situation, he professed to be greatly amused, and said it reminded him of the case of an old lady in " Punch" who had to pass a surveyor in the street, behind a theodolite. " Please, sir, don't shoot till I get past," she begged.

Before leaving England, I had made very elaboate arrangements, both with the Astronomer Royal and with the telegraph companies, to determine the longitude of Gibraltar by telegraphic signals. The most difficult part of the operation was the transfer of the signals from the end of the land line into the cable, which had to be done by hand, because the cable companies were not willing to trust to an automatic action of any sort between the land line and the cable. It was therefore necessary to show the operator at the point of junction how signals were to be transmitted. This required a journey to Port Curno, at the very end of the Land's End, several miles beyond the terminus of the railway. It was the most old-time place I ever saw; one might have imagined himself thrown back into the days of the Lancasters. The thatched inn had a hard stone floor, with a layer of loose sand scattered over it as a carpet in the bedroom. My linguistic qualities were put to a severe test in talking with the landlady. But the cable operators were pleasing and intelligent young gentlemen, and I had no difficulty in making them understand how the work was to be done.

The manager of the cable was Sir James Anderson, who had formerly commanded a Cunard steamship from Boston, and was well known to the Harvard professors, with whom he was a favorite. I had met him, or at least seen him, at a meeting of the American Academy ten years before, where he was introduced by one of his Harvard friends. After commanding the ship that laid the first Atlantic cable, he was made manager of the cable line from England to Gibraltar. He gave me a letter to the head operator at Gibraltar, the celebrated de Sauty.

I say " the celebrated," but may it not be that this appellation can only suggest the vanity of all human greatness? It just occurs to me that many of the present generation may not even have heard of the —

> Whispering Boanerges, son of silent thunder,
> Holding talk with nations,

immortalized by Holmes in one of his humorously scientific poems. During the two short weeks that the first Atlantic cable transmitted its signals, his fame spread over the land, for the moment obscuring by its brilliancy that of Thomson, Field, and all others who had taken part in designing and laying the cable. On the breaking down of the cable he lapsed into his former obscurity. I asked him if he had ever seen Holmes's production. He replied that he had received a copy of " The Atlantic Monthly " containing it from the poet himself, accompanied by a note saying that he might find in it something of interest. He had been overwhelmed with invitations to continue his journey from Newfoundland to the United States and lecture on the cable, but was sensible enough to decline them.

The rest of the story of the telegraphic longitude is short. The first news which de Sauty had to give me was that the cable was broken, — just where, he did not know, and would not be able soon to discover. After the break was located, an unknown period would be required to raise the cable, find the place, and repair the breach. The weather, on the day of the eclipse, was more than half cloudy, so that I did not succeed in making observations of such value as would justify my waiting indefinitely for the repair of the cable, and the project of determining the longitude had to be abandoned.

XI

MEN AND THINGS IN EUROPE

We went from Gibraltar to Berlin in January by way of Italy. The Mediterranean is a charming sea in summer, but in winter is a good deal like the Atlantic. The cause of the blueness of its water is not completely settled; but its sharing this color with Lake Geneva, which is tinged with detritus from the shore, might lead one to ascribe it to substances held in solution. The color is noticeable even in the harbor of Malta, to which we had a pleasant though not very smooth passage of five days.

Here was our first experience of an Italian town of a generation ago. I had no sooner started to take a walk than a so-called guide, who spoke what he thought was English, got on my track, and insisted on showing me everything. If I started toward a shop, he ran in before me, invited me in, asked what I would like to buy, and told the shopman to show the gentleman something. I could not get rid of him till I returned to the hotel, and then he had the audacity to want a fee for his services. I do not think he got it. Everything of interest was easily seen, and we only stopped to take the first Italian steamer to Messina. We touched at Syracuse and Catania, but did not land.

Ætna, from the sea, is one of the grandest sights I ever saw. Its snow-covered cone seems to rise on all sides out of the sea or the plain, and to penetrate the blue sky. In this it gives an impression like that of the Weisshorn seen from Randa, but gains by its isolation.

At Messina, of course, our steamer was visited by a commissionnaire, who asked me in good English whether I wanted a hotel. I told him that I had already decided upon a hotel, and therefore did not need his services. But it turned out that he belonged to the very hotel I was going to, and was withal an American, a native-born Yankee, in fact, and so obviously honest that I placed myself unreservedly in his hands, — something which I never did with one of his profession before or since. He said the first thing was to get our baggage through the custom-house, which he could do without any trouble, at the cost of a franc. He was as good as his word. The Italian custom-house was marked by primitive rigor, and baggage was commonly subjected to a very thorough search. But my man was evidently well known and fully trusted. I was asked to raise the lid of one trunk, which I did; the official looked at it, with his hands in his pockets, gave a nod, and the affair was over. My Yankee friend collected one franc for that part of the business. He told us all about the place, changed our money so as to take advantage of the premium on gold, and altogether looked out for our interests in a way to do honor to his tribe. I thought there might be some curious story of the way in which a New Englander of such qualities could have dropped into such a place, but it will have to be left to imagination.

We reached the Bay of Naples in the morning twilight, after making an unsuccessful attempt to locate Scylla and Charybdis. If they ever existed, they must have disappeared. Vesuvius was now and then lighting up the clouds with its intermittent flame. But we had passed a most uncomfortable night, and the morning was wet and chilly. A view requires something more than the objective to make it appreciated, and the effect of a rough voyage and bad weather was such as to deprive of all its beauty what is considered one of the finest views in the world. Moreover, the experience made me so ill-natured that I was determined that the custom-house officer at the landing should have no fee from me. The only article that could have been subject to duty was on top of everything in the trunk, except a single covering of some loose garment, so that only a touch was necessary to find it. When it came to the examination, the officer threw the top till contemptuously aside, and devoted himself to a thorough search of the bottom. The only unusual object he stumbled upon was a spy-glass inclosed in a shield of morocco. Perhaps a gesture and a remark on my part aroused his suspicions. He opened the glass, tried to take it to pieces, inspected it inside and out, and was so disgusted with his failure to find anything contraband in it that he returned everything to the trunk, and let us off.

It is commonly and quite justly supposed that the more familiar the traveler is with the language of the place he visits, the better he will get along. It is a common experience to find that even when you can pronounce the language, you cannot understand what is said. But there are exceptions to all rules, and circumstances now and then occur in which one thus afflicted has an advantage over the native. You can talk to him, while he cannot talk to you. There was an amusing case of this kind at Munich. The only train that would take us to Berlin before nightfall of the same day left at eight o'clock in the morning, by a certain route. There was at Munich what we call a union station. I stopped at the first ticket-office where I saw the word "Berlin" on the glass, asked for a ticket good in the train that was going to leave at eight o'clock the next morning for Berlin, and took what the seller gave me. He was a stupid-looking fellow, so when I got to my hotel I showed the ticket to a friend. "That is not the ticket that you want at all," said he; "it will take you by a circuitous route in a train that does not leave until after nine, and you will not reach Berlin until long after dark." I went directly back to the station and showed my ticket to the agent.

"I — asked — you — for — a — ticket — good

— in — the — train — which — leaves — at — eight — o' — clock. This — ticket — is — not — good — in — that — train. Sie — haben — mich — betrügen. I — want — you — to — take — the — ticket — back — and — return — me — the — money. What — you — say — can — I — not — understand."

He expostulated, gesticulated, and fumed, but I kept up the bombardment until he had to surrender. He motioned to me to step round into the office, where he took the ticket and returned the money. I mention the matter because taking back a ticket is said to be quite unusual on a German railway.

At Berlin, the leading astronomers then, as now, were Förster, director of the observatory, and Auwers, permanent secretary of the Academy of Sciences. I was especially interested in the latter, as we had started in life nearly at the same time, and had done much work on similar lines. It was several days before I made his acquaintance, as I did not know that the rule on the Continent is that the visitor must make the first call, or at least make it known by direct communication that he would be pleased to see the resident; otherwise it is presumed that he does not wish to see callers. This is certainly the more logical system, but it is not so agreeable to the visiting stranger as ours is. The art of making the latter feel at home is not brought to such perfection on the Continent as in England;

perhaps the French understand it less than any other people. But none can be pleasanter than the Germans, when you once make their acquaintance; and we shall always remember with pleasure the winter we passed in Berlin.

To-day, Auwers stands at the head of German astronomy. In him is seen the highest type of the scientific investigator of our time, one perhaps better developed in Germany than in any other country. The work of men of this type is marked by minute and careful research, untiring industry in the accumulation of facts, caution in propounding new theories or explanations, and, above all, the absence of effort to gain recognition by being the first to make a discovery. When men are ambitious to figure as Newtons of some great principle, there is a constant temptation to publish unverified speculations which are likely rather to impede than to promote the advance of knowledge. The result of Auwers's conscientiousness is that, notwithstanding his eminence in his science, there are few astronomers of note whose works are less fitted for popular exposition than his. His specialty has been the treatment of all questions concerning the positions and motions of the stars. This work has required accurate observations of position, with elaborate and careful investigations of a kind that offer no feature to attract public attention, and only in exceptional cases lead to conclusions that would interest the general reader. He considers no work as ready for publication until it is completed in every detail.

The old astronomical observations of which I was in quest might well have been made by other astronomers than those of Paris, so while awaiting the end of the war I tried to make a thorough search of the writings of the mediæval astronomers in the Royal Library. If one knew exactly what books he wanted, and had plenty of time at his disposal, he would find no difficulty in consulting them in any of the great Continental libraries. But at the time of my visit, notwithstanding the cordiality with which all the officials, from Professor Lepsius down, were disposed to second my efforts, the process of getting any required book was very elaborate. Although one could obtain a book on the same day he ordered it, if he went in good time, it was advisable to leave the order the day before, if possible. When, as in the present case, one book only suggests another, this a third, and so on, in an endless chain, the carrying on of an extended research is very tedious.

One feature of the library strongly impressed me with the comparatively backward state of mathematical science in our own country. As is usual in the great European libraries, those books which are most consulted are placed in the general reading-room, where any one can have access to them, at any moment. It was surprising to see amongst these books a set of Crelle's "Journal of Mathematics," and to find it well worn by constant use. At that time, so far as I could learn, there were not more than two or three sets of the Journal in the United States; and these were almost unused. Even the Library of Congress did not contain a set. There has been a great change since that time, — a change in which the Johns Hopkins University took the lead, by inviting Sylvester to this country, and starting a mathematical school of the highest grade. Other universities followed its example to such an extent that, to-day, an American student need not leave his own country to hear a master in any branch of mathematics.

I believe it was Dr. B. A. Gould who called the Pulkova Observatory the astronomical capital of the world. This institution was founded in 1839 by the Emperor Nicholas, on the initiative of his greatest astronomer. It is situated some twelve miles south of St. Petersburg, not far from the railway between that city and Berlin, and gets its name from a peasant village in the neighborhood. From its foundation it has taken the lead in exact measurements relating to the motion of the earth and the positions of the principal stars. An important part of its equipment is an astronomical library, which is perhaps the most complete in existence. This, added to all its other attractions, induced me to pay a visit to Pulkova. Otto Struve, the director, had been kind enough to send me a message, expressing the hope that I would pay him a visit, and giving directions about telegraphing in advance, so as to insure the delivery of the dispatch. The time from Berlin to St. Petersburg is about forty-eight hours, the only through train leaving and arriving in the evening. On the morning of the day that the train was due I sent the dispatch. Early in the afternoon, as the train was stopping at a way station, I saw an official running hastily from one car to another, looking into each with some concern. When he came to my door, he asked if I had sent a telegram to Estafetta. I told him I had. He then informed me that Estafetta had not received it. But the train was already beginning to move, so there was no further chance to get information. The comical part of the matter was that " Estafetta " merely means a post or postman, and that the directions, as Struve had given them, were to have the dispatch sent by postman from the station to Pulkova.

It was late in the evening when the train reached Zarsko-Selo, the railway station for Pulkova, which is about five miles away. The station-master told me that no carriage from Pulkova was waiting for me, which tended to confirm the fear that the dispatch had not been received. After making known my plight, I took a seat in the station and awaited the course of events, in some doubt what to do. Only a few minutes had elapsed when a good-looking peasant, well wrapped in a fur overcoat, with a whip in his hand, looked in at the door, and pronounced very distinctly the words, "Observatorio Pulkova." Ah! this is Struve's driver at last, thought I, and I followed the man to the door. But when I looked at the conveyance, doubt once more supervened. It was scarcely more than a sledge, and was drawn by a single horse, evidently more familiar with hard work than good feeding. This did not seem exactly the vehicle that the great Russian observatory would send out to meet a visitor; yet it was a far country, and I was not acquainted with its customs.

The way in which my doubt was dispelled shows that there is one subject besides love on which difference of language is no bar to the communication of ideas. This is the desire of the uncivilized man for a little coin of the realm. In South Africa, Zulu chiefs, who do not know one other word of English, can say " shilling " with unmistakable distinctness. My Russian driver did not know even this little English word, but he knew enough of the universal language. When we had made a good start on the snow-covered prairie, he stopped his horse for a moment, looked round at me inquiringly, raised his hand, and stretched out two fingers so that I could see them against the starlit sky.

I nodded assent.

Then he drew his overcoat tightly around him with a gesture of shivering from the cold, beat his hands upon his breast as if to warm it, and again looked inquiringly at me.

I nodded again.

The bargain was complete. He was to have two rubles for the drive, and a little something to warm up his shivering breast. So he could not be Struve's man.

There is no welcome warmer than a Russian one, and none in any country warmer than that which the visiting astronomer receives at an observatory. Great is the contrast between the winter sky of a clear moonless night and the interior of a dining-room, forty feet square, with a big blazing fire at one end and a table loaded with eatables in the middle. The fact that the visitor had never before met one of his hosts detracted nothing from the warmth of his reception.

The organizer of the observatory, and its first director, was Wilhelm Struve, father of the one who received me, and equally great as man and astronomer. Like many other good Russians, he was the father of a large family. One of his sons was for ten years the Russian minister at Washington, and as popular a diplomatist as ever lived among us. The instruments which Struve designed sixty years ago still do as fine work as any in the world; but one may suspect this to be due more to the astronomers who handle them than to the instruments themselves.

The air is remarkably clear; the entrance to St. Petersburg, ten or twelve miles north, is distinctly visible, and Struve told me that during the Crimean war he could see, through the great telescope, the men on the decks of the British ships besieging Kronstadt, thirty miles away.

One drawback from which the astronomers suffer is the isolation of the place. The village at the foot of the little hill is inhabited only by peasants, and the astronomers and employees have nearly all to be housed in the observatory buildings. There is no society but their own nearer than the capital. At the time of my visit the scientific staff was almost entirely German or Swedish, by birth or language. In the state, two opposing parties are the Russian, which desires the ascendency of the native Muscovites, and the German, which appreciates the fact that the best and most valuable of the Tsar's subjects are of German or other foreign descent. During the past twenty years the Russian party has gradually got the upper hand; and the result of this ascendency at Pulkova will be looked for with much solicitude by astronomers everywhere.

Once a year the lonely life of the astronomers is enlivened by a grand feast — that of the Russian New Year. One object of the great dining-room which I have mentioned, the largest room, I believe, in the whole establishment, was to make this feast possible. My visit took place early in March, so that I did not see the celebration; but from what I have heard, the little colony does what it can to make up for a year of ennui. Every twenty-five years it celebrates a jubilee; the second came off in 1889.

There is much to interest the visitor in a Russian peasant village, and that of Pulkova has features some of which I have never seen described. Above the door of each log hut is the name of the occupant, and below the name is a rude picture of a bucket, hook, or some other piece of apparatus used in extinguishing fire. Inside, the furniture is certainly meagre enough, yet one could not see why the occupants should be otherwise than comfortable. I know of no good reason why ignorance should imply unhappiness; altogether, there is some good room for believing that the less civilized races can enjoy themselves, in their own way, about as well as we can. What impressed me as the one serious hardship of the peasantry was their hours of labor. Just how many hours of the twenty-four these beings find for sleep was not clear to the visitor; they seemed to be at work all day, and at midnight many of them had to start on their way to St. Petersburg with a cartload for the market. A church ornamented with tinsel is a feature of every Russian village; so also are the priests. The only two I saw were sitting on a fence, wearing garments that did not give evidence of having known water since they were made. One great drawback to the growth of manufactures in Russia is the number of feast days, on which the native operators must one and all abandon their work, regardless of consequences.

The astronomical observations made at Pulkova are not published annually, as are those made at most of the other national observatories; but a volume relating to one subject is issued whenever the work is done. When I was there, the volumes containing the earlier meridian observations were in press. Struve and his chief assistant, Dr. Wagner, used to pore nightly over the proof sheets, bestowing on every word and detail a minute attention which less patient astronomers would have found extremely irksome.

Dr. Wagner was a son-in-law of Hansen, the astronomer of the little ducal observatory at Gotha, as was also our Bayard Taylor. My first meeting with Hansen, which occurred after my return to Berlin, was accompanied with some trepidation. Modest as was the public position that he held, he may now fairly be considered the greatest master of celestial mechanics since Laplace. In what order Leverrier, Delaunay, Adams, and Hill should follow him, it is not necessary to decide. To many readers it will seem singular to place any name ahead of that of the master who pointed out the position of Neptune before a human eye had ever recognized it. But this achievement, great as it was, was more remarkable for its boldness and brilliancy than for its inherent difficulty. If the work had to be done over again to-day, there are a number of young men who would be as successful as Leverrier; but there are none who would attempt to reinvent the methods of Hansen, or even to improve radically upon them. Their main feature is the devising of new and refined methods of computing the variations in the motions of a planet produced by the attraction of all the other planets. As Laplace left this subject, the general character of these variations could be determined without difficulty, but the computations could not be made with mathematical exactness. Hansen's methods led to results so precise that, if they were fully carried out, it is doubtful whether any deviation between the predicted and the observed motions of a planet could be detected by the most refined observation.

At the time of my visit Mrs. Wagner was suffering from a severe illness, of which the crisis passed while I was at Pulkova, and left her, as was supposed, on the road to recovery. I was, of course, very desirous of meeting so famous a man as Hansen. He was expected to preside at a session of the German commission on the transit of Venus, which was to be held in Berlin about the time of my return thither from Pulkova. The opportunity was therefore open of bringing a message of good news from his daughter. Apart from this, the prospect of the meeting might have been embarrassing. The fact is that I was at odds with him on a scientific question, and he was a man who did not take a charitable view of those who differed from him in opinion.

He was the author of a theory, current thirty or forty years ago, that the farther side of the moon is composed of denser materials than the side turned toward us. As a result of this, the centre of gravity of the moon was supposed to be farther from us than the actual centre of her globe. It followed that, although neither atmosphere nor water existed on our side of the moon, the other side might have both. Here was a very tempting field into which astronomical speculators stepped, to clothe the invisible hemisphere of the moon with a beautiful terrestrial landscape, and people it as densely as they pleased with beings like ourselves. If these beings should ever attempt to explore the other half of their own globe, they would find themselves ascending to a height completely above the limits of their atmosphere. Hansen himself never countenanced such speculations as these, but confined his claims to the simple facts he supposed proven.

In 1868 I had published a little paper showing what I thought a fatal defect, a vicious circle in fact, in Hansen's reasoning on this subject. Not long before my visit, Delaunay had made this paper the basis of a communication to the French Academy of Sciences, in which he not only indorsed my views, but sought to show the extreme improbability of Hansen's theory on other grounds.

When I first reached Germany, on my way from Italy, I noticed copies of a blue pamphlet lying on the tables of the astronomers. Apparently, the paper had been plentifully distributed; but it was not until I reached Berlin that I found it was Hansen's defense against my strictures, — a defense in which mathematics were not unmixed with scathing sarcasm at the expense of both Delaunay and myself. The case brought to mind a warm discussion between Hansen and Encke, in the pages of a scientific journal, some fifteen years before. At the time it had seemed intensely comical to see two enraged combatants — for so I amused myself by fancying them — hurling algebraic formulæ, of frightful complexity, at each other's heads. I did not then dream that I should live to be an object of the same sort of attack, and that from Hansen himself.

To be revised, pulled to pieces, or superseded, as science advances, is the common fate of most astronomical work, even the best. It does not follow that it has been done in vain; if good, it forms a foundation on which others will build. But not every great investigator can look on with philosophic calm when he sees his work thus treated, and Hansen was among the last who could.

Under these circumstances, it was a serious question what sort of reception Hansen would accord to a reviser of his conclusions who should venture to approach him. I determined to assume an attitude that would show no consciousness of offense, and was quite successful. Our meeting was not attended by any explosion; I gave him the pleasant message with which I was charged from his daughter, and, a few days later, sat by his side at a dinner of the German commission on the coming transit of Venus.

As Hansen was Germany's greatest master in mathematical astronomy, so was the venerable Argelander in the observational side of the science. He was of the same age as the newly crowned Emperor, and the two were playmates at the time Germany was being overrun by the armies of Napoleon. He was held in love and respect by the entire generation of young astronomers, both Germans and foreigners, many of whom were proud to have had him as their preceptor. Among these was Dr. B. A. Gould, who frequently related a story of the astronomer's wit. When with him as a student, Gould was beardless, but had a good head of hair. Returning some years later, he had become bald, but had made up for it by having a full, long beard. He entered Argelander's study unannounced. At first the astronomer did not recognize him.

" Do you not know me, Herr Professor? "

The astronomer looked more closely. " Mine Gott! It is Gould mit his hair struck through! "

Argelander was more than any one else the founder of that branch of his science which treats of variable stars. His methods have been followed by his successors to the present time. It was his policy to make the best use he could of the instruments at his disposal, rather than to invent new ones that might prove of doubtful utility. The results of his work seem to justify this policy.

We passed the last month of the winter in Berlin waiting for the war to close, so that we could visit Paris. Poor France had at length to succumb, and in the latter part of March, we took almost the first train that passed the lines.

Delaunay was then director of the Paris Observatory, having succeeded Leverrier when the emperor petulantly removed the latter from his position. I had for some time kept up an occasional correspondence with Delaunay, and while in England, the autumn before, had forwarded a message to him, through the Prussian lines, by the good offices of the London legation and Mr. Washburn. He was therefore quite prepared for our arrival. The evacuation of a country by a hostile army is rather a slow process, so that the German troops were met everywhere on the road, even in France. They had left Paris just before we arrived; but the French national army was not there, the Communists having taken possession of the city as fast as the Germans withdrew. As we passed out of the station, the first object to strike our eyes was a flaming poster addressed to " Citoyens," and containing one of the manifestoes which the Communist government was continually issuing.

Of course we made an early call on Mr. Washburn. His career in Paris was one of the triumphs of diplomacy; he had cared for the interests of German subjects in Paris in such a way as to earn the warm recognition both of the emperor and of Bismarck, and at the same time had kept on such good terms with the French as to be not less esteemed by them. He was surprised that we had chosen such a time to visit Paris; but I told him the situation, the necessity of my early return home, and my desire to make a careful search in the records of the Paris Observatory for observations made two centuries ago. He advised us to take up our quarters as near to the observatory as convenient, in order that we might not have to pass through the portions of the city which were likely to be the scenes of disturbance.

We were received at the observatory with a warmth of welcome that might be expected to accompany the greeting of the first foreign visitor, after a siege of six months. Yet a tinge of sadness in the meeting was unavoidable. Delaunay immediately began lamenting the condition of his poor ruined country, despoiled of two of its provinces by a foreign foe, condemned to pay an enormous subsidy in addition, and now the scene of an internal conflict the end of which no one could foresee.

While I was mousing among the old records of the Paris Observatory, the city was under the reign of the Commune and besieged by the national forces. The studies had to be made within hearing of the besieging guns; and I could sometimes go to a window and see flashes of artillery from one of the fortifications to the south. Nearly every day I took a walk through the town, occasionally as far as the Arc de Triomphe. The story of the Commune has been so often written that I cannot hope to add anything to it, so far as the main course of events is concerned. Looking back on a sojourn at so interesting a period, one cannot but feel that a golden opportunity to make observations of historic value was lost. The fact is, however, that I was prevented from making such observations not only by my complete absorption in my work, but by the consideration that, being in what might be described as a semi-official capacity, I did not want to get into any difficulty that would have compromised the position of an official visitor. I should not deem what we saw worthy of special mention, were it not that it materially modifies the impressions commonly given by writers on the history of the Commune. What an historian says may be quite true, so far as it goes, and yet may be so far from the whole truth as to give the reader an incorrect impression of the actual course of events. The violence and disease which prevail in the most civilized country in the world may be described in such terms as to give the impression of a barbarous community. The murder of the Archbishop of Paris and of the hostages show how desperate were the men who had seized power, yet the acts of these men constitute but a small part of the history of Paris during that critical period.

What one writes at the time is free from the suspicion that may attach to statements not recorded till many years after the events to which they relate. The following extract from a letter which I wrote to a friend, the day after my arrival, may therefore be taken to show how things actually looked to a spectator: —

DEAR CHARLIE, — Here we are, on this slumbering volcano. Perhaps you will hear of the burst-up long before you get this. We have seen historic objects which fall not to the lot of every generation, the barricades of the Paris streets. As we were walking out this morning, the pavement along one side of the street was torn up for some distance, and used to build a temporary fort. Said fort would be quite strong against musketry or the bayonet; but with heavy shot against it, I should think it would be far worse than nothing, for the flying stones would kill more than the balls.

The streets are placarded at every turn with all sorts of inflammatory appeals, and general orders of the Comité Central or of the Commune. One of the first things I saw last night was a large placard beginning "Citoyens!" Among the orders is one forbidding any one from placarding any orders of the Versailles government under the severest penalties; and another threatening with instant dismissal any official who shall recognize any order issuing from the said government.

I must do all hands the justice to say that they are all very well behaved. There is nothing like a mob anywhere, so far as I can find. I consulted my map this morning, right alongside the barricade and in full view of the builders, without being molested, and wife and I walked through the insurrectionary districts without being troubled or seeing the slightest symptoms of disturbance. The stores are all open, and every one seems to be buying and selling as usual. In all the cafés I have seen, the habitués seem to be drinking their wine just as coolly as if they had nothing unusual on their minds.

From this date to that of our departure I saw nothing suggestive of violence within the limited range of my daily walks, which were mostly within the region including the Arc de Triomphe, the Hôtel de Ville, and the observatory; the latter being about half a mile south of the Luxembourg. The nearest approach to a mob that I ever noticed was a drill of young recruits of the National Guard, or a crowd in the court of the Louvre being harangued by an orator. With due allowance for the excitability of the French nature, the crowd was comparatively as peaceable as that which we may see surrounding a gospel wagon in one of our own cities. A drill-ground for the recruits happened to be selected opposite our first lodgings, beside the gates of the Luxembourg. This was so disagreeable that we were glad to accept an invitation from Delaunay to be his guests at the observatory, during the remainder of our stay. We had not been there long before the spacious yard of the observatory was also used as a drill-ground; and yet later, two or three men were given *billets de logement* upon the observatory; but I should not have known of the latter occurrence, had not Delaunay told me. I believe he bought the men off, much as one pays an organ-grinder to move on. In one of our walks we entered the barricade around the Hôtel de Ville, and were beginning to make a close examination of a mitrailleuse, when a soldier (beg his pardon, *un citoyen membre de la Garde Nationale*) warned us away from the weapon. The densest crowd of Communists was along the Rue de Rivoli and in the region of the Colonne Vendôme, where some of the principal barricades were being erected. But even here, not only were the stores open as usual, but the military were doing their work in the midst of piles of trinkets exposed for sale on the pavement by the shopwomen. The order to destroy the Column was issued before we left, but not executed until later. I have no reason to suppose that the shopwomen were any more concerned while the Column was being undermined than they were before. To complete the picture, not a policeman did we see in Paris; in fact, I was told that one of the first acts of the Commune had been to drive the police away, so that not one dared to show himself.

An interesting feature of the sad spectacle was the stream of proclamations poured forth by the Communist authorities. They comprised not only decrees, but sensational stories of victories over the Versailles troops, denunciations of the Versailles government, and even elaborate legal arguments, including a not intemperate discussion of the ethical question whether citizens who were not adherents of the Commune should be entitled to the right of suffrage. The conclusion was that they should not. The lack of humor on the part of the authorities was shown by their commencing one of a rapid succession of battle stories with the words, "Citoyens! Vous avez soif de la vérité!" The most amusing decree I noticed ran thus: —

"Article I. All conscription is abolished.

"Article II. No troops shall hereafter be allowed in Paris, except the National Guard.

"Article III. Every citizen is a member of the National Guard."

We were in daily expectation and hope of the capture of the city, little imagining by what scenes it would be accompanied. It did not seem to my unmilitary eye that two or three batteries of artillery could have any trouble in demolishing all the defenses, since a wall of paving-stones, four or five feet high, could hardly resist solid shot, or prove anything but a source of destruction to those behind it if attacked by artillery. But the capture was not so easy a matter as I had supposed.

We took leave of our friend and host on May 5, three weeks before the final catastrophe, of which he wrote me a graphic description. As the barricades were stormed by MacMahon, the Communist line of retreat was through the region of the observatory. The walls of the building and of the yard were so massive that the place was occupied as a fort by the retreating forces, so that the situation of the few non-combatants who remained was extremely critical. They were exposed to the fire of their friends, the national troops, from without, while enraged men were threatening their lives within. So hot was the fusillade that, going into the great dome after the battle, the astronomer could imagine all the constellations of the sky depicted by the bullet-holes. When retreat became inevitable, the Communists tried to set the building on fire, but did not succeed. Then, in their desperation, arrangements were made for blowing it up; but the most violent man among them was killed by a providential bullet, as he was on the point of doing his work. The remainder fled, the place was speedily occupied by the national troops, and the observatory with its precious contents was saved.

The Academy of Sciences had met regularly through the entire Prussian siege. The legal quorum being three, this did not imply a large attendance. The reason humorously assigned for this number was that, on opening a session, the presiding officer must say, *Messieurs, la séance est ouverte*, and he cannot say *Messieurs* unless there are at least two to address. At the time of my visit a score of members were in the city. Among them were Elie de Beaumont, the geologist; Milne-Edwards, the zoölogist; and Chevreul, the chemist. I was surprised to learn that the latter was in his eighty-fifth year; he seemed a man of seventy or less, mentally and physically. Yet we little thought that he would be the longest-lived man of equal eminence that our age has known. When he died, in 1889, he was nearly one hundred and three years old. Born in 1786, he had lived through the whole French Revolution, and was seven years old at the time of the Terror. His scientific activity, from beginning to end, extended over some eighty years. When I saw him, he was still very indignant at a bombardment of the Jardin des Plantes by the German besiegers. He had made a formal statement of this outrage to the Academy of Sciences, in order that posterity might know what kind of men were besieging Paris. I suggested that the shells might have fallen in the place by accident; but he maintained that it was not the case, and that the bombardment was intentional.

The most execrated man in the scientific circle

at this time was Leverrier. He had left Paris before the Prussian siege began, and had not returned. Delaunay assured me that this was a wise precaution on his part; for had he ventured into the city he would have been mobbed, or the Communists would have killed him as soon as caught. Just why the mob should have been so incensed against one whose life was spent in the serenest fields of astronomical science was not fully explained. The fact that he had been a senator, and was politically obnoxious, was looked on as an all-sufficient indictment. Even members of the Academy could not suppress their detestation of him. Their language seemed not to have words that would fully express their sense of his despicable meanness, not to say turpitude.

Four years later I was again in Paris, and attended a meeting of the Academy of Sciences. In the course of the session a rustle of attention spread over the room, as all eyes were turned upon a member who was entering rather late. Looking toward the door, I saw a man of sixty, a decided blond, with light chestnut hair turning gray, slender form, shaven face, rather pale and thin, but very attractive, and extremely intellectual features. As he passed to his seat hands were stretched out on all sides to greet him, and not until he sat down did the bustle caused by his entrance subside. He was evidently a notable.

"Who is that?" I said to my neighbor.

"Leverrier."

Delaunay was one of the most kindly and attractive men I ever met. We spent our evenings walking in the grounds of the observatory, discussing French science in all its aspects. His investigation of the moon's motion is one of the most extraordinary pieces of mathematical work ever turned out by a single person. It fills two quarto volumes, and the reader who attempts to go through any part of the calculations will wonder how one man could do the work in a lifetime. His habit was to commence early in the morning, and work with but little interruption until noon. He never worked in the evening, and generally retired at nine. I felt some qualms of conscience at the frequency with which I kept him up till nearly ten. I found it hopeless to expect that he would ever visit America, because he assured me that he did not dare to venture on the ocean. The only voyage he had ever made was across the Channel, to receive the gold medal of the Royal Astronomical Society for his work. Two of his relatives — his father and, I believe, his brother — had been drowned, and this fact gave him a horror of the water. He seemed to feel somewhat like the clients of the astrologists, who, having been told from what agencies they were to die, took every precaution to avoid them. I remember, as a boy, reading a history of astrology, in which a great many cases of this sort were described; the peculiarity being that the very measures which the victim took to avoid the decree of fate became the engines that executed it. The death of Delaunay was not exactly a case of this kind, yet it could not but bring it to mind. He was at Cherbourg in the autumn of 1872. As he was walking on the beach with a relative, a couple of boatmen invited them to take a sail. Through what inducement Delaunay was led to forget his fears will never be known. All we know is that he and his friend entered the boat, that it was struck by a sudden squall when at some distance from the land, and that the whole party were drowned.

There was no opposition to the reappointment of Leverrier to his old place. In fact, at the time of my visit, Delaunay said that President Thiers was on terms of intimate friendship with the former director, and he thought it not at all unlikely that the latter would succeed in being restored. He kept the position with general approval till his death in 1877.

The only occasion on which I met Leverrier was after the incident I have mentioned, in the Academy of Sciences. I had been told that he was incensed against me on account of an unfortunate remark I had made in speaking of his work which led to the discovery of Neptune. I had heard this in Germany as well as in France, yet the matter was so insignificant that I could hardly conceive of a man of philosophic mind taking any notice of it. I determined to meet him, as I had met Hansen, with entire unconsciousness of offense. So I called on him at the observatory, and was received with courtesy, but no particular warmth. I suggested to him that now, as he had nearly completed his work on the tables of the planets, the question of the moon's motion would be the next object worthy of his attention. He replied that it was too large a subject for him to take up.

To Leverrier belongs the credit of having been the real organizer of the Paris Observatory. His work there was not dissimilar to that of Airy at Greenwich; but he had a much more difficult task before him, and was less fitted to grapple with it. When founded by Louis XIV. the establishment was simply a place where astronomers of the Academy of Sciences could go to make their observations. There was no titular director, every man working on his own account and in his own way. Cassini, an Italian by birth, was the best known of the astronomers, and, in consequence, posterity has very generally supposed he was the director. That he failed to secure that honor was not from any want of astuteness. It is related that the monarch once visited the observatory to see a newly discovered comet through the telescope. He inquired in what direction the comet was going to move. This was a question it was impossible to answer at the moment, because both observations and computations would be necessary before the orbit could be worked out. But Cassini reflected that the king would not look at the comet again, and would very soon forget what was told him; so he described its future path in the heavens quite at random, with entire confidence that any deviation of the actual motion from his prediction would never be noted by his royal patron.

One of the results of this lack of organization has been that the Paris Observatory does not hold an historic rank correspondent to the magnificence of the establishment. The go-as-you-please system works no better in a national observatory than it would in a business institution. Up to the end of the last century, the observations made there were too irregular to be of any special importance. To remedy this state of things, Arago was appointed director early in the present century; but he was more eminent in experimental physics than in astronomy, and had no great astronomical problem to solve. The result was that while he did much to promote the reputation of the observatory in the direction of physical investigation, he did not organize any well-planned system of regular astronomical work.

When Leverrier succeeded Arago, in 1853, he had an extremely difficult problem before him. By a custom extending through two centuries, each astronomer was to a large extent the master of his own work. Leverrier undertook to change all this in a twinkling, and, if reports are true, without much regard to the feelings of the astronomers. Those who refused to fall into line either resigned or were driven away, and their places were filled with men willing to work under the direction of their chief. Yet his methods were not up to the times; and the work of the Paris Observatory, so far as observations of precision go, falls markedly behind that of Greenwich and Pulkova.

In recent times the institution has been marked by an energy and a progressiveness that go far to atone for its former deficiencies. The successors of Leverrier have known where to draw the line between routine, on the one side, and initiative on the part of the assistants, on the other. Probably no other observatory in the world has so many able and well-trained young men, who work partly on their own account, and partly in a regular routine. In the direction of physical astronomy the observatory is especially active, and it may be expected in the future to justify its historic reputation.

XII

THE OLD AND THE NEW WASHINGTON

A FEW features of Washington as it appeared during the civil war are indelibly fixed in my memory. An endless train of army wagons ploughed its streets with their heavy wheels. Almost the entire southwestern region, between the War Department and the Potomac, extending west on the river to the neighborhood of the observatory, was occupied by the Quartermaster's and Subsistence Departments for storehouses. Among these the astronomers had to walk by day and night, in going to and from their work. After a rain, especially during winter and spring, some of the streets were much like shallow canals. Under the attrition of the iron-bound wheels the water and clay were ground into mud, which was at first almost liquid. It grew thicker as it dried up, until perhaps another rainstorm reduced it once more to a liquid condition. In trying first one street and then another to see which offered the fewest obstacles to his passage, the wayfarer was reminded of the assurance given by a bright boy to a traveler who wanted to know the best road to a certain place: " Whichever road you take, before you get halfway there you'll wish you had taken t' other." By night swarms of rats, of a size proportional to their ample food supply, disputed the right of way with the pedestrian.

Across the Potomac, Arlington Heights were whitened by the tents of soldiers, from which the discharges of artillery or the sound of the fife and drum became so familiar that the dweller almost ceased to notice it. The city was defended by a row of earthworks, generally not far inside the boundary line of the District of Columbia, say five or six miles from the central portions of the city. One of the circumstances connected with their plans strikingly illustrates the exactness which the science or art of military engineering had reached. Of course the erection of fortifications was one of the first tasks to be undertaken by the War Department. Plans showing the proposed location and arrangements of the several forts were drawn up by a board of army engineers, at whose head, then or afterward, stood General John G. Barnard. When the plans were complete, it was thought advisable to test them by calling in the advice of Professor D. H. Mahan of the Military Academy at West Point. He came to Washington, made a careful study of the maps and plans, and was then driven around the region of the lines to be defended to supplement his knowledge by personal inspection. Then he laid down his ideas as to the location of the forts. There were but two variations from the plans proposed by the Board of Engineers, and these were not of fundamental importance.

Willard's Hotel, then the only considerable one

in the neighborhood of the executive offices, was a sort of headquarters for arriving army officers, as well as for the thousands of civilians who had business with the government, and for gossip generally. Inside its crowded entrance one could hear every sort of story, of victory or disaster, generally the latter, though very little truth was ever to be gleaned.

The newsboy flourished. He was a bright fellow too, and may have developed into a man of business, a reporter, or even an editor. "Another great battle!" was his constant cry. But the purchaser of his paper would commonly read of nothing but a skirmish or some fresh account of a battle fought several days before — perhaps not even this. On one occasion an officer in uniform, finding nothing in his paper to justify the cry, turned upon the boy with the remark, —

"Look here, boy, I don't see any battle here."

"No," was the reply, "nor you won't see one as long as you hang around Washington. If you want to see a battle you must go to the front."

The officer thought it unprofitable to continue the conversation, and beat a retreat amid the smiles of the bystanders. This story, I may remark, is quite authentic, which is more than one can say of the report that a stick thrown by a boy at a dog in front of Willard's Hotel struck twelve brigadier generals during its flight.

The presiding genius of the whole was Mr. Edwin M. Stanton, Secretary of War. Before the actual outbreak of the conflict he had been, I believe, at least a Democrat, and, perhaps, to a certain extent, a Southern sympathizer so far as the slavery question was concerned. But when it came to blows, he espoused the side of the Union, and after being made Secretary of War he conducted military operations with a tireless energy, which made him seem the impersonation of the god of war. Ordinarily his character seemed almost savage when he was dealing with military matters. He had no mercy on inefficiency or lukewarmness. But his sympathetic attention, when a case called for it, is strikingly shown in the following letter, of which I became possessed by mere accident. At the beginning of the war Mr. Charles Ellet, an eminent engineer, then resident near Washington, tendered his services to the government, and equipped a fleet of small river steamers on the Mississippi under the War Department. In the battle of June 6, 1862, he received a wound from which he died some two weeks later. His widow sold or leased his house on Georgetown Heights, and I boarded in it shortly afterward. Amongst some loose rubbish and old papers lying around in one of the rooms I picked up the letter which follows.

War Department,
Washington City, D. C., June 9, 1862.

Dear Madam, — I understand from Mr. Ellet's dispatch to you that as he will be unfit for duty for some time it will be agreeable to him for you to visit him, traveling slowly so as not to expose your own health.

With this view I will afford you every facility within

the control of the Department, by way of Pittsburg and Cincinnati to Cairo, where he will probably meet you.

Yours truly,

EDWIN M. STANTON,
Secretary of War.

The interesting feature of this letter is that it is entirely in the writer's autograph, and bears no mark of having been press copied. I infer that it was written out of office hours, after all the clerks had left the Department, perhaps late at night, while the secretary was taking advantage of the stillness of the hour to examine papers and plans.

Only once did I come into personal contact with Mr. Stanton. A portrait of Ferdinand R. Hassler, first superintendent of the Coast Survey, had been painted about 1840 by Captain Williams of the Corps of Engineers, U. S. A., a son-in-law of Mr. G. W. P. Custis, and therefore a brother-in-law of General Lee. The picture at the Arlington house was given to Mrs. Colonel Abert, who loaned it to Mr. Custis. When the civil war began she verbally donated it to my wife, who was Mr. Hassler's granddaughter, and was therefore considered the most appropriate depositary of it, asking her to get it if she could. But before she got actual possession of it, the Arlington house was occupied by our troops and Mr. Stanton ordered the picture to be presented to Professor Agassiz for the National Academy of Sciences. On hearing of this, I ventured to mention the matter to Mr. Stanton, with a brief statement of our claims upon the picture.

"Sir," said he, "that picture was found in the house of a rebel in arms [General Robert E. Lee], and was justly a prize of war. I therefore made what I considered the most appropriate disposition of it, by presenting it to the National Academy of Sciences."

The expression "house of a rebel in arms" was uttered with such emphasis that I almost felt like one under suspicion of relations with the enemy in pretending to claim the object in question. It was clearly useless to pursue the matter any further at that time. Some years later, when the laws were no longer silent, the National Academy decided that whoever might be the legal owner of the picture, the Academy could have no claim upon it, and therefore suffered it to pass into the possession of the only claimant.

Among the notable episodes of the civil war was the so-called raid of the Confederate general, Early, in July, 1864. He had entered Maryland and defeated General Lew Wallace. This left nothing but the well-designed earthworks around Washington between his army and our capital. Some have thought that, had he immediately made a rapid dash, the city might have fallen into his hands.

All in the service of the War and Navy departments who were supposed capable of rendering efficient help, were ordered out to take part in the defense of the city, among them the younger professors of the observatory. By order of Captain Gilliss I became a member of a naval brigade, organized in the most hurried manner by Admiral Goldsborough, and including in it several officers of high and low rank. The rank and file was formed of the workmen in the Navy Yard, most of whom were said to have seen military service of one kind or another. The brigade formed at the Navy Yard about the middle of the afternoon, and was ordered to march out to Fort Lincoln, a strong earthwork built on a prominent hill, half a mile southwest of the station now known as Rives. The Reform School of the District of Columbia now stands on the site of the fort. The position certainly looked very strong. On the right the fort was flanked by a deep intrenchment running along the brow of the hill, and the whole line would include in the sweep of its fire the region which an army would have to cross in order to enter the city. The naval brigade occupied the trench, while the army force, which seemed very small in numbers, manned the front.

I was not assigned to any particular duty, and simply walked round the place in readiness to act whenever called upon. I supposed the first thing to be done was to have the men in the trench go through some sort of drill, in order to assure their directing the most effective fire on the enemy should he appear. The trench was perhaps six feet deep; along its bottom ran a little ledge on which the men had to step in order to deliver their fire, stepping back into the lower depth to load again. Along the edge was a sort of rail fence, the bottom rail of which rested on the ground. In order to fire on an enemy coming up the hill, it would be necessary to rest the weapon on this bottom rail. It was quite evident to me that a man not above the usual height, standing on the ledge, would have to stand on tiptoe in order to get the muzzle of his gun properly directed down the slope. If he were at all flurried he would be likely to fire over the head of the enemy. I called attention to this state of things, but did not seem to make any impression on the officers, who replied that the men had seen service and knew what to do.

We bivouacked that night, and remained all the next day and the night following awaiting the attack of the enemy, who was supposed to be approaching Fort Stevens on the Seventh Street road. At the critical moment, General H. G. Wright arrived from Fort Monroe with his army corps. He and General A. McD. McCook both took their stations at Fort Lincoln, which it was supposed would be the point of attack. A quarter or half a mile down the hill was the mansion of the Rives family, which a passenger on the Baltimore and Ohio Railway can readily see at the station of that name. A squad of men was detailed to go to this house and destroy it, in case the enemy should appear. The attack was expected at daybreak, but General Early, doubtless hearing of the arrival of reinforcements, abandoned any project he might have entertained and had beat a retreat the day before.

Whether the supposition that he could have taken the city with great celerity has any foundation, I cannot say; I should certainly greatly doubt it, remembering the large loss of life generally suffered during the civil war by troops trying to storm intrenchments or defenses of any sort, even with greatly superior force.

I was surprised to find how quickly one could acquire the stolidity of the soldier. During the march from the Navy Yard to the fort I felt extremely depressed, as one can well imagine, in view of the suddenness with which I had to take leave of my family and the uncertainty of the situation, as well as its extreme gravity. But this depression wore off the next day, and I do not think I ever had a sounder night's sleep in my life than when I lay down on the grass, with only a blanket between myself and the sky, with the expectation of being awakened by the rattle of musketry at daybreak.

I remember well how kindly we were treated by the army. The acquaintance of Generals Wright and McCook, made under such circumstances, was productive of a feeling which has never worn off. It has always been a matter of sorrow to me that the Washington of to-day does not show a more lively consciousness of what it owes to these men.

One of the entertainments of Washington during the early years of the civil war was offered by President Lincoln's public receptions. We used to go there simply to see the people and the cos-

THE OLD AND THE NEW WASHINGTON 343

tumes, the latter being of a variety which I do not think was ever known on such occasions before or since. Well-dressed and refined ladies and gentlemen, men in their working clothes, women arrayed in costumes fanciful in cut and brilliant in color, mixed together in a way that suggested a convention of the human race. Just where the oddly dressed people came from, or what notion took them at this particular time to don an attire like that of a fancy-dress ball, no one seemed to know.

Among the never-to-be-forgotten scenes was that following the news of the fall of Richmond. If I described it from memory, a question would perhaps arise in the reader's mind as to how much fancy might have added to the picture in the course of nearly forty years. I shall therefore quote a letter written to Chauncey Wright immediately afterwards, of which I preserved a press copy.

OBSERVATORY, April 7, 1865.

DEAR WRIGHT, — Yours of the 5th just received. I heartily reciprocate your congratulations on the fall of Richmond and the prospective disappearance of the S. C. alias C. S.

You ought to have been here Monday. The observatory is half a mile to a mile from the thickly settled part of the city. At 11 A. M. we were put upon the qui vive by an unprecedented commotion in the city. From the barracks near us rose a continuous stream of cheers, and in the city was a hubbub such as we had never before heard. We thought it must be Petersburg or Richmond, but hardly dared to hope which. Miss Gilliss sent us word that it was really Richmond. I went down to the city. All the bedlams in creation broken loose could not have made such a scene. The stores were half closed, the clerks given a holiday, the streets crowded, every other man drunk, and drums were beating and men shouting and flags waving in every direction. I never felt prouder of my country than then, as I compared our present position with our position in the numerous dark days of the contest, and was almost ashamed to think that I had ever said that any act of the government was not the best possible.

Not many days after this outburst, the city was pervaded by an equally intense and yet deeper feeling of an opposite kind. Probably no event in its history caused such a wave of sadness and sympathy as the assassination of President Lincoln, especially during the few days while bands of men were scouring the country in search of the assassin. One could not walk the streets without seeing evidence of this at every turn. The slightest bustle, perhaps even the running away of a dog, caused a tremor.

I paid one short visit to the military court which was trying the conspirators. The court itself was listening with silence and gravity to the reading of the testimony taken on the day previous. General Wallace produced on the spectators an impression a little different from the other members, by exhibiting an artistic propensity, which subsequently took a different direction in " Ben Hur." The most impressive sight was that of the conspirators, all heavily manacled; even Mrs. Surratt, who kept her irons partly concealed in the folds of her gown. Payne, the would-be assassin of Seward, was a powerful-looking man, with a face that showed him ready for anything; but the other two conspirators were such simple-minded, mild-looking youths, that it seemed hardly possible they could have been active agents in such a crime, or capable of any proceeding requiring physical or mental force.

The impression which I gained at the time from the evidence and all the circumstances, was that the purpose of the original plot was not the assassination of the President, but his abduction and transportation to Richmond or some other point within the Confederate lines. While Booth himself may have meditated assassination from the beginning, it does not seem likely that he made this purpose known to his fellows until they were ready to act. Then Payne alone had the courage to attempt the execution of the programme.

Two facts show that a military court, sitting under such circumstances, must not be expected to reach exactly the verdict that a jury would after the public excitement had died away. Among the prisoners was the man whose business it was to assist in arranging the scenery on the stage of the theatre where the assassination occurred. The only evidence against him was that he had not taken advantage of his opportunity to arrest Booth as the latter was leaving, and for this he was sentenced to twenty years penal servitude. He was pardoned out before a great while.

The other circumstance was the arrest of Surratt, who was supposed to stand next to Booth in the conspiracy, but who escaped from the country and was not discovered until a year or so later, when he was found to have enlisted in the papal guards at Rome. He was brought home and tried twice. On the first trial, notwithstanding the adverse rulings and charge of the judge, only a minority of the jury were convinced of his guilt. On the second trial he was, I think, acquitted.

One aftermath of the civil war was the influx of crowds of the newly freed slaves to Washington, in search of food and shelter. With a little training they made fair servants if only their pilfering propensities could be restrained. But religious fervor did not ensure obedience to the eighth commandment. "The good Lord ain't goin' to be hard on a poor darky just for takin' a chicken now and then," said a wench to a preacher who had asked her how she could reconcile her religion with her indifference as to the ownership of poultry.

In the seventies I had an eight-year-old boy as help in my family. He had that beauty of face very common in young negroes who have an admixture of white blood, added to which were eyes of such depth and clearness that, but for his color, he would have made a first-class angel for a mediæval painter.

One evening my little daughters had a children's party, and Zeke was placed as attendant in charge of the room in which the little company met. Here he was for some time left alone. Next morning a gold pen was missing from its case in a drawer. Suspicion rested on Zeke as the only person who could possibly have taken it, but there was no positive proof. I thought so small and innocent-looking a boy could be easily cowed into confessing his guilt; so next morning I said to him very solemnly, —

"Zeke, come upstairs with me."

He obeyed with alacrity, following me up to the room.

"Zeke, come into this room."

He did so.

"Now, Zeke," I said sternly, "look here and see what I do."

I opened the drawer, took out the empty case, opened it, and showed it to him.

"Zeke, look into my eyes!"

He neither blinked nor showed the slightest abashment or hesitation as his soft eyes looked steadily into mine with all the innocence of an angel.

"Zeke, where is the pen out of that case?"

"Missr Newcomb," he said quietly, "I don't know nothin' about it."

I repeated the question, looking into his face as sternly as I could. As he repeated the answer with the innocence of childhood, "Deed, Missr Newcomb, I don't know what was in it," I felt almost like a brute in pressing him with such severity. Threats were of no avail, and I had to give the matter up as a failure.

On coming home in the afternoon, the first news was that the pen had been found by Zeke's mother hidden in one corner of her room at home, where the little thief had taken it. She, being an honest woman, and suspecting where it had come from, had brought it back.

There was a vigorous movement, having its origin in New England, for the education of the freedmen. This movement was animated by the most philanthropic views. Here were several millions of blacks of all ages, suddenly made citizens, or eligible to citizenship, and yet savage so far as any education was concerned. A small army of teachers, many, perhaps most of them, young women, were sent south to organize schools for the blacks. It may be feared that there was little adaptation of the teaching to the circumstances of the case. But one method of instruction widely adopted was, so far as I can learn, quite unique. It was the "loud method" of teaching reading and spelling. The whole school spelled in unison. The passer-by on the street would hear in chorus from the inside of the building, "B-R-E-A-D — BREAD!" all at the top of the voice of the speakers. Schools in which this method was adopted were known as "loud schools."

A queer result of this movement once fell under my notice. I called at a friend's house in Georgetown. In the course of the conversation, it came out that the sable youngster who opened the door for me filled the double office of scullion to the household and tutor in Latin to the little boy of the family.

Probably the Senate of the United States never had a member more conscientious in the discharge of his duties than Charles Sumner. He went little into society outside the circles of the diplomatic corps, with which his position as chairman of the Foreign Affairs Committee placed him in intimate relations. My acquaintance with him arose from the accident of his living for some time almost opposite me. I was making a study of some historic subject, pertaining to the feeling in South Carolina before the civil war, and called at his rooms to see if he would favor me with the loan of a book, which I was sure he possessed. He received me so pleasantly that I was, for some time, an occasional visitor. He kept bachelor quarters on a second floor, lived quite alone, and was accessible to all comers without the slightest ceremony.

One day, while I was talking with him, shortly after the surrender of Lee, a young man in the garb of a soldier, evidently fresh from the field, was shown into the room by the housemaid, unannounced, as usual. Very naturally, he was timid and diffident in approaching so great a man, and the latter showed no disposition to say anything that would reassure him. He ventured to tell the senator that he had come to see if he could recommend him for some public employment. I shall never forget the tone of the reply.

"But *I* do not know *you*." The poor fellow was completely dumfounded, and tried to make some excuses, but the only reply he got was, "I cannot do it; I do not know you at all." The visitor had nothing to do but turn round and leave.

At the time I felt some sympathy with the poor fellow. He had probably come, thinking that the great philanthropist was quite ready to become a friend to a Union soldier without much inquiry into his personality and antecedents, and now he met with a stinging rebuff. But it must be confessed that subsequent experience has diminished my sympathy for him, and probably it would be better for the country if the innovation were introduced of having every senator of the United States dispose of such callers in the same way.

Foreign men of letters, with whom Sumner's acquaintance was very wide, were always among his most valued guests. A story is told of Thackeray's visit to Washington, which I distrust only for the reason that my ideas of Sumner's make-up do not assign him the special kind of humor which the story brings out. He was, however, quoted as saying, "Thackeray is one of the most perfect gentlemen I ever knew. I had a striking illustration of that this morning. We went out for a walk together and, thoughtlessly, I took him through Lafayette

Square. Shortly after we entered it, I realized with alarm that we were going directly toward the Jackson statue. It was too late to retrace our steps, and I wondered what Thackeray would say when he saw the object. But he passed straight by without seeming to see it at all, and did not say one word about it."

Sumner was the one man in the Senate whose seat was scarcely ever vacant during a session. He gave the closest attention to every subject as it arose. One instance of this is quite in the line of the present book. About 1867, an association was organized in Washington under the name of the "American Union Academy of Literature, Science, and Art." Its projectors were known to few, or none, but themselves. A number of prominent citizens in various walks of life had been asked to join it, and several consented without knowing much about the association. It soon became evident that the academy was desirous of securing as much publicity as possible through the newspapers and elsewhere. It was reported that the Secretary of the Treasury had asked its opinion on some instrument or appliance connected with the work of his department. Congress was applied to for an act of incorporation, recognizing it as a scientific adviser of the government by providing that it should report on subjects submitted to it by the governmental departments, the intent evidently being that it should supplant the National Academy of Sciences.

The application to Congress satisfied the two requirements most essential to favorable consideration. These are that several respectable citizens want something done, and that there is no one to come forward and say that he does not want it done. Such being the case, the act passed the House of Representatives without opposition, came to the Senate, and was referred to the appropriate committee, that on education, I believe. It was favorably reported from the committee and placed on its passage. Up to this point no objection seems to have been made to it in any quarter. Now, it was challenged by Mr. Sumner.

The ground taken by the Massachusetts senator was comprehensive and simple, though possibly somewhat novel. It was, in substance, that an academy of literature, science, and art, national in its character, and incorporated by special act of Congress, ought to be composed of men eminent in the branches to which the academy related. He thought a body of men consisting very largely of local lawyers, with scarcely a man of prominence in either of the three branches to which the academy was devoted, was not the one that should receive such sanction from the national legislature.

Mr. J. W. Patterson, of New Hampshire, was the principal advocate of the measure. He claimed that the proposed incorporators were not all unscientific men, and cited as a single example the name of O. M. Poe, which appeared among them.

This man, he said, was a very distinguished meteorologist.

This example was rather unfortunate. The fact is, the name in question was that of a well-known officer of engineers in the army, then on duty at Washington, who had been invited to join the academy, and had consented out of good nature without, it seems, much if any inquiry. It happened that Senator Patterson had, some time during the winter, made the acquaintance of a West Indian meteorologist named Poey, who chanced to be spending some time in Washington, and got him mixed up with the officer of engineers. The senator also intimated that the gentleman from Massachusetts had been approached on the subject and was acting under the influence of others. This suggestion Mr. Sumner repelled, stating that no one had spoken to him on the subject, that he knew nothing of it until he saw the bill before them, which seemed to him to be objectionable for the very reasons set forth. On his motion the bill was laid on the table, and thus disposed of for good. The academy held meetings for some time after this failure, but soon disappeared from view, and was never again heard of.

In the year 1862, a fine-looking young general from the West became a boarder in the house where I lived, and sat opposite me at table. His name was James A. Garfield. I believe he had come to Washington as a member of the court in the case of General Fitz John Porter. He left after a short time and had, I supposed, quite forgotten me. But, after his election to Congress, he one evening visited the observatory, stepped into my room, and recalled our former acquaintance.

I soon found him to be a man of classical culture, refined tastes, and unsurpassed eloquence, — altogether, one of the most attractive of men. On one occasion he told me one of his experiences in the State legislature of Ohio, of which he was a member before the civil war. A bill was before the House enacting certain provisions respecting a depository. He moved, as an amendment, to strike out the word " depository " and insert " depositary." Supposing the amendment to be merely one of spelling, there was a general laugh over the house, with a cry of " Here comes the schoolmaster ! " But he insisted on his point, and sent for a copy of Webster's Dictionary in order that the two words might be compared. When the definitions were read, the importance of right spelling became evident, and the laughing stopped.

It has always seemed to me that a rank injustice was done to Garfield on the occasion of the Credit Mobilier scandal of 1873, which came near costing him his position in public life. The evidence was of so indefinite and flimsy a nature that the credence given to the conclusion from it can only illustrate how little a subject or a document is exposed to searching analysis outside the precincts of a law court. When he was nominated for the presidency this scandal was naturally raked up and much made of it. I was so strongly impressed with the injustice as to write for a New York newspaper, anonymously of course, a careful analysis of the evidence, with a demonstration of its total weakness. Whether the article was widely circulated, or whether Garfield ever heard of it, I do not know; but it was amusing, a few days after it appeared, to see a paragraph in an opposition paper claiming that its contemporary had gone to the trouble of hiring a lawyer to defend Garfield.

No man better qualified as a legislator ever occupied a seat in Congress. A man cast in the largest mould, and incapable of a petty sentiment, his grasp of public affairs was rarely equaled, and his insight into the effects of legislation was of the deepest. But on what the author of the Autocrat calls the arithmetical side, — in the power of judging particular men and not general principles; in deciding who were the good men and who were not, he fell short of the ideal suggested by his legislative career. The brief months during which he administered the highest of offices were stormy enough, perhaps stormier than any president before him had ever experienced, and they would probably have been outdone by the years following, had he lived. But I believe that, had he remained in the Senate, his name would have gone into history among those of the greatest of legislators.

Sixteen years after the death of Lincoln public feeling was again moved to its depth by the assassination of Garfield. The cry seemed to pass from mouth to mouth through the streets faster than a messenger could carry the news, "The President has been shot." It chanced to reach me just as I was entering my office. I at once summoned my messenger and directed him to go over to the White House, and see if anything unusual had happened, but gave him no intimation of my fears. He promptly returned with the confirmation of the report. The following are extracts from my journal at the time: —

"July 2, Saturday: At 9.20 this morning President Garfield was shot by a miserable fellow named Guiteau, as he was passing through the Baltimore and Potomac R. R. station to leave Washington. One ball went through the upper arm, making a flesh wound, the other entered the right side on the back and cannot be found; supposed to have lodged in the liver. In the course of the day President rapidly weakened, and supposed to be dying from hemorrhage."

"Sunday morning: President still living and rallied during the day. Small chance of recovery. At night alarming symptoms of inflammation were exhibited, and at midnight his case seemed almost hopeless."

"Monday: President slightly better this morning, improving throughout the day."

"July 6. This P. M. sought an interview with Dr. Woodward at the White House, to talk of an apparatus for locating the ball by its action in retarding a rapidly revolving el. magnet. I hardly think the plan more than theoretically practical, owing to the minuteness of the action."

"The President still improving, but great dangers are

yet to come, and nothing has been found of the ball, which is supposed to have stayed in the liver because, were it anywhere else, symptoms of irritation by its presence would have been shown."

"July 9. This is Saturday evening. Met Major Powell at the Cosmos Club, who told me that they would like to have me look at the air-cooling projects at the White House. Published statement that the physicians desired some way to cool the air of the President's room had brought a crowd of projects and machines of all kinds. Among other things, a Mr. Dorsey had got from New York an air compressor such as is used in the Virginia mines for transferring power, and was erecting machinery enough for a steamship at the east end of the house in order to run it."

Dr. Woodward was a surgeon of the army, who had been on duty at Washington since the civil war, in charge of the Army Medical Museum. Among his varied works here, that in micro-photography, in which he was a pioneer, gave him a wide reputation. His high standing led to his being selected as one of the President's physicians. To him I wrote a note, offering to be of any use I could in the matter of cooling the air of the President's chamber. He promptly replied with a request to visit the place, and see what was being done and what suggestions I could make. Mr. Dorsey's engine at the east end was dispensed with after a long discussion, owing to the noise it would make and the amount of work necessary to its final installation and operation.

Among the problems with which the surgeons

had to wrestle was that of locating the ball. The question occurred to me whether it was not possible to do so by the influence produced by the action of a metallic conductor in retarding the motion of a rapidly revolving magnet, but the effect would be so small, and the apparatus to be made so delicate, that I was very doubtful about the matter. If there was any one able to take hold of the project successfully, I knew it would be Alexander Graham Bell, the inventor of the telephone. When I approached him on the subject, he suggested that the idea of locating the ball had also occurred to him, and that he thought the best apparatus for the purpose was a telephonic one which had been recently developed by Mr. Hughes. As there could be no doubt of the superiority of his project, I dropped mine, and he went forward with his. In a few days an opportunity was given him for actually trying it. The result, though rather doubtful, seemed to be that the ball was located where the surgeons supposed it to be. When the autopsy showed that their judgment had been at fault, Mr. Bell admitted his error to Dr. Woodward, adding some suggestion as to its cause. "Expectant attention," was Woodward's reply.

I found in the basement of the house an apparatus which had been brought over by a Mr. Jennings from Baltimore, which was designed to cool the air of dairies or apartments. It consisted of an iron box, two or three feet square, and some five feet long. In this box were suspended cloths, kept cool and damp by the water from melting ice contained in a compartment on top of the box. The air was driven through the box by a blower, and cooled by contact with the wet cloths. But no effect was being produced on the temperature of the room.

One conversant with physics will see one fatal defect in this appliance. The cold of the ice, if I may use so unscientific an expression, went pretty much to waste. The air was in contact, not with the ice, as it should have been, but with ice-water, which had already absorbed the latent heat of melting.

Evidently the air should be passed over the unmelted ice. The question was how much ice would be required to produce the necessary cooling? To settle this, I instituted an experiment. A block of ice was placed in an adjoining room in a current of air with such an arrangement that, as it melted, the water would trickle into a vessel below. After a certain number of minutes the melted water was measured, then a simple computation led to a knowledge of how much heat was absorbed from the air per minute by a square foot of the surface of the ice. From this it was easy to calculate from the known thermal capacity of air, and the quantity of the latter necessary per minute, how many feet of cooling surface must be exposed. I was quite surprised at the result. A case of ice nearly as long as an ordinary room, and large enough for men to walk about in it, must be provided. This was speedily done, supports were erected for the blocks of ice, the case was placed at the end of Mr. Jennings's box, and everything gotten in readiness for directing the air current through the receptacle, and into the room through tubes which had already been prepared.

It happened that Mr. Jennings's box was on the line along which the air was being conducted, and I was going to get it out of the way. The owner implored that it should be allowed to remain, suggesting that the air might just as well as not continue to pass through it. The surroundings were those in which one may be excused for not being harsh. Such an outpouring of sympathy on the part of the public had never been seen in Washington since the assassination of Lincoln. Those in charge were overwhelmed with every sort of contrivance for relieving the sufferings of the illustrious patient. Such disinterested efforts in behalf of a public and patriotic object had never been seen. Mr. Jennings had gone to the trouble and expense of bringing his apparatus all the way from Baltimore to Washington in order to do what in him lay toward the end for which all were striving. To leave his box in place could not do the slightest harm, and would be a gratification to him. So I let it stand, and the air continued to pass through it on its way to the ice chest.

While these arrangements were in progress three officers of engineers of the navy reported under orders at the White House, to do what they could toward the cooling of the air. They were Messrs. William L. Baillie, Richard Inch, and W. S. Moore. All four of us coöperated in the work in a most friendly way, and when we got through we made our reports to the Navy Department. A few weeks later these reports were printed in a pamphlet, partly to correct a wrong impression about the Jennings cold-box. Regular statements had appeared in the local evening paper that the air was being cooled by this useless contrivance. Their significance first came out several months later, on the occasion of an exhibition of mechanical or industrial implements at Boston. Among these was Mr. Jennings's cold-box, which was exhibited as the instrument that had cooled the air of President Garfield's chamber.

More light yet was thrown on the case when the question of rewarding those who had taken part in treating the President, or alleviating his sufferings in any way, came before Congress. Mr. Jennings was, I believe, among the claimants. Congress found the task of making the proper awards to each individual to be quite beyond its power at the time, so a lump sum was appropriated, to be divided by the Treasury Department according to its findings in each particular case. Before the work of making the awards was completed, I left on the expedition to the Cape of Good Hope to observe the transit of Venus, and never learned what had been done with the claims of Mr. Jennings. It might naturally be supposed that when an official report to the Navy Department showed that he had no claims whatever except those of a patriotic citizen who had done his best, which was just nothing at all, to promote the common end, the claim would have received little attention. Possibly this may have been the case. But I do not know what the outcome of the matter was.

Shortly after the death of the President, I had a visit from an inventor who had patented a method of cooling the air of a room by ice. He claimed that our work at the Executive Mansion was an infringement on his patent. I replied that I could not see how any infringement was possible, because we had gone to work in the most natural way, without consulting any previous process whatever, or even knowing of the existence of a patent. Surely the operation of passing air over ice to cool it could not be patentable.

He invited me to read over the statement of his claims. I found that although this process was not patented in terms, it was practically patented by claiming about every possible way in which ice could be arranged for cooling purposes. Placing the ice on supports was one of his claims; this we had undoubtedly done, because otherwise the process could not have been carried out. In a word, the impression I got was that the only sure way of avoiding an infringement would have been to blindfold the men who put the ice in the box, and ask them to throw it in pellmell. Every method of using judgment in arranging the blocks of ice he had patented.

I had to acknowledge that his claim of infringement might have some foundation, and inquired what he proposed to do in the case. He replied that he did not wish to do more than have his priority recognized in the matter. I replied that I had no objection to his doing this in any way he could, and he took his leave. Nothing more, so far as I am aware, was done in his case. But I was much impressed by this as by other examples I have had of the same kind, of the loose way in which our Patent Office sometimes grants patents.

I do not think the history of any modern municipality can show an episode more extraordinary or, taken in connection with its results, more instructive than what is known as the " Shepherd régime" in Washington. What is especially interesting about it is the opposite views that can be taken of the same facts. As to the latter there is no dispute. Yet, from one point of view, Shepherd made one of the most disastrous failures on record in attempting to carry out great works, while, from another point of view, he is the author of the beautiful Washington of to-day, and entitled to a public statue in recognition of his services. As I was a resident of the city and lived in my own house, I was greatly interested in the proposed improvements, especially of the particular street on which I lived. I was also an eye-witness to so much of the whole history as the public was cognizant of. The essential facts of the case, from the two opposing points of view, are exceedingly simple.

One fact is the discreditable condition of the streets of Washington during and after the civil war. The care of these was left entirely to the local municipality. Congress, so far as I know, gave no aid except by paying its share of street improvements in front of the public buildings. It was quite out of the power of the residents, who had but few men of wealth among them, to make the city what it ought to be. Congress showed no disposition to come to the help of the citizens in this task.

In 1871, however, some public-spirited citizens took the matter in hand and succeeded in having a new government established, which was modeled after that of the territories of the United States. There was a governor, a legislature, and a board of public works. The latter was charged with the improvements of the streets, and the governor was *ex officio* its president. The first governor was Henry D. Cooke, the banker, and Mr. Shepherd was vice-president of the board of public works and its leading member. Mr. Cooke resigned after a short term, and Mr. Shepherd was promoted to his place. He was a plumber and gas-fitter by trade, and managed the leading business in his line in Washington. Through the two or three years of his administration the city directory still contained the entry —

Shepherd, Alex. R. & Co., plumbers and gas-fitters, 910 Pa. Ave. N. W.

In recent years he had added to his plumbing business that of erecting houses for sale. He had had no experience in the conduct of public business, and, of course, was neither an engineer nor a financier. But such was the energy of his character and his personal influence, that he soon became practically the whole government, which he ran in his own way, as if it were simply his own business enlarged. Of the conditions which the law imposes on contracts, of the numerous and complicated problems of engineering involved in the drainage and street systems of a great city, of the precautions to be taken in preparing plans for so immense a work, and of the legal restraints under which it should be conducted, he had no special knowledge. But he had in the highest degree a quality which will bear different designations according to the point of view. His opponents would call it unparalleled recklessness; his supporters, boldness and enterprise.

Such were the preliminaries. Three years later the results of his efforts were made known by an investigating committee of Congress, with Senator Allison, a political friend, at its head. It was found that with authority to expend $6,000,000 in the improvement of the streets, there was an actual or supposed expenditure of more than $18,000,000, and a crowd of additional claims which no man could estimate, based on the work of more than one thousand principal contractors and an unknown number of purchasers and sub-contractors. Chaos reigned supreme. Some streets were still torn up and impassable; others completely paved, but done so badly that the pavements were beginning to rot almost before being pressed by a carriage. A debt had been incurred which it was impossible for the local municipality to carry and which was still piling up.

For all this Congress was responsible, and manfully shouldered its responsibility. Mr. Shepherd was legislated out of office as an act of extreme necessity, by the organization of a government at the head of which were three commissioners. The feeling on the subject may be inferred from the result when President Grant, who had given Shepherd his powerful support all through, nominated him as one of the three commissioners. The Senate rejected the nomination, with only some half dozen favorable votes.

The three commissioners took up the work and carried it on in a conservative way. Congress came to the help of the municipality by bearing one half the taxation of the District, on the very sound basis that, as it owned about one half of the property, it should pay one half the taxes.

The spirit of the time is illustrated by two little episodes. The reservation on which the public library founded by Mr. Carnegie is now built, was then occupied by the Northern Liberties Market, one of the three principal markets of the city. Being a public reservation, it had no right to remain there except during the pleasure of the authorities. Due notice was given to the marketmen to remove the structures. The owners were dilatory in doing so, and probably could not see why they should be removed when the ground was not wanted for any other purpose, and before they had time to find a new location. It was understood that, if an attempt was made to remove the buildings, the marketmen would apply to the courts for an injunction. To prevent this, an arrangement was made by which the destruction of the buildings was to commence at dinner-time. At the same time, according to current report, it was specially arranged that all the judges to whom an application could be made should be invited out to dinner. However this may have been, a large body of men appeared upon the scene in the course of the evening and spent the night in destroying the buildings. With such energy was the work carried on that one marketman was killed and another either wounded or seriously injured in trying to save their wares from destruction. The indignation against Shepherd was such that his life was threatened, and it was even said that a body-guard of soldiers had to be supplied by the War Department for his protection.

The other event was as comical as this was tragic. It occurred while the investigating committee of Congress was at its work. The principal actors in the case were Mr. Harrington, secretary of the local government and one of Mr. Shepherd's assistants, the chief of police, and a burglar. Harrington produced an anonymous letter, warning him that an attempt would be made in the course of a certain night to purloin from the safe in which they were kept, certain government papers, which the prosecutors of the case against Shepherd were anxious to get hold of. He showed this letter to the chief of police, who was disposed to make light of the matter. But on Harrington's urgent insistence the two men kept watch about the premises on the night in question. They were in the room adjoining that in which the records were kept, and through which the robber would have to pass. In due time the latter appeared, passed through the room and proceeded to break into the safe. The chief wanted to arrest him immediately, but Harrington asked him to wait, in order that they might see what the man was after, and especially what he did with the books. So they left and took their stations outside the door. The burglar left the building with the books in a satchel, and, stepping outside, was confronted by the two men.

I believe every burglar of whom history or fiction has kept any record, whether before or after this eventful night, when he broke open a safe and, emerging with his booty, found himself confronted by a policeman, took to his heels. Not so this burglar. He walked up to the two men, and with the utmost unconcern asked if they could tell him where Mr. Columbus Alexander lived. Mr. Alexander, it should be said, was the head man in the prosecu-

THE OLD AND THE NEW WASHINGTON 369

tion. The desired information being conveyed to the burglar, he went on his way to Mr. Alexander's house, followed by the two agents of the law. Arriving there, he rang the bell.

In the ordinary course of events, Mr. Alexander or some member of his family would have come to the door and been informed that the caller had a bundle for him. A man just awakened from a sound sleep and coming downstairs rubbing his eyes, would not be likely to ask any questions of such a messenger, but would accept the bundle and lock the door again. Then what a mess the prosecution would have been in! Its principal promoter detected in collusion with a burglar in order to get possession of the documents necessary to carry on his case!

It happened, however, that Mr. Alexander and the members of his household all slept the sleep of the just and did not hear the bell. The patience of the policeman was exhausted and the burglar was arrested and lodged in jail, where he was kept for several months. Public curiosity to hear the burglar's story was brought to a high pitch, but never gratified. Before the case came to trial the prisoner was released on straw bail and never again found. I do not think the bottom facts, especially those connected with the anonymous letter, were ever brought to light. So every one was left to form his own theory of what has since been known as the " Safe Burglary Conspiracy."

What seems at present the fashionable way of

looking at the facts is this: Shepherd was the man who planned the beautiful Washington of to-day, and who carried out his project with unexampled energy until he was stopped through the clamor of citizens who did not want to see things go ahead so fast. Other people took the work up, but they only carried out Shepherd's ideas. The latter, therefore, should have all the credit due to the founder of the new Washington.

The story has always seemed to me most interesting as an example of the way in which public judgment of men and things is likely to be influenced. Public sentiment during the thirty years which have since elapsed has undergone such a revolution in favor of Shepherd that a very likely outcome will be a monument to commemorate his work. But it is worth while to notice the mental processes by which the public now reaches this conclusion. It is the familiar and ordinarily correct method of putting this and that together.

This is one of the most beautiful cities in the United States, of which Americans generally are proud when they pay it a visit.

That is the recollection of the man who commenced the work of transforming an unsightly, straggling, primitive town into the present Washington, and was condemned for what he did.

These two considerations form the basis of the conclusion, all intermediate details dropping out of sight and memory. The reckless maladministration of the epoch, making it absolutely necessary to introduce a new system, has no place in the picture.

There is also a moral to the story, which is more instructive than pleasant. The actors in the case no doubt believed that if they set about their work in a conservative and law-abiding way, spending only as much money as could be raised, Congress would never come to their help. So they determined to force the game, by creating a situation which would speedily lead to the correct solution of the problem. I do not think any observant person will contest the proposition that had Shepherd gone about his work and carried it to a successful conclusion in a peaceable and law-abiding way, — had he done nothing to excite public attention except wisely and successfully to administer a great public work, — his name would now have been as little remembered in connection with what he did as we remember those of Ketchem, Phelps, and the other men who repaired the wreck he left and made the city what it is to-day.

In my mind one question dominates all others growing out of the case: What will be the moral effect on our children of holding up for their imitation such methods as I have described?

XIII

MISCELLANEA

IF the "Great Star-Catalogue Case" is not surrounded with such mystery as would entitle it to a place among *causes célèbres*, it may well be so classed on account of the novelty of the questions at issue. It affords an instructive example of the possibility of cases in which strict justice cannot be done through the established forms of legal procedure. It is also of scientific interest because, although the question was a novel one to come before a court, it belongs to a class which every leader in scientific investigation must constantly encounter in meting out due credit to his assistants.

The plaintiff, Christian H. F. Peters, was a Dane by birth, and graduated at the University of Berlin in 1836. During the earlier years of his manhood he was engaged in the trigonometrical survey of the kingdom of Naples, where, for a time, he had charge of an observatory or some other astronomical station. It is said that, like many other able European youth of the period, he was implicated in the revolution of 1848, and had to flee the kingdom in consequence. Five years later, he came to the United States. Here his first patron was Dr. B. A. Gould, who procured for him first a position on the Coast Survey, and then one as his assistant at the Dudley Observatory in Albany. He was soon afterward appointed professor of astronomy and director of the Litchfield Observatory at Hamilton College, where he spent the remaining thirty years of his life. He was a man of great learning, not only in subjects pertaining to astronomy, but in ancient and modern languages. The means at his disposal were naturally of the slenderest kind; but he was the discoverer of some forty asteroids, and devoted himself to various astronomical works and researches with great ability.

Of his personality it may be said that it was extremely agreeable so long as no important differences arose. What it would be in such a case can be judged by what follows. Those traits of character which in men like him may be smoothed down to a greater or less extent by marital discipline were, in the absence of any such agency, maintained in all their strength to his latest years.

The defendant, Charles A. Borst, was a graduate of the college and had been a favorite pupil of Peters. He was a man of extraordinary energy and working capacity, ready to take hold in a business-like way of any problem presented to him, but not an adept at making problems for himself. His power of assimilating learning was unusually developed; and this, combined with orderly business habits, made him a most effective and valuable assistant. The terms of his employment were of the first importance in the case. Mr. Litchfield of New York was the patron of the observatory; he had given the trustees of Hamilton College a capital for its support, which sufficed to pay the small salary of the director and some current expenses, and he also, when the latter needed an assistant, made provision for his employment. It appears that, in the case of Borst, Peters frequently paid his salary for considerable periods at a time, which sums were afterward reimbursed to him by Mr. Litchfield.

I shall endeavor to state the most essential facts involved as they appear from a combination of the sometimes widely different claims of the two parties, with the hope of showing fairly what they were, but without expecting to satisfy a partisan of either side. Where an important difference of statement is irreconcilable, I shall point it out.

In his observations of asteroids Peters was continually obliged to search through the pages of astronomical literature to find whether the stars he was using in observation had ever been catalogued. He long thought that it would be a good piece of work to search all the astronomical journals and miscellaneous collections of observations with a view of making a complete catalogue of the positions of the thousands of stars which they contained, and publishing it in a single volume for the use of astronomers situated as he was. The work of doing this was little more than one of routine search and calculation, which any well-trained youth could take up; but it was naturally quite without the power of Peters to carry it through with his own hand. He had employed at least one former assistant on the work, Professor John G. Porter, but very little progress was made. Now, however, he had a man with the persistence and working capacity necessary to carry out the plan.

There was an irreconcilable difference between the two parties as to the terms on which Borst went to work. According to the latter, Peters suggested to him the credit which a young man would gain as one of the motives for taking up the job. But plaintiff denied that he had done anything more than order him to do it. He did not, however, make it clear why an assistant at the Litchfield Observatory should be officially ordered to do a piece of work for the use of astronomy generally, and having no special connection with the Litchfield Observatory.

However this may be, Borst went vigorously to work, repeating all the calculations which had been made by Peters and former assistants, with a view of detecting errors, and took the work home with him in order that his sisters might make a great mass of supplementary calculations which, though not involved in the original plan, would be very conducive to the usefulness of the result. One or two of these bright young ladies worked for about a year at the job. How far Peters was privy to what they did was not clear; according to his claim he did not authorize their employment to do anything but copy the catalogue.

By the joint efforts of the assistant and his two sisters, working mostly or entirely at their own home, the work was brought substantially to a conclusion about the beginning of 1888. Borst then reported the completion to his chief and submitted a proposed title-page, which represented that the work was performed by Charles A. Borst under the direction of Christian H. F. Peters, Professor of Astronomy, etc. According to Borst's account, Peters tore up the paper, opened the stove door, put the fragments into the fire, and then turned on the assistant with the simple order, "Bring me the catalogue!"

This was refused, and a suit in replevin was immediately instituted by Peters. The ablest counsel were engaged on both sides. That of the plaintiff was Mr. Elihu Root, of New York, afterward Secretary of War, one of the leading members of the New York bar, and well known as an active member of the reform branch of the Republican party of that city. For the defendant was the law firm of an ex-senator of the United States, the Messrs. Kernan of Utica.

I think the taking of evidence and the hearing of arguments occupied more than a week. One claim of the defendant would, if accepted, have brought the suit to a speedy end. Peters was an employee of the corporation of Hamilton College, and by the terms of his appointment all his work at the Litchfield Observatory belonged to that institution. Borst was summoned into the case as an official employee of the Litchfield Observatory. Therefore the corporation of the college was the only authority which had power to bring the suit. But this point was disposed of by a decision of the judge that it was not reasonable, in view of the low salary received by the plaintiff, to deprive him of the right to the creations of his own talent. He did not, however, apply this principle of legal interpretation to the case of the defendant, and not only found for the plaintiff, but awarded damages based on the supposed value of the work, including, if I understand the case aright, the value of the work done by the young ladies. It would seem, however, that in officially perfecting the details of his decision he left it a little indefinite as to what papers the plaintiff was entitled to, it being very difficult to describe in detail papers many of which he had never seen. Altogether it may be feared that the decision treated the catalogue much as the infant was treated by the decision of Solomon.

However this might be, the decision completely denied any right of the defendant in the work. This feature of it I thought very unjust, and published in a Utica paper a review of the case in terms not quite so judicial as I ought to have chosen. I should have thought such a criticism quite a breach of propriety, and therefore would never have ventured upon it but for an eminent example then fresh in my mind.

Shortly after the Supreme Court of the United States uttered its celebrated decision upholding the constitutionality of the Legal Tender Act, I happened to be conversing at an afternoon reception with one of the judges, Gray, who had sustained the decision. Mr. George Bancroft, the historian, stepped up, and quite surprised me by expressing to the judge in quite vigorous language his strong dissent from the decision. He soon afterward published a pamphlet reviewing it adversely. I supposed that what Mr. Bancroft might do with a decision of the Supreme Court of the United States, a humbler individual might be allowed to do with the decision of a local New York judge.

The defense appealed the case to a higher court of three judges, where the finding of the lower court was sustained by a majority of two to one. It was then carried to the Court of Appeals, the highest in the State. Here the decision was set aside on what seemed to me the common sense ground that the court had ignored the rights of the defendant in the case, who certainly had some, and it must therefore be remanded for a new trial.

Meantime Peters had died; and it is painful to think that his death may have been accelerated by the annoyances growing out of the suit. One morning, in the summer of 1890, he was found dead on the steps of his little dwelling, having apparently fallen in a fit of apoplexy or heart failure as he was on his way to the observatory the night before. His heirs had no possible object in pushing the suit; probably his entire little fortune was absorbed in the attendant expenses.

When the difference with Borst was first heard of it was, I think, proposed to Peters by several of his friends, including myself, that the matter should be submitted to an arbitration of astronomers. But he would listen to nothing of the sort. He was determined to enforce his legal rights by legal measures. A court of law was, in such a case, at an enormous disadvantage, as compared with an astronomical board of arbitration. To the latter all the circumstances would have been familiar and simple, while the voluminous evidence, elucidated as it was by the arguments of counsel on the two sides, failed to completely enlighten the court on the points at issue. One circumstance will illustrate this. Some allusion was made during the trial to Peters's work while he was abroad, in investigating the various manuscripts of the Almagest of Ptolemy and preparing a commentary and revised edition of Ptolemy's Catalogue of Stars. This would have been an extremely important and original work, most valuable in the history of ancient astronomy. But the judge got it mixed up in his mind with the work before the court, and actually supposed that Peters spent his time in Europe in searching ancient manuscripts to get material for the catalogue in question. He also attributed great importance to the conception of the catalogue, forgetting that, to use the simile of a writer in the "New York Evening Post," such a conception was of no more value than the conception of a railroad from one town to another by a man who had no capital to build it. No original investigation was required on one side or the other. It was simply a huge piece of work done by a young man with help from his sisters, suggested by Peters, and now and then revised by him in its details. It seemed to me that the solution offered by Borst was eminently proper, and I was willing to say so, probably at the expense of Peters's friendship, on which I set a high value.

I have always regarded the work on Ptolemy's catalogue of stars, to which allusion has just been made, as the most important Peters ever undertook. It comprised a critical examination and comparison of all the manuscripts of the Almagest in the libraries of Europe, or elsewhere, whether in Arabic or other languages, with a view of learning what light might be thrown on the doubtful questions growing out of Ptolemy's work. At the Litchfield Observatory I had an opportunity of examining the work, especially the extended commentaries on special points, and was so impressed by the learning shown in the research as to express a desire for its speedy completion and publication. In fact, Peters had already made one or more communications to the National Academy of Sciences on the subject, which were supposed to be equivalent to presenting the work to the academy for publication. But before the academy put in any claim for the manuscript, Mr. E. B. Knobel of London, a well-known member of the Royal Astronomical Society, wrote to Peters's executors, stating that he was a collaborator with Peters in preparing the work, and as such had a claim to it, and wished to complete it. He therefore asked that the papers should be sent to him. This was done, but during the twelve years which have since elapsed, nothing more has been heard of the work. No one, so far as I know, ever heard of Peters's making any allusion to Mr. Knobel or any other collaborator. He seems to have always spoken of the work as exclusively his own.

Among the psychological phenomena I have witnessed, none has appeared to me more curious than a susceptibility of certain minds to become imbued with a violent antipathy to the theory of gravitation. The anti-gravitation crank, as he is commonly called, is a regular part of the astronomer's experience. He is, however, only one of a large and varied class who occupy themselves with what an architect might consider the drawing up of plans and specifications for a universe. This is, no doubt, quite a harmless occupation; but the queer part of it is the seeming belief of the architects that the actual universe has been built on their plans, and runs according to the laws which they prescribe for it. Ether, atoms, and nebulæ are the raw material of their trade. Men of otherwise sound intellect, even college graduates and lawyers, sometimes engage in this business. I have often wondered whether any of these men proved that, in all the common schools of New York, the power which conjugates the verbs comes, through some invisible conduit in the earth, from the falls of Niagara. This would be quite like many of the theories propounded.

Babbage's "Budget of Paradoxes" is a goodly volume descriptive of efforts of this sort. It was supplemented a year or two ago by a most excellent and readable article on eccentric literature, by Mr. John Fiske, which appeared in the "Atlantic Monthly." Here the author discussed the subject so well that I do not feel like saying much about it, beyond giving a little of my own experience.

Naturally the Smithsonian Institution was, and I presume still is, the great authority to which these men send their productions. It was generally a rule of Professor Henry always to notice these communications and try to convince the correspondents of their fallacies. Many of the papers were referred to me; but a little experience showed that it was absolutely useless to explain anything to these "paradoxers." Generally their first communication was exceedingly modest in style, being evidently designed to lead on the unwary person to whom it was addressed. Moved to sympathy with so well-meaning but erring an inquirer, I would point out wherein his reasoning was deficient or his facts at fault. Back would come a thunderbolt demonstrating my incapacity to deal with the subject in terms so strong that I could not have another word to say.

The American Association for the Advancement of Science was another attraction for such men. About thirty years ago there appeared at one of its meetings a man from New Jersey who was as much incensed against the theory of gravitation as if it had been the source of all human woe. He got admission to the meetings, as almost any one can, but the paper he proposed to read was refused by the committee. He watched his chance, however, and when discussion on some paper was invited, he got up and began with the words, " It seems to me that the astronomers of the present day have gravitation on the brain." This was the beginning of an impassioned oration which went on in an unbroken torrent until he was put down by a call for the next paper. But he got his chance at last. A meeting of Section Q was called; what this section was the older members will recall and the reader may be left to guess. A programme of papers had been prepared, and on it appeared Mr. Joseph Treat, on Gravitation. Mr. Treat got up with great alacrity, and, amid the astonishment and laughter of all proceeded to read his paper with the utmost seriousness.

I remember a visit from one of these men with great satisfaction, because, apparently, he was an exception to the rule in being amenable to reason. I was sitting in my office one morning when a modest-looking gentleman opened the door and looked in.

" I would like to see Professor Newcomb."

" Well, here he is."

"You Professor Newcomb?"

"Yes."

"Professor, I have called to tell you that I don't believe in Sir Isaac Newton's theory of gravitation!"

"Don't believe in gravitation! Suppose you jump out of that window and see whether there is any gravitation or not."

"But I don't mean that. I mean"—

"But that is all there is in the theory of gravitation; if you jump out of the window you'll fall to the ground."

"I don't mean that. What I mean is I don't believe in the Newtonian theory that gravitation goes up to the moon. It does n't extend above the air."

"Have you ever been up there to see?"

There was an embarrassing pause, during which the visitor began to look a little sheepish.

"N-no-o," he at length replied.

"Well, I have n't been there either, and until one of us can get up there to try the experiment, I don't believe we shall ever agree on the subject."

He took his leave without another word.

The idea that the facts of nature are to be brought out by observation is one which is singularly foreign not only to people of this class, but even to many sensible men. When the great comet of 1882 was discovered in the neighborhood of the sun, the fact was telegraphed that it might be seen with the naked eye, even in the sun's neighborhood. A news reporter came to my office with this statement, and wanted to know if it was really true that a comet could be seen with the naked eye right alongside the sun.

"I don't know," I replied; "suppose you go out and look for yourself; that is the best way to settle the question."

The idea seemed to him to be equally amusing and strange, and on the basis of that and a few other insipid remarks, he got up an interview for the "National Republican" of about a column in length.

I think there still exists somewhere in the Northwest a communistic society presided over by a genius whose official name is Koresh, and of which the religious creed has quite a scientific turn. Its fundamental doctrine is that the surface of the earth on which we live is the inside of a hollow sphere, and therefore concave, instead of convex, as generally supposed. The oddest feature of the doctrine is that Koresh professes to have proved it by a method which, so far as the geometry of it goes, is more rigorous than any other that science has ever applied. The usual argument by which we prove to our children the earth's rotundity is not purely geometric. When, standing on the seashore, we see the sails of a ship on the sea horizon, her hull being hidden because it is below, the inference that this is due to the convexity of the surface is based on the idea that light moves in a straight line. If a ray of light is curved toward the surface, we should have the same appearance, although the earth might be perfectly flat. So the Koresh people professed to have determined the figure of the earth's surface by the purely geometric method of taking long, broad planks, perfectly squared at the two ends, and using them as a geodicist uses his base apparatus. They were mounted on wooden supports and placed end to end, so as to join perfectly. Then, geometrically, the two would be in a straight line. Then the first plank was picked up, carried forward, and its end so placed against that of the second as to fit perfectly; thus the continuation of a straight line was assured. So the operation was repeated by continually alternating the planks. Recognizing the fact that the ends might not be perfectly square, the planks were turned upside down in alternate settings, so that any defect of this sort would be neutralized. The result was that, after they had measured along a mile or two, the plank was found to be gradually approaching the sea sand until it touched the ground.

This quasi-geometric proof was to the mind of Koresh positive. A horizontal straight line continued does not leave the earth's surface, but gradually approaches it. It does not seem that the measurers were psychologists enough to guard against the effect of preconceived notions in the process of applying their method.

It is rather odd that pure geometry has its full share of paradoxers. Runkle's "Mathematical Monthly" received a very fine octavo volume, the printing of which must have been expensive, by Mr. James Smith, a respectable merchant of Liverpool. This gentleman maintained that the circumference of a circle was exactly $3\frac{1}{5}$ times its diameter. He had pestered the British Association with his theory, and come into collision with an eminent mathematician whose name he did not give, but who was very likely Professor DeMorgan. The latter undertook the desperate task of explaining to Mr. Smith his error, but the other evaded him at every point, much as a supple lad might avoid the blows of a prize-fighter. As in many cases of this kind, the reasoning was enveloped in a mass of verbiage which it was very difficult to strip off so as to see the real framework of the logic. When this was done, the syllogism would be found to take this very simple form: —

The ratio of the circumference to the diameter is the same in all circles. Now, take a diameter of 1 and draw round it a circumference of $3\frac{1}{5}$. In that circle the ratio is $3\frac{1}{5}$; therefore, by the major premise, that is the ratio for all circles.

The three famous problems of antiquity, the duplication of the cube, the quadrature of the circle, and the trisection of the angle, have all been proved by modern mathematics to be insoluble by the rule and compass, which are the instruments assumed in the postulates of Euclid. Yet the problem of the trisection is frequently attacked by men of some mathematical education. I think it was about 1870 that I received from Professor

Henry a communication coming from some institution of learning in Louisiana or Texas. The writer was sure he had solved the problem, and asked that it might receive the prize supposed to be awarded by governments for the solution. The construction was very complicated, and I went over the whole demonstration without being able at first to detect any error. So it was necessary to examine it yet more completely and take it up point by point. At length I found the fallacy to be that three lines which, as drawn, intersected in what was to the eye the same point on the paper, were assumed to intersect mathematically in one and the same point. Except for the complexity of the work, the supposed construction would have been worthy of preservation.

Some years later I received, from a teacher, I think, a supposed construction, with the statement that he had gone over it very carefully and could find no error. He therefore requested me to examine it and see whether there was anything wrong. I told him in reply that his work showed that he was quite capable of appreciating a geometric demonstration; that there was surely something wrong in it, because the problem was known to be insoluble, and I would like him to try again to see if he could not find his error. As I never again heard from him, I suppose he succeeded.

One of the most curious of these cases was that of a student, I am not sure but a graduate, of the University of Virginia, who claimed that geometers were in error in assuming that a line had no thickness. He published a school geometry based on his views, which received the endorsement of a well-known New York school official and, on the basis of this, was actually endorsed, or came very near being endorsed, as a text-book in the public schools of New York.

From my correspondence, I judge that every civilized country has its share of these paradoxers. I am almost constantly in receipt of letters not only from America, but from Europe and Asia, setting forth their views. The following are a few of these productions which arrived in the course of a single season.

BALTIMORE, Sept. 29, 1897.
104 Collington Ave.

PROF. SIMON NEWCOMB:

Dear Sir, — Though a stranger to you, Sir, I take the liberty to enlist your interest in a Cause, — so grand, so beautiful, as to eclipse anything ever presented to the highest tribunal of human intellect and intuition.

Trusting you to be of liberal mind, Sir, I have mailed you specimen copy of the "Banner of Light," which will prove somewhat explanatory of my previous remarks.

Being a student of Nature and her wonderful laws, as they operate in that subtle realm of human life, — the soul, for some years, I feel well prepared to answer inquiries pertaining to this almost unknown field of scientific research, and would do so with much pleasure, as I am desirous to contribute my mite to the enlightenment of mankind upon this most important of all subjects.

Yours very truly, ———

P. S. — Would be pleased to hear from you, Sir.

MEXICO, 16 Oct. 1897.

DEAR SIR, — I beg to inform you that I have forwarded by to days mail to your adress a copy of my 20th Century planetary spectacle with a clipping of a german newspaper here. Thirty hours for 3000 years is to day better accepted than it was 6 years ago when I wrote it, although it called even then for some newspaper comment, especially after President Cleveland's election, whose likeness has been recognized on the back cover, so has been my comet, which was duly anounced by an Italian astronomer 48 hours before said election. A hint of Jupiters fifth satelite and Mars satelites is also to be found in my planetary spectacle but the most striking feature of such a profetic play is undoubtedly the Allegory of the Paris fire my entire Mercury scene and next to it is the Mars scene with the wholesale retreat of the greecs that is just now puzzling some advanced minds. Of cours the musical satelites represent at the same time the european concert with the disgusted halfuroons face in one corner and Egypt next to it and there can be no doubt that the world is now about getting ready to applaud such a grand realistic play on the stage after even the school children of Chicago adopted a great part of my moral scuol-club (act II) as I see from the Times Herald Oct. 3d. and they did certainly better than the Mars Fools did in N. Y. 4 years ago with that Dire play, A trip to Mars. The only question now is to find an enterprising scientist to not only recomend my play but put some 1500$ up for to stage it at once perhaps you would be able to do so.

   Yours truly
    G. A. KASTELIC, Hotel Buenavista.

In the following Dr. Diaforus of the *Malade Imaginaire* seems to have a formidable rival.

CHICAGO, Oct. 31, 1897.

MR. NEWCOMBE:

Dear Sir, — I forwarded you photographs of several designs which demonstrate by illustrations in physics, metaphysics, phrenology, mechanics, Theology, Law magnetism Astronomy etc — the only true form and principles of universal government, and the greatest life sustaining forces in this universe, I would like to explain to you and to some of the expert government detectives every thing in connection with those illustrations since 1881; I have traveled over this continent; for many years I have been persecuted. my object in sending you those illustrations is to see if you could influence some Journalist in this City, or in Washington to illustrate and write up the interpretation of those designs, and present them to the public through the press.

You know that very few men can grasp or comprehend in what relation a plumb line stands to the sciences, or to the nations of this earth, at the present time, by giving the correct interpretation of Christian, Hebrew, & Mohammedian prophesy, this work presents a system of international law which is destined to create harmony peace and prosperity.

sincerely yours

1035 Monadnock Bld
Chicago Ill

C/o L. L. SMITH.

P. S. The very law that moulds a tear; and bids it trickel from its source; that law preserves this earth a sphere, and guides the planets in their course.

ORD NEB Nove 18, 1897.

PROFESSOR SIMON NEWCOMB
    Washington D C

Dear Sir, — As your labors have enabled me to protect my honor And prove the Copernican Newton Keplar and Gallileo theories false I solicit transportation to your department so that I can come and explain the whole of Nature and so enable you to obtain the true value of the Moon from both latitudes at the same instant.

My method of working does not accord with yours Hence will require more time to comprehend I have asked Professor James E Keeler to examine the work and forward his report with this application for transportation
                      Yours truly   —— ——

One day in July, 1895, I was perplexed by the receipt of a cable dispatch from Paris in the following terms: —

        Will you act?   Consult Gould.   Furber.

The dispatch was accompanied by the statement that an immediate answer was requested and prepaid. Dr. Gould being in Cambridge, and I in Washington, it was not possible to consult him immediately as to what was meant. After consultation with an official of the Coast Survey, I reached the conclusion that the request had something to do with the International Metric Commission, of which Dr. Gould was a member, and that I was desired to act on some committee. As there could be no doubt of my willingness to do this, I returned an affirmative answer, and wrote to Dr. Gould to know exactly what was required. Great was my surprise to receive an answer stating that he knew nothing of the subject, and could not imagine what was meant. The mystery was dispelled a few days later by a visit from Dr. E. R. L. Gould, the well-known professor of economics, who soon after extended his activities into the more practical line of the presidency of the Suburban Homes and Improvement Company of New York. He had just arrived from Paris, where a movement was on foot to induce the French government to make such modifications in the regulations governing the instruction and the degrees at the French universities as would make them more attractive to American students, who had hitherto frequented the German universities to the almost entire exclusion of those of France. It was desired by the movers in the affair to organize an American committee to act with one already formed at Paris; and it was desired that I should undertake this work.

I at first demurred on two grounds. I could not see how, with propriety, Americans could appear as petitioners to the French government to modify its educational system for their benefit. Moreover, I did not want to take any position which would involve me in an effort to draw American students from the German universities.

He replied that neither objection could be urged in the case. The American committee would act only as an adviser to the French committee, and its sole purpose was to make known to the latter what arrangements as regarded studies, examinations, and degrees would be best adapted to meet the views and satisfy the needs of American students. There was, moreover, no desire to draw American students from the German universities; it was only desired to give them greater facilities in Paris.

The case was fortified by a letter from M. Michel Bréal, member of the Institute of France, and head of the Franco-American committee, as it was called in Paris, expressing a very flattering desire that I should act.

I soon gave my consent, and wrote to the presidents of eight or ten of our leading universities and several Washington officials interested in education, to secure their adhesion. With a single exception, the responses were unanimous in the affirmative, and I think the exception was due to a misapprehension of the objects of the movement. The views of all the adhering Americans were then requested, and a formal meeting was held, at which they were put into shape. It is quite foreign to my present object to go into details, as everything of interest in connection with the matter will be found in educational journals. One point may, however, be mentioned. The French committee was assured that whatever system of instruction and of degrees was offered, it must be one in which no distinction was made between French and foreigners. American students would not strive for a degree which was especially arranged for them alone.

I soon found that the movement was a much more complex one than it appeared at first sight, and that all the parties interested in Paris did not belong to one and the same committee. Not long after we had put our suggestions into shape, I was gratified by a visit from Dom de la Tremblay, prior of the Benedictine Convent of Santa Maria, in Paris, a most philanthropic and attractive gentleman, who desired to promote the object by establishing a home for the American students when they should come. Knowing the temptations to which visiting youth would be exposed, he was desirous of founding an establishment where they could live in the best and most attractive surroundings. He confidently hoped to receive the active support of men of wealth in this country in carrying out his object.

It was a somewhat difficult and delicate matter to explain to the philanthropic gentleman that American students were not likely to collect in a home specially provided for them, but would prefer to find their own home in their own way. I tried to do it with as little throwing of cold water as was possible, but, I fear, succeeded only gradually. But after two or three visits to New York and Washington, it became evident to him that the funds necessary for his plan could not be raised.

The inception of the affair was still not clear to me. I learned it in Paris the year following. Then

I found that the movement was started by Mr. Furber, the sender of the telegram, a citizen of Chicago, who had scarcely attained the prime of life, but was gifted with that indomitable spirit of enterprise which characterizes the metropolis of the West. What he saw of the educational institutions of Paris imbued him with a high sense of their value, and he was desirous that his fellow-countrymen should share in the advantages which they offered. To induce them to do this, it was only necessary that some changes should be made in the degrees and in the examinations, the latter being too numerous and the degrees bearing no resemblance to those of Germany and the United States. He therefore addressed a memorial to the Minister of Public Instruction, who was much impressed by the view of the case presented to him, and actively favored the formation of a Franco-American committee to carry out the object. Everything was gotten ready for action, and it only remained that the prime mover should submit evidence that educators in America desired the proposed change, and make known what was wanted.

Why I should have been selected to do this I do not know, but suppose it may have been because I had just been elected a foreign associate of the Institute, and was free from trammels which might have hindered the action of men who held official positions in the government or at the heads of universities. The final outcome of the affair was the establishment in the universities of France of the degree of Doctor of the University, which might be given either in letters or in science, and which was expected to correspond as nearly as possible to the degree of Doctor of Philosophy in Germany and America.

One feature of the case was brought out which may be worthy of attention from educators. In a general way it may be said that our Bachelor's degree does not correspond to any well-defined stage of education, implying, as it does, something more than that foundation of a general liberal education which the degree implies in Europe, and not quite so much as the Doctor's degree. I found it very difficult, if not impossible, to make our French friends understand that our American Bachelor's degree was something materially higher than the Baccalaureate of the French Lycée, which is conferred at the end of a course midway between our high school and our college.

From education at the Sorbonne I pass to the other extreme. During a stay in Harper's Ferry in the autumn of 1887, I had an object lesson in the state of primary education in the mountain regions of the South. Accompanied by a lady friend, who, like myself, was fond of climbing the hills, I walked over the Loudon heights into a sequestered valley, out of direct communication with the great world. After visiting one or two of the farmhouses, we came across a school by the roadside. It was the hour of recess, and the teacher was taking an active part in promoting the games in which the children were engaged. It was suggested by one of us that it would be of interest to see the methods of this school; so we approached the teacher on the subject, who very kindly offered to call his pupils together and show us his teaching.

First, however, we began to question him as to the subjects of instruction. The curriculum seemed rather meagre, as he went over it. I do not think it went beyond the three R's.

"But do you not teach grammar as well as reading?" I asked.

"No, I am sorry to say, I do not. I did want to teach grammar, but the people all said that they had not been taught grammar, and had got along very well without it, and did not see why the time of the children should be taken up by it."

"If you do not teach grammar from the book, you could at least teach it by practice in composition. Do you not exercise them in writing compositions?"

"I did try that once, and let me tell you how it turned out. They got up a story that I was teaching the children to write love letters, and made such a clamor about it that I had to stop."

He then kindly offered to show us what he did teach. The school was called together and words to spell were given out from a dictionary. They had got as far as "patrimony," and went on from that word to a dozen or so that followed it. The words were spelled by the children in turn, but nothing was said about the definition or meaning of the word. He did not explain whether, in the opinion of the parents, it was feared that disastrous events might follow if the children knew what a "patrimony" was, but it seems that no objections were raised to their knowing how to spell it.

We thanked him and took our leave, feeling that we were well repaid for our visit, however it might have been with the teacher and his school.

I have never been able to confine my attention to astronomy with that exclusiveness which is commonly considered necessary to the highest success in any profession. The lawyer finds almost every branch of human knowledge to be not only of interest, but of actual professional value, but one can hardly imagine why an astronomer should concern himself with things mundane, and especially with sociological subjects. But there is very high precedent for such a practice. Quite recently the fact has been brought to light that the great founder of modern astronomy once prepared for the government of his native land a very remarkable paper on the habit of debasing the currency, which was so prevalent during the Middle Ages.[1] The paper of Copernicus is, I believe, one of the strongest expositions of the evil of a debased currency that had ever appeared. Its tenor may be judged by the opening sentence, of which the following is a free translation: —

[1] Prowe : Nicolaus Copernicus, Bd. ii. (Berlin, 1884), p. 33.

Innumerable though the evils are with which kingdoms, principalities, and republics are troubled, there are four which in my opinion outweigh all others, — war, death, famine, and debasement of money. The three first are so evident that no one denies them, but it is not thus with the fourth.

A certain interest in political economy dates with me from the age of nineteen, when I read Say's work on the subject, which was at that time in very wide circulation. The question of protection and free trade was then, as always, an attractive one. I inclined towards the free trade view, but still felt that there might be another side to the question which I found myself unable fully to grasp. I remember thinking it quite possible that Smith's "Wealth of Nations" might be supplemented by a similar work on the strength of nations, in which not merely wealth, but everything that conduces to national power should be considered, and that the result of the inquiry might lead to practical conclusions different from those of Smith. Very able writers, among them Henry C. Carey, had espoused the side of protection, but for some years I had not time to read their works, and therefore reserved my judgment until more light should appear.

Thus the matter stood until an accident impelled me to look into the subject. About 1862 or 1863 President Thomas Hill, of Harvard University, paid a visit to Washington. I held him in very high esteem. He was a mathematician, and had been the favorite student of Professor Benjamin Peirce;

but I did not know that he had interested himself in political economy until, on the occasion in question, I passed an evening with him at the house where he was a guest. Here he told me that in a public lecture at Philadelphia, a few evenings before, he had informed his hearers that they had amongst them one of the greatest philosophers of the time, Henry C. Carey. He spoke of his works in such enthusiastic terms, describing especially his law of the tendency of mankind to be attracted towards the great capitals or other centres of population, that I lost no time in carefully reading Carey's "Principles of Social Science."

The result was much like a slap in the face. With every possible predisposition to look favorably on its teachings, I was unable to find anything in them but the prejudiced judgments of a one-sided thinker, fond of brilliant general propositions which really had nothing serious to rest upon either in fact or reason. The following parody on his method occurred to me : —

The physicians say that quinine tends to cure intermittent fever. If this be the case, then where people use most quinine, they will have least intermittent fever. But the facts are exactly the opposite. Along the borders of the lower Mississippi, where people take most quinine, they suffer most from fever; therefore the effect of quinine is the opposite of that alleged.

I earnestly wished for an opportunity to discuss the matter further with Mr. Hill, but it was never offered.

During the early years of the civil war, when the country was flooded with an irredeemable currency, I was so much disturbed by what seemed to me the unwisdom of our financial policy, that I positively envied the people who thought it all right, and therefore were free from mental perturbation on the subject. I at length felt that I could keep silent no longer, and as the civil war was closing, I devoted much time to writing a little book, " Critical Examination of Our Financial Policy during the Southern Rebellion." I got this published by the Appletons, but had to pay for the production. It never yielded enough to pay the cost of printing, as is very apt to be the case with such a book when it is on the unpopular side and by an unknown author. It had, however, the pleasant result of bringing me into friendly relations with two of the most eminent financiers of the country, Mr. Hugh McCulloch and Mr. George S. Coe, the latter president of one of the principal banks of New York. The compliments which these men paid to the book were the only compensation I got for the time and money expended upon it.

In 1876 the "North American Review" published a centennial number devoted to articles upon our national progress during the first century of our existence. I contributed the discussion of our work in exact science. Natural science had been cultivated among us with great success, but I was obliged to point out our backward condition in every branch of exact science, which was more marked the more mathematical the character of the scientific work. In pure mathematics we seemed hopelessly behind in the race.

I suppose that every writer who discusses a subject with a view of influencing the thought of the public, must be more or less discouraged by the small amount of attention the best he can say is likely to receive from his fellow-men. No matter what his own opinion of the importance of the matters he discusses, and the results that might grow out of them if men would only give them due attention, they are lost in the cataract of utterances poured forth from the daily, weekly, and monthly press. I was therefore much pleased, soon after the article appeared, to be honored with a visit from President Gilman, who had been impressed with my views, and wished to discuss the practicability of the Johns Hopkins University, which was now being organized, doing something to promote the higher forms of investigation among us.

One of the most remarkable mathematicians of the age, Professor J. J. Sylvester, had recently severed his connection with the Royal Military Academy at Woolich, and it had been decided to invite him to the chair of mathematics at the new university. It was considered desirable to have men of similar world-wide eminence in charge of the other departments in science. But this was found to be impracticable, and the policy adopted was to find young men whose reputation was yet to be made, and who would be the leading men of the future, instead of belonging to the past.

All my experience would lead me to say that the selection of the coming man in science is almost as difficult as the selection of youth who are to become senators of the United States. The success of the university in finding the young men it wanted, has been one of the most remarkable features in the history of the Johns Hopkins University. Of this the lamented Rowland affords the most striking, but by no means the only instance. Few could have anticipated that the modest and scarcely known youth selected for the chair of physics would not only become the leading man of his profession in our country, but one of the chief promoters of scientific research among us. Mathematical study and research of the highest order now commenced, not only at Baltimore, but at Harvard, Columbia, and other centres of learning, until, to-day, we are scarcely behind any nation in our contributions to the subject.

The development of economic study in our country during the last quarter of the last century is hardly less remarkable than that of mathematical science. A great impulse in this direction was given by Professor R. T. Ely, who, when the Johns Hopkins University was organized, became its leading teacher in economics. He had recently come from Germany, where he had imbibed what was supposed to be a new gospel in economics, and he now appeared as the evangelist of what was termed the historical school. My own studies were of course too far removed from this school to be a factor in it. But, so far as I was able, I fought the idea of there being two schools, or of any necessary antagonism between the results of the two methods. It was true that there was a marked difference in form between them. Some men preferred to reach conclusions by careful analysis of human nature and study of the acts to which men were led in seeking to carry out their own ends. This was called the old-school method. Others preferred to study the problem on a large scale, especially as shown in the economic development of the country. But there could be no necessary difference between the conclusions thus reached.

One curious fact, which has always been overlooked in the history of economics in our country, shows how purely partisan was the idea of a separation of the two schools. The fact is that the founder of the historic school among us, the man who first introduced the idea, was not Ely, but David A. Wells. Up to the outbreak of the civil war, Mr. Wells had been a writer on scientific subjects without any special known leaning toward economics; but after it broke out he published a most noteworthy pamphlet, setting forth the resources of our country for carrying on war and paying a debt, in terms so strong as to command more attention than any similar utterance at the time. This led to his appointment as Special

Commissioner of Revenue, with the duty of collecting information devising the best methods of raising revenue. His studies in this line were very exhaustive, and were carried on by the methods of the historic school of economics. I was almost annoyed to find that, if any economic question was presented to him, he rushed off to the experience of some particular people or nation — it might be Sweden or Australia — instead of going down to fundamental principles. But I could never get him interested in this kind of analysis.

One of Professor Ely's early movements resulted in the organization of the American Economic Association. His original plan was that this society should have something like a creed to which its members were expected to subscribe. A discussion of the whole subject appeared in the pages of "Science," a number of the leading economists of the country being contributors to it. The outcome of the whole matter has been a triumph for what most men will now consider reason and good sense. The Economic Association was scarcely more than organized when it broke loose from all creeds and admitted into its ranks investigators of the subject belonging to every class. I think the last discussion on the question of two schools occurred at the New York meeting, about 1895, after which the whole matter was dropped and the association worked together as a unit.

As Professor Ely is still a leader on the stage, I desire to do him justice in one point. I am able to do so because of what I have always regarded as one of the best features of the Johns Hopkins University — the unity of action which pervaded its work. There is a tendency in such institutions to be divided up into departments, not only independent of each other, but with little mutual help or sympathy. Of course every department has the best wishes of every other, and its coöperation when necessary, but the tendency is to have nothing more than this. In 1884, after the resignation of Professor Sylvester, I was invited by President Gilman to act as head of the department of mathematics. I could not figure as the successor of Sylvester, and therefore suggested that my title should be professor of mathematics and astronomy. The examinations of students for the degree of Doctor of Philosophy were then, as now, all conducted by a single "Board of University Studies," in which all had equal powers, although of course no member of the board took an active part in cases which lay entirely outside of his field. But the general idea was that of mutual coöperation and criticism all through. Each professor was a factor in the department of another in a helpful and not an antagonistic way, and all held counsel on subjects where the knowledge of all was helpful to each. I cannot but think that the wonderful success of the Johns Hopkins University is largely due to this feature of its activity, which tended to broaden both professors and students alike.

In pursuance of this system I for several years took part in the examinations of students of economics for their degrees. I found that Professor Ely's men were always well grounded in those principles of economic theory which seemed to me essential to a comprehension of the subject on its scientific side.

Being sometimes looked upon as an economist, I deem it not improper to disclaim any part in the economic research of to-day. What I have done has been prompted by the conviction that the greatest social want of the age is the introduction of sound thinking on economic subjects among the masses, not only of our own, but of every other country. This kind of thinking I have tried to promote in our own country by such books as "A Plain Man's Talk on the Labor Question," and "Principles of Political Economy."

My talks with Professor Henry used to cover a wide field in scientific philosophy. Adherence to the Presbyterian church did not prevent his being as uncompromising an upholder of modern scientific views of the universe as I ever knew. He was especially severe on the delusions of spiritualism. To a friend who once told him that he had seen a "medium" waft himself through a window, he replied, "Judge, you never saw that; and if you think you did, you are in a dangerous mental condition and need the utmost care of your family and your physician."

Among the experiences which I heard him relate more than once, I think, was one with a noted medium. Henry was quite intimate with President Lincoln, who, though not a believer in spiritualism, was from time to time deeply impressed by the extraordinary feats of spiritualistic performers, and naturally looked to Professor Henry for his views and advice on the subject. Quite early in his administration one of these men showed his wonderful powers to the President, who asked him to show Professor Henry his feats.

Although the latter generally avoided all contact with such men, he consented to receive him at the Smithsonian Institution. Among the acts proposed was that of making sounds in various quarters of the room. This was something which the keen senses and ready experimental faculty of the professor were well qualified to investigate. He turned his head in various positions while the sounds were being emitted. He then turned toward the man with the utmost firmness and said, "I do not know how you make the sounds, but this I perceive very clearly: they do not come from the room but from your person." It was in vain that the operator protested that they did not, and that he had no knowledge how they were produced. The keen ear of his examiner could not be deceived.

Sometime afterward the professor was traveling in the east, and took a seat in a railway car beside a young man who, finding who his companion was, entered into conversation with him, and informed him that he was a maker of telegraph and electrical instruments. His advances were received in so friendly a manner that he went further yet, and confided to Henry that his ingenuity had been called into requisition by spiritual mediums, to whom he furnished the apparatus necessary for the manifestations. Henry asked him by what mediums he had been engaged, and was surprised to find that among them was the very man he had met at the Smithsonian. The sounds which the medium had emitted were then described to the young man, who in reply explained the structure of the apparatus by which they were produced, which apparatus had been constructed by himself. It was fastened around the muscular part of the upper arm, and was so arranged that clicks would be produced by a simple contraction of the muscle, unaccompanied by any motion of the joints of the arm, and entirely invisible to a bystander.

During the Philadelphia meeting of the American Association for the Advancement of Science, held in 1884, a few members were invited by one of the foreign visitors, Professor Fitzgerald of Dublin, I think, to a conference on the subject of psychical research. The English society on this subject had been organized a few years before, and the question now was whether there was interest enough among us to lead to the organization of an American Society for Psychical Research. This was decided in the affirmative; the society was soon after formed, with headquarters in Boston, and I was elected its first president, a choice which Powell, of Washington, declared to be ridiculous in the highest degree.

On accepting this position, my first duty was to make a careful study of the publications of the parent society in England, with a view of learning their discoveries. The result was far from hopeful. I found that the phenomena brought out lacked that coherence and definiteness which is characteristic of scientific truths. Remarkable effects had been witnessed; but it was impossible to say, Do so and so, and you will get such an effect. The best that could be said was, perhaps you will get an effect, but more likely you will not. I could not feel any assurance that the society, with all its diligence, had done more than add to the mass of mistakes, misapprehensions of fact, exaggerations, illusions, tricks, and coincidences, of which human experience is full. In the course of a year or two I delivered a presidential address, in which I pointed out the difficulties of the case and the inconclusiveness of the supposed facts gathered. I suggested further experimentation, and called upon the English society to learn, by trials, whether the mental influences which they had observed to pass from mind to mind under specially arranged conditions, would still pass when a curtain or a door separated the parties. Fifteen years have since elapsed, and neither they nor any one else has settled this most elementary of all the questions involved. The only conclusion seems to be that only in exceptional cases does any effect pass at all; and when it does, it is just as likely to be felt halfway round the world as behind a curtain in the same room.

Shortly after the conference in Philadelphia I had a long wished-for opportunity to witness and investigate what, from the descriptions, was a wonder as great as anything recorded in the history of psychic research or spiritualism. Early in 1885 a tall and well-built young woman named Lulu Hurst, also known as the "Georgia magnetic girl," gave exhibitions in the eastern cities which equaled or exceeded the greatest feats of the Spiritualists. On her arrival in Washington invitations were sent to a number of our prominent scientific men to witness a private exhibition which she gave in advance of her public appearance. I was not present, but some who attended were so struck by her performance that they arranged to have another exhibition in Dr. Graham Bell's laboratory. I can give the best idea of the case if I begin with an account of the performance as given by the eye-witnesses at the first trial. We must remember that this was not the account of mere wonder-seekers, but of trained scientific men. Their account was in substance this: —

A light rod was firmly held in the hands of the tallest and most muscular of the spectators. Miss Lulu had only to touch the rod with her fingers when it would begin to go through the most extraordinary manœuvres. It jerked the holder around the room with a power he was unable to resist, and finally threw him down into a corner completely discomfited. Another spectator was then asked to take hold of the rod, and Miss Lulu extended her arms and touched each end with the tip of her finger. Immediately the rod began to whirl around on its central axis with such force that the skin was nearly taken off the holder's hands in his efforts to stop it.

A heavy man being seated in a chair, man and chair were lifted up by the fair performer placing her hands against the sides. To substantiate the claim that she herself exerted no force, chair and man were lifted without her touching the chair at all. The sitter was asked to put his hands under the chair; the performer put her hands around and under his in such a way that it was impossible for her to exert any force on the chair except through his hands. The chair at once lifted him up without her exerting any pressure other than the touch upon his hands.

Several men were then invited to hold the chair still. The performer then began to deftly touch it with her finger, when the chair again began to jump about in spite of the efforts of three or four men to hold it down.

A straw hat being laid upon a table crown downwards, she laid her extended hands over it. It was lifted up by what seemed an attractive force similar to that of a magnet upon an armature, and was in danger of being torn to pieces in the effort of any one holding it to keep it down, though she could not possibly have had any hold upon the object.

Among the spectators were physicians, one or more of whom grasped Miss Lulu's arms while the motions were going on, without finding any symptoms of strong muscular action. Her pulse remained normal throughout. The objects which she touched seemed endowed with a force which was wholly new to science.

So much for the story. Now for the reality. The party appeared at the Volta Laboratory, according to arrangement. Those having the matter in charge were not professional mystifiers of the public, and showed no desire to conceal anything. There was no darkening of rooms, no putting of hands under tables, no fear that spirits would refuse to act because of the presence of some skeptic, no trickery of any sort.

We got up such arrangements as we could for a scientific investigation of the movements. One of these was a rolling platform on which Miss Lulu was requested to stand while the forces were exerted. Another device was to seat her on a platform scale while the chair was lifting itself.

These several experiments were tried in the order in which I have mentioned them. I took the wonderful staff in my hands, and Miss Lulu placed the palms of her hands and extended them against the staff near the ends, while I firmly grasped it with my two hands in the middle. Of course this gave her a great advantage in the leverage. I was then asked to resist the staff with all my force, with the added assurance from Mrs. Hurst, the mother, that the resistance would be in vain.

Although the performer began with a delicate touch of the staff, I noticed that she changed the position of her hands every moment, sometimes seizing the staff with a firm grip, and that it never moved in any direction unless her hands pressed it in that direction. As nearly as I could estimate, the force which she exerted might have been equal to forty pounds, and this exerted first in one way and then in another was enough to upset the equilibrium of any ordinary man, especially when the jerks were so sudden and unexpected that it was impossible for one to brace himself against them. After a scene of rather undignified contortion I was finally compelled to retire in defeat, but without the slightest evidence of any other force than that exerted by a strong, muscular young woman. I asked that the rod might be made to whirl in my hands in the manner which has been described, but there was clearly some mistake in this whirl, for Miss Lulu knew nothing on the subject.

Then we proceeded to the chair performance, which was repeated a number of times. I noticed that although, at the beginning, the sitter held his fingers between the chair and the fingers of the performer, the chair would not move until Miss Lulu had the ball of her hand firmly in connection with it. Even then it did not actually lift the sitter from the ground, but was merely raised up behind, the front legs resting on the ground, whereupon the sitter was compelled to get out. This performance was repeated a number of times without anything but what was commonplace.

In order to see whether, as claimed, no force was exerted on the chair, the performer was invited to stand on the platform of the scales while making the chair move. The weights had been so adjusted as to balance a weight of forty pounds above her own. The result was that after some general attempts to make the chair move the lever clicked, showing that a lifting force exceeding forty pounds was being exerted by the young woman on the platform. The click seemed to demoralize the operator, who became unable to continue her efforts.

The experiment of raising a hat turned out equally simple, and the result of all the trials was only to increase my skepticism as to the whole doctrine of unknown forces and media of communication between one mind and another. I am now likely to remain a skeptic as to every branch of "occult science" until I find some manifestation of its reality more conclusive than any I have yet been able to find.

INDEX

INDEX

INDEX

Absence of mind, examples of, 73, 169.
Academy of Science, a would-be, 351.
Academy of Sciences, Paris, 327.
Adams, Prof. John C., 220; intellectual capacity, 282; politics, 283.
Agnesi, Donna Maria, 294.
Agassiz, Louis, discusses Origin of Species, 70.
Airy, Sir George B., Observations of Transit of Venus, 166; hospitality, 285; poetic taste, 286; executive ability, 286; methods of work, 289.
Alexander, Columbus, 368.
Anderson, Sir James, 300.
Angle, trisection of, 387.
Argelander, Prof., master of observational astronomy, 318, 319.
Atlantic Cable, the first, 300.
Auwers, the great astronomer, 306.

Bacon, Mr., teacher at Bedeque, 9.
Baillie, William, U. S. engineer, 361.
Baird, Spencer F., 240.
Bancroft, George, reviews judicial decision of Star Catalogue case, 378.
Barnard, E. E., 190.
Barnard, Gen. John G., 335.
Bartlett, William P. G., 83.
Belknap, Admiral G. H., 228.
Bell, Alexander Graham, tries to locate ball in Garfield's body, 358.
Black, Jeremiah, 168, 169.
Blackie, Prof. J. S., 294.
Bond, George P., 250.
Booth, Edwin, 157.
Borst, Charles A., 373.
Boss, Prof. Lewis, 124, 230.

Bowditch, Nathaniel, 1.
Bradford, Isaac, 74.
Brewster, Elder, 3.
Brown, Prof. S. J., 125.
Burnham, S. W., 188.

Campbell, William W., 190.
Carey, Henry C., 400.
Casey, Thomas L., Jr., 174.
Casserly, Eugene, 128.
Cassini, astronomer, of Paris Observatory, 331.
Cayley, Prof. Arthur, 280.
Chandler, Captain Ralph, U. S. N., 171.
Chandler, W. E., 126.
Chauvenet, William, 111.
Chevreul, M., his remarkable age, 327.
Circle, quadrature of, 387.
Clark, Alvan, 129, 144.
Clark, Alvan, & Sons, character of the firm, 147.
Cleveland, Keith, 224.
Cobbett, William, 7, 53.
Coe, George S., financier, 402.
Coffin, J. H. C., 111.
Combe, George, 4, 16.
Commune of Paris, 321–326.
Comstock, G. C., 126.
Cooke, Thomas, & Sons, 133.
Cox, Jacob D., 258.
Crank, the anti-gravitation, 381; a reasonable, 383.
Cranks, specimen letters from, 389.

Darwin's "Origin of Species," discussion of, 70.
Dawes, Henry L., 82.
Dawes, Rev. W. R., 148.
Davis, Charles H., 63; becomes superintendent at Naval Observatory, 107.

INDEX

Dayton, A. G., 126.
Delauney, Charles, indorses Prof. Newcomb, 317; director of Paris Observatory, 319; attractive personality, 329, 330.
Draper, Dr. Henry, expert in astronomical photography, 171, 223.
Draper, Dr. John W., 250.
Dudley Observatory troubles, 80.

Early, Gen. Jubal A., raid of, 339.
Eastman, John R., 107, 274.
Eclipse, solar, of 1860, journey to observe, 88.
Economics, studies in, 399; alleged schools of, 405.
Education in mountain regions of South, 397.
Eggleston, Edward, 89.
Eliot, Charles W., 74.
Elkin, Dr. W. L., 176.
Elliott, Benjamin S., 50.
Ely, Prof. R. T., as economist, 404; organizes American Economic Association, 406; merits as teacher, 408.
Evarts, William M., 241.
Eveleth, G. W., 55.

Feil, maker of optical discs, 185.
Ferguson, James, 111.
Ferrell, William, 72, 88.
Field, Cyrus W., 128.
Fiske, John, on eccentric literature, 382.
Fixed stars, Paris conference regarding, 230.
Floyd, Richard S., 186.
France, universities of, 392.
Franklin, Admiral, 122.
Furber, Mr., starts movement for admission of American students in French universities, 396.

Garfield, James A., first acquaintance with, 353; his early life, 354; injustice done him, 354; his intellectual gifts, 355; assassination of, 356.
Geological Survey, circumstances leading to origin of, 252-255; attacks on, 261.
Gibraltar, determination of the longitude of, 284, 299.
Gill, Sir David, 176.
Gillis, Capt. J. M., superintendent of Naval Observatory, 99; obtains new transit circle, 105.
Gilman, Daniel C., 403.
Gladstone, William Ewart, meeting with, 273, 276.
Glaisher, J. W. L., 72.
Goldsborough, Admiral, 340.
Gould, Benjamin A., personality, 78; Dudley Observatory directorship, 80; candidate for Naval Observatory director, 111.
Gould, Dr. E. R. L., 393.
Gravitation detestable to some minds, 381.
Green, Capt. F. M., 284.
Greenwich Observatory, situation, 285; value of observations at, 288.
Grubb, Sir Howard J., 156, 185.

Hagar, Judge, 189.
Hale, Eugene P., 123.
Hale, George E., 126.
Hall, Asaph, 107; discovers satellites of Mars, 141.
Hamlin, Hannibal, 128.
Harkness, William, appointed to Naval Observatory, 107; shares honor of discovering brightest line in spectrum of sun's corona, 113; director of Observatory, 180.
Harrington, attorney, 367.
Harvard Observatory, Prof. Newcomb called to directorship of, 211; Pickering's directorship, 212.
Hassler, J. J. S., 264.
Hansen, Prof., greatest master of celestial mechanics, 315, 316.
Hayden, Prof. F. V., 253.
Hayes, Rutherford B., 242, 259.
Hedrick, Prof., 73.
Hell, Father Maximilian, his alleged forgery, 154.
Henry, Prof. Joseph, Prof. Newcomb's relations with, 1, 54, 58, 161; characteristics, 234-237; on spiritualism, 408.
Herbert, Hilary A., 231.
Hewitt, A. S., 255.
Hilgard, J. E., 1, 59; in charge of Coast Survey, 65, 128.
Hill, George W., 218, 219, 221.
Hill, Thomas Prescott, 400.
Holcombe, Lieut. J. H. L., 174.

INDEX 421

Holden, Prof. E. S., 184–194.
Horsford, E. N., 74.
Hubbard, Prof. J. S., head astronomer of Naval Observatory, 98; in charge of mural circle, 102.
Huggins, Sir William, 279.
Hughes, Thomas, 272.
Humphreys, Gen., chief of engineers, 256.
Hurst, Lulu, the "Georgia magnetic girl," exhibitions of, 412–416.

Illusion, an astronomical, 137.
Inch, Richard, United States engineer, 361.

Jennings, Mr., cooling device of, 358.
Jewett, C. C., 237.

Keeler, James E., 191.
Kelvin, Lord, 248.
Kerr, Prof., 73.
King, Clarence, 258, 259.
Knobel, E. B., 380.
Koresh, his theory, 385.

Lamar, Judge Lucius, 264.
Langley, Prof. Samuel P., 240.
Language, advantage of not knowing a, 306.
Laplace, the "Méchanique Céleste" of, 1.
Lardner's "Popular Lectures on Science and Art," 19.
Lawrence, Prof. Smith J., 56.
Lee, Gen. Robert E., 339.
Lee's "Tables and Formulæ," 56.
Leverrier, M., two views of, 328; meeting with, 330; his merits, 331.
Leverrier and Hansen's systems of planetary computation, 219.
Lick, James, 182.
Lick Observatory, origin of, 182; location discussed, 187; telescope at, 185; Holden's administration, 192; Keeler's administration, 194; Campbell's administration, 194.
Lincoln, Pres., his war-time receptions, 342; assassination of, 344; trial of assassins, 345.
Lister, Lord, 278.
Litchfield Observatory, founder of, 374.
Loomis, E. J., 74.

Lowe, Mr. (Viscount Sherbrooke), 276.

Mahan, Prof. D. H., 335.
Mars, discovery of the satellites of, 141.
Marsh, Prof. O. C., exposure of Indian ring, 263; relation to "Wild West," 265; exposure of Cardiff giant, 266; his modern fossil, 269.
Maskelyne, Rev. Nevil, 152.
"Mathematical Monthly," foundation of, 84.
Mathematics and exact sciences, state of, in America, 402.
Maury, Matthew F., work of, 103.
McCook, Gen. A. D., 341.
McCormick, L. J., 132.
McCulloch, Hugh, 244, 402.
McMickan, Captain, of Cunard Line, 271.
McTavish, Governor, 91.
"Méchanique Céleste," first sight of, 56.
Meier, John, 223.
Meridian conference of 1884, 226.
Mill, John Stuart, 272.
Mills, D. O., 183.
Miner and Tully's "Fevers of the Connecticut Valley," 33.
Monroe, Rev. Alexander H., 36 n.
Moore, Capt. W. S., 361.
Moore's Navigator, 17.
Morrill, Justin S., 124.

National Academy of Science, early proceedings, 251; report of Geological Survey, 255; report of Forestry System, 261.
"National Intelligencer," letter in, 55.
Natural Philosophy, Mrs. Marcet's Conversations on, 18.
Nautical Almanac, assistants on, 66; in charge of, 120.
Naval Observatory, early history of, 102; work at, 109; conditions at, 110; civilian head proposed, 111; views of administration in regard to, 112; reports of eclipse of 1870, 113; visit of Emperor Dom Pedro, 117; efforts to improve, 122; Board of Visitors appointed, 126; telescope of, 128; Congressional action regarding new telescope, 131; observations of satel-

lites of Neptune, 136, 141; search for companion of Procyon, 138.
Negro, characteristics of, 346; education of, 348.
Neptune, observations of the satellites of, 136, 141.
Newall, R. S., 133.
Newcomb, John, father of Simon, characteristics and marriage, 4.
Newcomb, Simon, the first, 2.
Newcomb, Judge Simon B., 2.
Newcomb, Prof. Simon, ancestry, 2, 3; parentage, 6; early education at Bedeque, 9; begins study of arithmetic, 10; influence of books, 14–22; winter spent with farmer Jefferson, 18; residence at Yarmouth, 23; ancestral home, 23; begins study of medicine, 27; manufacture of botanic medicine under Dr. Foshay, 31, 32; joins temperance lodge, 37; intimacy with Parkin family, 39; first sight of Smithsonian, 52; reading in political economy, 53; study of Newton's "Principia," 54; first attempt at mathematical paper, 54; letter in "National Intelligencer," 55; Colonel Abert sends Lee's "Tables and Formulæ," 56; letter from Prof. L. J. Smith, 56; teaching in a planter's family, 56; first sight of "Méchanique Céleste," 56; assistant on staff of Nautical Almanac, 66; discussion of Darwin's "Origin of Species," 70; student in Lawrence Scientific School, 74; acquaintance with Dr. B. A. Gould, 78; friendship with William P. G. Bartlett, 83; journey in 1860 to observe solar eclipse, 88; meets Governor Ramsey and Edward Eggleston, 89; received by Governor McTavish, 91; Saskatchewan journey, 92; candidate for professorship in Washington University, 95; application for professorship in Naval Observatory, 97; early experience at Observatory, 101; edits Yarnall's observations, 105; in charge of mural circle, 107; journey to observe 1869 eclipse, 113; new transit circle, 114; investigation of moon's motion, 115; visit of Dom Pedro to Observatory, 117; assumes charge of Nautical Almanac Office, 120; verification of satellites of Mars, 141; transit of Venus expedition to Europe, 167; expedition to Cape of Good Hope, 174; agent of Lick Observatory trustees, 184; first meeting with Schaeberle, 190; study of orbits of asteroids, 195; problems of astronomy, 198; motion of moon, 202; occultations of stars, 207; offered Harvard Observatory directorship, 211; head of Nautical Almanac Office, 214; policy of office, 216, 233; computations for Planet Tables, 216; assistants, 218; suggestions to Meridian Conference, 226; computations regarding fixed stars, 230; member Yale Alumni Association, 241; member Washington Scientific Club, 244; first trip to Europe, 271; meets Thomas Hughes, 272; John Stuart Mill, 272; William Ewart Gladstone, 273; General Burnside, 273; attends banquet of Royal Society, 276; visit to Lord Lister, 278; meets Prof. Cayley, 280; Prof. J. C. Adams calls, 281; determination of Gibraltar longitude, 284; visits Greenwich, 285; friendship with Sir George Airy, 285–289; visits Edinburgh, 292; meets Prof. Blackie, 294; joins party of English astronomers bound for Algeria, 295; stormy voyage, 296; at Gibraltar, 297; Sir James Anderson, an old acquaintance, 300; Mediterranean trip, 302–305; Wilhelm Förster, a Berlin acquaintance, 306; meets great astronomer Auwers, 306; visits Pulkova Observatory, 309; winter ride in Russia, 310; first meeting with Hansen, 315; arrives in Paris during German evacuation, 319; visits Paris Observatory, 321; meets Leverrier, 330; Washington during Civil War and after, 334–371; two days military service, 339; assassination of Lincoln, 344; attends trial of conspirators, 345; acquaintance with Sumner, 349; with President Garfield, 353; asked to devise means for cooling

INDEX

his sick chamber, 357; suggestions for location of bullet, 358; experience with eccentric theorists, 381–389; assists in obtaining entrance of American students to French universities, 396; object lesson in regard to education in mountain regions of South, 397; studies in economics, 399; publishes "Critical Examination of our Financial Policy during the Southern Rebellion," 402; contribution to "North American Review," 402; conference with Prof. Daniel C. Gilman, 403; contributions to economic literature: "A Plain Man's Talk on the Labor Question," "Principles of Political Economy," 408; "Psychical Research," 410–412.
Nixon, Thomas, 37, 41.

Occultism, 93.
Old Peake, janitor of the Smithsonian, 58.
Oldright, Mr., 53.
Oliver, James E., 72.
Ommaney, Sir Erastus, 295.

Paine, Thomas, 3.
Paradoxers, experience with, 382.
Paris Conference, conclusions of, 230; attacked by Prof. Boss and S. C. Chandler, 230.
Paris Observatory, 321, 332.
Parkin, George R., 39.
Patent claim, a curious, 361.
Patterson, J. W., 352.
Peirce, Benjamin, professor of mathematics, 75; personality, 77, 78; chairman of committee on methods of observing transit of Venus, 161; director of solar eclipse expedition, 274; presence in England valuable to British astronomers, 277.
Peters, C. H. F., heads Transit of Venus expedition, 139; Star Catalogue Case, 372; work on Ptolemy's Catalogue, 380.
Photoheliograph, horizontal, 164.
Phrenology, study of, 14, 34.
Pickering, E. C., 126.
Pistor and Martin's transit circle, 105.
Poe, Gen. O. M., 352.

Powell, John W., 240; during Garfield's illness, 357.
"Principia," Newton's, 54.
Procyon, search for companion of, 138; at Lick Observatory, 140.
Professors in Navy, origin of corps of, 101.
"Psychical Research," 410.
Ptolemy's Star Catalogue, Peters's work on, 380.
Pulkova Observatory, object glass made by Alvan Clark & Sons, 144, 145; foundation and situation, 309–313.

Reed, Thomas B., 125.
Rhodes scholarships, 37.
Rodgers, Admiral John, 120.
Rogers, William B., 250.
Royal Society, banquet of, 275.
Runkle, John D., 1, 66.

Safe burglary conspiracy, 367.
Safford, Truman H., 67.
Sampson, Admiral W. T., 121.
Sands, Admiral, superintendent of Naval Observatory, 112; retirement, 116; assists in obtaining new telescope, 130.
Sauty, de, cable operator at Gibraltar, 300.
Schaeberle, assistant to Prof. Holden, 190.
Schofield, J. M., 96.
Schurman, Caleb, 11.
Schurman, Jacob Gould, 11 n.
Scientific Club, 244.
Scudder, Samuel H., 88.
Shepherd, Alexander H., career, 364–371.
Sherman, Gen. W. T., 243.
Sibley, J. Langdon, 76.
Smith, James, circle squarer, 387.
Smithson, James, 235.
Smithsonian Institution, policy of, 235, 236; difficulties in administration, 237; expansion of scope, 240.
Smyth, Prof. C. Piazzi, 293.
Smyth, Admiral W. H., 152.
Sophocles, Evangelinus Apostolides, 75.
Standard time, adoption of, 225, 226.
Stanton, Edwin M., 336; his tireless energy, 337; his law of war, 338.

INDEX

Star Catalogue case, the great, 372.
Steeves, Isaac, 38.
Struve, Otto, 144, 309.
Struve, Wilhelm, 312.
Struve, Russian minister at Washington, 312.
Sudler, Dr. Arthur E., 50.
Sumner, Charles, characteristics, 349, 350; kills an incipient "Academy," 352.
Sylvester, Prof. J. J., 403.

Telescope, horizontal, planned by Prof. Winlock, 163.
Thomson, Sir William, 248.
Tilley, Sir Leonard, 40.
Tracy, Benjamin, 123.
Transit of Venus, early observations of, 151; observed by Mason and Dixon, 153; Hell's alleged forgeries, 157; preparation for observation of, 160; Committee of National Academy of Sciences to consider subject, 161; transit commission, 163; appropriation for observation station, 170, 171, 174; value of observations, 173; observations at Cape Town, 177; publication of observations, 178.
Tremblay, Dom de la, 395.
Tuttle, H. P., 192.
Tyndall, Prof., 296.

Van Vleck, Prof., 73.

Wagner, Dr., 315.
Wallace, Gen. Lew, 339.
Washburn, Mr., minister to Paris, 320.
Washington, during the Civil War, 334; newsboys of, 336; Early's raid on, 339; after the fall of Richmond, 343; Shepherd régime, 363; the new city, 366.
Weiss, director of Vienna Observatory, 157.
Welles, Gideon, 111.
Wells, David A., 405.
White House, incidents at, during Garfield's illness, 357.
Whitney, William C., 123.
Williams, Sir Fenwick, 298.
Wilson, Henry, 250.
Winlock, Prof. Joseph, superintendent Nautical Almanac, 59, 61; personality, 65; constructs instrument for astronomical photography, 163.
Wolf, Prof. Charles, 144.
Woodward, Dr. J. J., 357.
Wright, Chauncey, 70.
Wright, Gen. H. G., 341.

Yale Alumni Association, 241.
Yarnall, Prof. M., characteristics, 101; observations of, 105.

A STATISTICAL INQUIRY INTO THE PROBABILITY OF CAUSES OF THE PRODUCTION OF SEX IN HUMAN OFFSPRING

Simon Newcomb

PREFATORY NOTE.

The present paper is an attempt to apply a rigorous theory of probable inference to a question of genetic biology, taking statistical data as the basis of the inquiry. If, in making such an investigation, the author may seem to stray outside his professional field, he would reply that the discussion of a biological question was by no means the sole object with which the work has been undertaken. It has appeared to him that the treatment of statistical data generally on a large scale, by the rigorous methods of probable induction, leads one into a field the cultivation of which promises important results to the science of the future; and he hopes the work will show how it is possible by such methods to reach conclusions on questions which elude all direct investigation.

The author has to acknowledge his indebtedness to the Trustees of the Bache fund, who made him a grant to pay the expenses of the necessary examination of genealogical data, and to the Census Bureau, through Mr. W. C. Hunt, chief statistician for population, who supplied statistical data relating to several thousand families culled from the Census records.

WASHINGTON, *November, 1903.*

A STATISTICAL INQUIRY INTO THE PROBABILITY OF CAUSES OF THE PRODUCTION OF SEX IN HUMAN OFFSPRING.

By Simon Newcomb.

1. Introductory 5
2. Preponderance of male births 6
3. Is the ratio of male to female births the same in all races?.. 7
4. Inquiry whether any unisexual tendency exists among parents... 9
5. Unisexual tendency in multiple births 16
6. Processes suggested by the statistics of multiple births 21
7. Influence of the age of the parent on sex 22
8. Supposed influence of other conditions 26
9. Summary of conclusions....................................... 28
    Appendix. Mathematical theory of the effect of unisexual tendency.... 29

1. INTRODUCTORY.

The object of the present investigation is to make an application of certain hitherto undeveloped statistical methods to a class of cases in which the causes elude all direct inquiry. The subject of such methods, although its general principles are well understood, is one which is susceptible of being worked out in detail to a far greater extent than has yet been seriously attempted. The present paper may therefore be regarded as having a double object: one to illustrate and apply certain methods; the other to draw conclusions on a physiological question of the widest scientific and human interest.

The idea that the sex of offspring may depend, to a greater or less extent, on discoverable causes, perhaps even on causes within the control of the parents, is a very natural one. Examples of what might be possible causes are the respective ages of the parents, their vigor or condition of health, their conjugal habits, which may be infinitely varied, or the relation of the time of conception to the periods of menstruation. Besides this, it is possible that some parents, male or female, may be endued with some special faculty for producing children of one sex rather than another, owing to constitutional or other causes which, not being apparent on the surface, might entirely elude direct investigation.

The method of investigating this tendency in the case of recognizable possible causes is well known. The question whether parents having a certain characteristic, A, are more likely than parents of the class not-A to have children of either sex is determined by counting a sufficient number of the offspring of parents of each class and comparing the ratio of male and female children in each class with the average ratio. If we find this ratio to be markedly different in the two classes, we conclude that the characteristic A is associated with some cause having a definite effect upon the production of sex.

Although the author proposes to apply this simple and obvious method to certain cases in the following investigations, his main object is to go farther, and inquire whether there are any causes or conditions whatever, known or unknown, which, in a decided degree, affect the production of sex. When we see one family consisting mainly or wholly of male children, and another consisting mainly or wholly of female children, it is very natural to suspect that the excess, in each case, may be due to some characteristic or faculty of the parents, or some peculiarity of their constitution, which may or may not admit of discovery and investigation. The mere fact of this inequality, taken in itself, does not, however, prove anything, because it may be the natural result of those accidents which determine sex, but of which we know nothing. Granting the existence of constitutional or other tendencies of the kind supposed, we may apply the term *unisexual* to them. We may then make the hypothesis that there is something in the constitution or habits of parents which results in some having a unisexual tendency toward the production of male children, and others toward the production of females, which tendencies nevertheless elude investigation otherwise than by statistical investigation of their effects. The desideratum is to discover a criterion by which we may distinguish between inequalities in the division of a family between the two sexes which are simply the result of chance and those which are the result of a unisexual tendency on the part of the parents. Such a criterion is pointed out in the first four sections of the present paper, and its mathematical theory developed in the Appendix.

2. THE PREPONDERANCE OF MALE BIRTHS.

Certain facts preliminary to this inquiry may be set forth which will serve as points of comparison in coordinating our conclusions. The first of these is the well-known general fact that, in the entire Semitic race, there is a small but well-marked preponderance of male over female births. This preponderance is remarkably uniform in all European and American countries where complete statistics of births are available. Mulhall finds the ratio to be 1052 male to 1000 female births in Europe generally. For our present purpose it will be convenient to express the excess in a different form from this. The result just cited may be expressed by saying that, out of 2052 births, 1052 are males and 1000 females. We may also say that 51.3 per cent are males and 48.7 per cent are females. Altogether, it seems to the writer that the best expression for the sexes is the excess of male over female in 100 births. Since Mulhall's conclusion implies that in 100 births 51.3 per cent are males and 48.7 per cent are females, the number expressing the excess would be 2.6 per cent. The excess thus expressed is represented by the symbol E_m. That is, we put E_m = the excess of male over female births in a total of 100 births.

The slight variations in this excess found in different countries are no greater than may be the result of accident. Although its value is so nearly the same for different countries, there are still some circumstances to be considered in connection with this definition. The most important of these is that it is decidedly larger when still-births are included. In France the male excess of such births is between 4 and 5 per cent of the whole number, a proportion which is probably substantially the same in most European countries. If we desire E_m to express the physiological probability of the production of a male child, still-births should be included. We should then have, in France,
$$E^m = 2.93.$$
Although this expresses the proper physiological probability of the production of male offspring, it will better conduce to our present operations to consider only living births. For these we may take, as a normal ratio, Mulhall's value, which will give
$$E_m = 2.6$$

From the statistics of Massachusetts the ratios are found to be:

Living children, $E_m = 2.8$ All births, $E_m = 3.3$

It appears, therefore, that there is no material difference on the two sides of the Atlantic.

3. **IS THE RATIO OF MALE TO FEMALE BIRTHS THE SAME IN ALL RACES?**

So far as the author is aware the only races besides the Semitic for which we have statistical data on which to base a conclusion are the Mongolian race of Japan and the negro race in America. In the latter we have only a limited registration of births, but sufficient to at least point to a conclusion. The census of 1900 gives a registration record, and a total record of births of children of the colored race as follows:

	Births.	Registration.	Total.
Male		13,526	136,350
Female		13,244	136,625

The registration record shows a preponderance of males less than one-half that of the Semitic race; the census enumeration an actual preponderance of female births.

The census tables also give the number of negro children of each of certain ages at a number of very early ages.

From the census of 1900 we find negro children:

Age.	Male.	Female.
Under one month	10,200	10,322
Under three months	32,840	33,353
Under one year	121,329	123,181

It is difficult to conceive of any cause why the data given to a census taker should be systematically in error as to sex. The number under the age of one month can be nothing else than the number born during that month, less the deaths during the month. If, then, the preponderance of male deaths is no greater in the negro than in the white race, it would seem that an excess of male births does not characterize the negro race, but rather the contrary. Moreover, the uniformity of the excess throughout the first year seems to confirm this conclusion. I find it also to be confirmed by all the statistics of previous censuses since 1870. There is, in general, in all these cases, a slight excess of female negro or colored children under one month of age, which is greater than would be due to the excess of male over female deaths during the early period of life.

On the other hand, the recent Japanese census shows, in a number of births exceeding one million, an excess of males practically the same as in European countries. It would seem therefore that there is no difference in this respect between the Semitic and the Mongolian races.

As the numbers relating to the negro race in America are not beyond the possibility of doubt, and especially as those of actual births registered show a male excess, the suspicion, sometimes expressed, that this excess may run through the whole order of mammals is at least worthy of examination. But the statistics of horses kept with much detail in England and Germany show, on the whole, an almost equal division between the sexes, the general tendency being toward a preponderancy on the female side.

It is also a curious fact that in European countries where complete statistics are available, the excess of male births is smaller for illegitimate than for legitimate children. The problem of explaining this difference, which we can scarcely believe to be real, is one which the writer must leave to others.

4. INQUIRY WHETHER ANY UNISEXUAL TENDENCY, PERMANENT IN THE INDIVIDUAL IN EITHER DIRECTION, EXISTS AMONG PARENTS.

The well-known fact being that inequalities in the proportion of the two sexes are almost the universal rule, some families consisting mainly or entirely of male children, others mainly or entirely of female children, the question before us is whether these inequalities are simply the result of chance, or show unisexual tendencies on the part of the respective parents. What we want is a criterion for distinguishing between these two cases. To make clear the principle of the proposed criterion, we begin with an illustration of its application. Let us suppose that we select at random 100 families of two children each. Granting that this selection corresponds to the general average, and leaving out of consideration the small preponderance of male births, we shall have, in these families, 50 cases in which the first-born was a male, and 50 in which it was a female. Of these two sets of 50 each, if there is no unisexual tendency, the second child will be a male in one-half, or 25 cases, and a female in the other cases. If there is no tendency of the kind sought, the final result will be:

25 families of 2 females each;
25 families of 2 males each;
50 of 1 male and 1 female.

That is to say, in the great mass the number of families comprising two children of the same sex will be the same as that of families comprising two children of opposite sexes.

Now let us suppose that, in consequence of some cause not known in advance, some parents have a tendency toward the production of male and others toward the production of female children. To fix the ideas, let us suppose that in the case of one-half of the parents, which we call Class A, there is a probability of three-fifths in favor of a male child, and in the remaining half, called Class B, a similar preponderance in favor of a female child. The method presupposes that we have no clue to a decision as to which of these classes any given parents belong. The first-born will still be a male in one-half and a female in the other half of the cases. But, in case of the second child, the 50 parents of Class A will have 30 male and only 20 female children. The 50 of Class B will have 20 male and 30 female children. The probable distribution of children between the two classes will be as follows: Of the 50 parents of Class A, 30 will have male first-born children, and 20 will have female first-born. The preponderance being the same in the cases of the second child, the 30 males will be followed by 18 males and 12 females. The 20 females will be followed by 12 males and 8 females.

The numbers will be the same for Class B, only substituting female for male preponderance. Thus the total outcome will be:

	Class A	Class B	Total
2 males	18	8	26
Male-female	12	12	24
Female-male	12	12	24
2 females	8	18	26

That is, in 100 families of 2 children each, we shall have 52 with children of one sex, and 48 with children of different sexes, instead of 50 of each class.

The result will be yet more decisive when we consider families of a greater number of children. Whatever the number of the family, the theory of probabilities and combinations gives a certain distribution of male and female children, which would be the most likely result of pure chance. For example, in families of 3, the most likely chance distribution would be three cases in which the children were of different sexes for one in which all three were of the same sex. But, were there any tendency among special parents to a production of children of one sex, the proportion of families in which all three were of the same sex would be greater than that given by the law of chance distribution. To show the preponderance, let us suppose the tendency to be the same as in the preceding example, and the number of families of 3 children each to be 500, or 250 of each class. The result will be:

	Class A	Class B	Total
3 males	54	16	70
2 males, 1 female	108	72	180
1 female, 2 males	72	108	180
3 females	16	54	70

Assuming no unisexual tendency on the part of parents, the probable result would be a proportion of three families not all of the same sex to one of the same sex. The results would then compare thus:

Families	Actual	Probable
3 children of 1 sex	140	125
3 children of 2 sexes	360	375

The result is an excess of 15 unisexual families.

Let us now pass on to the general problem of which the preceding examples are special cases. We have cited the well-known fact of a general or average unisexual tendency in the male direction among parents of the Semitic race. The question before us is whether this tendency of this race is the same among all parents. What we know to start with is that, if some parents have a tendency greater than the normal to produce male children, then there must be a corresponding tendency among other parents to produce female children. It is this combined tendency toward the production of children of one sex in some cases and the other sex in other cases that, for the present purpose, I term unisexual.

The data for the investigation in question have been derived from two sources. Mr. Hunt, chief of the Division of Population Statistics in the Census Office, very courteously made for me a count of 2000 families in which the parents were of various nationalities, enumerated in the census of 1900. I have also had counts made from a genealogy including all the known families descended from one Andrew Newcomb, who died about the year 1650, and which served a valuable purpose as probably including a wider range of conditions than those affecting the families enumerated in the Census. This was further extended to include a great number of other family genealogies. The entire list may therefore be taken to include the widest possible range of ordinary conditions which might affect the sex of offspring.

In the following summary of families, the first column of figures gives numbers for the white families as supplied by the Census Office. The second gives the corresponding numbers for families taken from the genealogies. The third and fourth columns give the data for the negro and Indian families, as supplied by the Census. The fifth gives the sum for all the families.

This is followed by the probable numbers given by the theory of chances, in case that there is no unisexual tendency among parents.

In each column the first line is the total number of families, the second the number of those families of which the children are all of the same sex, whether male or female. The following lines give the number of bisexual families of each class, each division of the number between the two classes being combined. For example, in families of 4 children, the line marked 3 and 1 gives the combined number of families comprising 3 males and 1 female, together with those comprising 1 male and 3 females. The totality of the families enumerated is too small to give any value to the separate enumeration of males and females. A combination is therefore made in order to reduce the results to the smallest number of distinct data. Thus the reader can see at a glance to what extent, if any, a bisexual tendency can be found in the families enumerated.

FAMILIES OF 2 CHILDREN.

Class	White Census	White Genealogy	Negro	Indian	Total	Probable
Number of families	670	1051	56	6	1783	
Same sex	322	547	30	2	901	892
Opposite sexes	348	504	26	4	882	892

In the case of the Census families, there are fewer pairs of children of the same than of opposite sexes, which result is the opposite of that of a unisexual tendency. In the case of the genealogical families, the excess is in the unisexual direction, but is in part due to an excess of male offspring recorded in the genealogies.

FAMILIES OF 3 CHILDREN.

	White Census	White Genealogy	Negro	Indian	Total	Probable
Number	436	914	38	4	1392	
Same sex	112	219	13	2	346	348
Different sexes	324	695	25	2	1046	1044

In a chance distribution the unisexual families should be one-fourth the entire number. The actual number is slightly below this, so that no unisexual tendency is shown.

FAMILIES OF 4 CHILDREN.

	White Census	White Genealogy	Negro	Indian	Total	Probable
Number	322	729	24	5	1080	
4 of same sex	40	110	4	0	154	135
3 and 1	174	336	14	3	527	540
2 and 2	108	283	6	2	399	405

In a chance distribution the numbers of the three classes should be in the proportion 1:4:3. The actual number of unisexual families in the entire list is 19 in excess of the probable number. This excess is, however, no greater than might well be the result of chance. The number of families having 3 out of 4 children of the same sex shows the opposite of a unisexual tendency.

FAMILIES OF 5 CHILDREN.

	White Census	White Genealogy	Negro	Indian	Total	Probable
Number of families	191	620	18	3	832	
5 of same sex	10	49	1	0	60	52
4 and 1	56	177	4	1	238	260
3 and 2	125	394	13	2	534	520

Here again there is an excess of 8 unisexual families over the normal number of 52. The effect of a unisexual tendency would be to produce a number smaller than the probable one of families consisting of 3 children of one sex and 2 of another, but there is an excess in the case of these families. We can, therefore, only attribute the deviation to chance.

Families of 6 Children.

	White Census	White Genealogy	Negro	Indian	Total	Probable
Number of families	94	626	8	2	730	
6 of same sex	6	16	0	0	22	23
5 and 1	18	141	1	2	162	137
4 and 2	43	287	6	0	336	342
3 and 3	27	182	1	0	210	228

The deviations from the normal are in all cases rather less than we might expect as a result of chance deviation.

Families of 7 Children.

	White Census	White Genealogy	Negro	Indian	Total	Probable
Number of families	62	517	10	0	589	
7 of one sex	1	11	0	0	12	9
6 and 1	4	59	0	0	63	65
5 and 2	23	165	3	0	191	193
4 and 3	34	282	7	0	323	322

Families of 8 Children.

	White Census	White Genealogy	Negro	Indian	Total	Probable
Number of families	24	417	6	0	447	
8 of one sex	0	3	1	0	4	3
7 and 1	1	27	0	0	28	28
6 and 2	7	85	3	0	95	98
5 and 3	12	185	1	0	198	196
4 and 4	4	117	1	0	122	122

Families of 9 Children.

	White Census	White Genealogy	Negro	Indian	Total	Probable
Number of families	8	306	3	0	317	
9 of one sex	0	1	0	0	1	1
8 and 1	1	9	0	0	10	11
7 and 2	0	52	1	0	53	45
6 and 3	3	99	0	0	102	104
5 and 4	4	145	2	0	151	156

Families of 10 Children.

	White Census	White Genealogy	Negro	Indian	Total	Probable
Number of families	5	295	0	0	300	
10 of 1 sex	0	0	0	0	0	0.6
9 and 1	0	7	0	0	7	5.9
8 and 2	1	37	0	0	38	26.3
7 and 3	3	67	0	0	70	70.3
6 and 4	0	117	0	0	117	123.0
5 and 5	1	67	0	0	68	73.9

Families of 11 to 16 Children.

209 families of 11 children:

	Geneal.	Prob.
11 of one sex	0	0.2
10 and 1	6	2.2
9 and 2	16	11.2
8 and 3	32	33.7
7 and 4	68	67.3
6 and 5	87	94.2

124 families of 12 children:

12 of one sex	0	0.1
11 and 1	2	0.7
10 and 2	10	4.1
9 and 3	18	13.3
8 and 4	39	30.0
7 and 5	35	48.0
6 and 6	20	28.0

50 families of 13 children:

13 of one sex	0	0.0
13 and 1	0	0.2
11 and 2	1	0.9
10 and 3	4	3.4
9 and 4	9	8.7
8 and 5	15	15.7
7 and 6	21	21.0

25 families of 14 children:

14 of one sex	0	0.0
13 and 1	0	0.0
12 and 2	0	0.3
11 and 3	3	1.1
10 and 4	5	3.0
9 and 5	9	6.1
8 and 6	6	9.1
7 and 7	2	5.1

11 families of 15 children:

15 of one sex	0	0.0
14 and 1	0	0.0
13 and 2	0	0.1
12 and 3	0	0.3
11 and 4	1	0.9
10 and 5	2	2.0
9 and 6	5	3.4
8 and 7	3	4.4

7 families of 16 children:

16 of one sex	0	0.0
15 and 1	0	0.0
14 and 2	0	0.0
13 and 3	0	0.1
12 and 4	0	0.4
11 and 5	1	1.0
10 and 6	1	1.7
9 and 7	3	2.4
8 and 8	2	1.4

In the case of families of 7 or more, a unisexual tendency would be shown by a deficiency in the numbers of families with a nearly equal number of males and females. These are given in the last line of each series of families. We find such a deficiency to be actually shown. The numbers are:

        Actual: 101, 35, 46, 19, 25, 4, 7, 2
        Probable: 107, 39, 51, 23, 23, 7, 5, 1
        Actual   sum = 239
        Probable sum = 256

        Deficit.......... 17

This deviation is not large enough to base a conclusion upon. It is partly due to a deficit in the reports of female children in the genealogies, the respective recorded numbers of male and female being:

Number of male children..............................3,339
Number of female children............................3,020

Excess of male children......................... 319

It thus appears that about 110 males are reported against 100 females; in other words, 52.5 per cent of the whole number are males. This excess over the normal is to be attributed to the greater difficulty of tracing female than male children, owing to the greater liability of the former to be omitted from a record, especially when they die young.

We shall therefore expect to see a slight unisexual indication in the numbers from this cause alone, so that we may regard the deviation as explained without supposing any actual tendency of the kind sought.

The general result of the count of 2838 families embracing 13,257 children is that the distribution of male and female offspring follows the statistical laws of chance within the limits of probable deviation, the actual deviations being as great in one direction as in the opposite one. Consequently, there are no unisexual tendencies on the part of parents sufficiently great to be of practical importance. We are not, however, justified in concluding from these numbers alone that there can be absolutely no unisexual differences in the human race, nor that no possible conditions are productive of such a difference. Our conclusions only preclude conditions affecting the sex of a child which may occur with a certain frequency. For example, if one-tenth of the parents in the whole list practiced any habit, or possessed any characteristic, which would lead to two-thirds of their children being of one sex, the effect would show in the statistics. But if only a single pair of parents in the whole list had a unisexual tendency, or if, in the great majority of cases, this tendency was exceedingly small, the effect would not be shown.

As to this latter case, the fact of any unisexual tendency, however minute, would be of great scientific interest, but of no practical importance. It would hardly be worth while for any parent or any community to lay great stress on any cause which would result only in increasing the chances of male or female offspring by 3 or 4 per cent. It seems highly improbable that any very rare or highly artificial cause would produce a unisexual tendency, if no ordinary cause produced it. The absence of any strongly marked unisexual tendency in the families we have examined, therefore, justifies us in concluding, at least with a high degree of probability, that the causes of sex are beyond artificial control.

5. THE UNISEXUAL TENDENCY IN MULTIPLE BIRTHS.

We have next to consider a sort of family in which the conditions are peculiarly favorable for drawing conclusions on the general question of the cause of sex. These are families consisting of children of a single birth in twins or triplets. Considering first the case of twins, we begin with the effects which would result on two extreme hypotheses as to the cause of sex.

HYPOTHESIS I. *The distinction of male and female exists in original germs, antecedent to conception, presumably supplied by the father.*

In this case we should find the same random distribution of twin children between the two sexes that we find to exist in families of two. In four births of twins we should have one of two males, one of two females, and two bisexual. Only in two ways could this conclusion be avoided. One is by supposing that a germ of either sex is more likely than not to have one of the same sex in physical juxtaposition with it. This view seems inadmissible because, even if such juxtaposition did exist in any case at any moment, it would not be permanent. The other supposition is that at certain periods there is an abnormal excess of male germs and at other periods a similar excess of female germs in the same father. In the absence of permanent unisexualism on the part of any one father, which was shown in the preceding section, such an inequality can not be permanent. It therefore seems to me that this supposition is also too artificial and unlikely to be considered. We may therefore consider the statistical distribution of twin children between the sexes to afford a test of the above hypothesis.

HYPOTHESIS II. *Sex is entirely determined by the conditions to which the germ is subject during the early stages of its development.*

All these conditions are the same *ab initio* for the two members of the pair. The result of the hypothesis would therefore be that twin children would always be of the same sex.

The statistics show that neither of these hypotheses is correct taken singly; the actual result being an intermediate one between those of the two hypotheses. A child of one sex is more likely than not to have a twin of the same sex; but there is only a certain preponderance of probability for this.

The negation of the first hypothesis leads to the conclusion that the original germs supplied by the father, if not completely asexual, can at most have no other sexual quality than a slightly greater tendency to develop into one sex than into the other. While the existence of such a tendency is not out of the question, the more likely conclusion would seem to be that the part of the father is completely asexual, and that the determination of sexes is entirely the function of the mother. It is true that this contravenes certain supposed conclusions from statistics which will be considered in the next section. But these seem to me open to misconstruction.

We pass next to the actual numbers shown by the statistics of France and Germany. In the following table the first line shows the number of births giving rise to two males; the second to the number of bisexual births; the third of births of two females. In the next two lines are the total number of male and female children resulting from all the births. The preponderance of males is, on the whole, the normal one, showing that, in the production of twins, there are no causes affecting sex which act in any way differently from those in the ordinary cases. Then is given the normal number of bisexual births as it would have been were the determination of sex, in the case of the two children, completely independent, as required by hypothesis I. The proportional deficiency of the actual number of bisexual births over this probable number may be used to define the unisexual tendency in the case of twins. The observed fact may be set forth thus: The probable proportions of unisexual and bisexual twins, if the sex of each child were determined independently of the other, and the percentage of pairs of the several classes would be:

2 m., 0.260; m. and f., 0.500; 2 f., 0.240,

while the actual proportions are:

2 m., 0.332; m. and f., 0.354; 2 f., 0.314.

Sex of Twins—France and Berlin.

	France 1898-1900	Berlin 1853-1880	Total
2 males	9,537	2,968	12,505
Bisexual	9,826	3,489	13,315
2 females	8,949	2,852	11,801
Total males	28,900	9,425	38,325
Total females	27,724	9,193	36,917
M. to 100 F.	104.2	102.5	103.8
E_m	2.1	1.2	1.9
Normal bisexual	14,149	4,603	18,752
Deficit	4,323	1,114	5,437

The formal discussion, by algebraic methods, of the unisexual tendency implied in these numbers will be found in the Appendix. These methods are not, however, necessary to convey an idea of the principles by which the results are to be explained. What makes a discussion of these principles of interest is that the numbers derived from the statistics of twins may be applied to the case of triplets, and a comparison of the actual statistics of triplets with those derived from the statistics of twins will be of interest.

The processes which we presuppose are these: During an unknown period of time, commencing with the moment of conception, the two germs are exposed to a series of common influences, either in the male or female direction, tending to make them of the same sex. As we can, without appreciable error, make abstraction of the small normal preponderance toward the male sex, we may say that, in the general average, this unisexual tendency will be as often in one direction as in the other. Thus, in one-half the cases, which we term group A, the common influences preponderate in the male direction, and in other cases, which we call group B, in the female direction.

But these preponderating influences do not completely determine the sexes. There are accidental causes operating differently on the two growing organisms, which may result in their becoming of opposite sexes. Numerically stated, the conclusion to which we are led is that the statistics of twins may be explained by supposing that, in Group A there is a probability of 0.77 in favor of either of the organisms taken at random becoming male, and therefore 0.23 in favor of its becoming female; while in group B there are similar probabilities in the opposite direction. This, I say, is a conclusion from the statistics of twins, assuming that there is no interaction between the two organisms tending to make them of the same sex. I may remark, however, that this assumption in the case of twins would have no special significance. The combination of probabilities would lead to the same result whether we supposed it or not. The main point is that there is some preponderating tendency of the pair of organisms towards one sex in some cases and the opposite sex in the remaining cases. Stating probable results in percentages, they would be:

In group A, probability of 2 males..........0.77 = 59.3 per cent.
In group A, probability of 2 females........0.23 = 5.3 per cent.

Total unisexual percentage....................64.6
Bisexual 35.4

In group B the results would be the same, interchanging male and female.

These numbers, it will be noticed, show the percentages actually given by the statistics. The mathematical method developed in the Appendix shows that the results may be accounted for by assuming a certain unisexual tendency represented by a fraction a having the value

$$a = 0.27$$

This coefficient may be considered to express the efficiency of all the causes tending to produce sex which are common to the two twin members of the family. In other words, assuming this unisexual coefficient, the result will be 77 per cent of twins of the same sex, and 23 per cent of twins of different sexes, these being the actual results of observation.

A most interesting fact is that, by the methods developed in the Appendix, we may apply this coefficient a to determine how the sexes in families of triplets should be divided. There are two ways of proceeding. We may assume the unisexual tendency to be the same in triplets as we have found it to be in twins, as naturally ought to be the case. From this we can determine what proportion of triplets should be unisexual, and compare the result with statistics. The other method consists in determining the value of the unisexual tendency from the statistics of the triplets in order to see how much it differs from that determined from the statistics of twins. Adopting the first method the problem is: We have three organisms subject to such conditions that, in the case of each, there is a probability of 0.77 that it will prove of one sex and of 0.23 that it will prove of the other. What are the respective probabilities that the three organisms will be unisexual; that is, all three of the same sex; and that one shall be of one sex and two of the opposite? These probabilities are found in the Appendix to be:

Unisexual: percentage, 46.9
Bisexual: "   53.1

To compare the statistics I have collected the sexes of triplets found in the French and German tables of births with the following results:

	France 1858-1900	Berlin 1853-1880	Total
3 males	342	32	374
3 females	304	28	332
Total unisexual	646	60	706
Bisexual	667	43	710

Observed proportion of unisexual triplets..........49.9 per cent.

It appears from these numbers that when we compare the probabilities derived from the case of twins with the actual facts in the case of triplets, there is a discrepancy. From the facts in the case of twins, we should conclude that 46.9 per cent of all triplets should be unisexual; we actually find that 49.9 per cent is unisexual, an excess of 3 per cent.

The discrepancy may take the other form by determining the amount of unisexual preponderance in the case of triplets as we have done in the case of twins. This preponderance is 0.79 instead of 0.77. That is to say, grouping the triplets as we have the twins, there is a probability of 0.79 that any one organism of a triplet of group A will develop into a male, and that one of group B will develop into a female. The coefficient of unisexual tendency is, therefore, for triplets,

$$a = 0.29$$

Now, we should suppose, *a priori,* that the ratio of the unisexual preponderance to the effects of the accidental causes which finally determine the sex would be the same with twins and triplets. It is true that the discrepancy between 0.27 and 0.29, or between 46.9 and 49.9 per cent is not greater than might easily have been the result of fortuitous deviation. Still it is larger than we should expect. If we may regard it as expressing a real law, we may suppose that, besides the independent causes at action tending toward one sex or the other, there is an interaction between the two organisms, by which the sex of one influences that of the other in its own direction.

Apart from this, the general conclusions from triplets confirm that from twins $\frac{1}{m}$ —there are not male and female germs. It would seem that we have in this a practically conclusive negation of the theory of completely determined sex in the original germs and may provisionally accept that of complete asexuality on the part of such germs, subject, however, to farther statistical tests.

6. PROCESSES IN THE DETERMINATION OF SEX SUGGESTED BY THE STATISTICS OF MULTIPLE BIRTHS.

The view that, if the sex is not completely determined in the original formation of a germ, it must be determined at some definite moment of development—that there can be no intermediate state between complete asexuality and complete sexuality—is one which, at first sight, seems almost axiomatic. And yet, the preceding statistics of multiple births seem to show that such is not the case, and that there may be a series of causes acting first in one direction and then in the other, each of which tends to make one sex or the other more probable until, gradually, the sex is definitely determined. An analogue to this determination by a succession of accidental causes may be constructed in the following way: Let A be a large pipe or aqueduct, from the mouth B of which a stream flows into a gradually widening river V. At a certain distance below

the exit B the river is divided into two branches by a promontory P. On one side of this promontory, which we may call the male side, the river is slightly broader than on the other. Between the exit and the promontory, the river flows over a rough bottom with many eddies, but the ultimate result must be that every drop of water which comes from the conduit ultimately passes on one side of the promontory or the other. But the side on which it shall pass is not determined at any one moment. As a drop, or, to give the analogy a more complete form, a small particle suspended in the water, leaves the conduit, it is equally likely to pass into one branch of the river or the other. If it chance to incline to the right after leaving the conduit, there will be a greater probability of its passing into the right branch, but this will be only a probability until a certain point of the course is reached. A particle reaching the point M, for example, will be likely to go into the female branch, but yet may be carried by an eddy across to the opposite side before it reaches it. One at N, although farther down, will still be uncertain; possibly its course may not be decided until it almost reaches P. A particle on one bank or the other will be more and more likely to pass into the corresponding branch the farther down it is found. When the particle once crosses one of the dotted lines PR and PS the branch it will take will be completely determined.

A case of twins or of triplets has its analogue in the case of two or three particles emerging from the conduit in contiguity. They are more likely to keep together and enter the same eddies than if they were widely separated in the beginning. To speak with numerical exactness there is a probability of 0.77 that they will pass on the same side of the promontory and of 0.23 that they will separate. In the case of triplets the corresponding probability would be 0.79; but these are only probabilities. At any moment any two particles may widen their distance and be drawn into different parts of the stream, never to reunite.

We may thus say that the question, which branch of the river a particle, emerging from the conduit, is to flow into, will be determined by a series of accidents tending in one direction or the other; and the most plausible conclusion from the statistics of twins is that sex is determined in an analogous way.

7. INFLUENCE OF THE AGE OF THE PARENT ON SEX.

The changes produced by age in the human system are such that we may most plausibly look to them as causes affecting the sex of offspring. The question of the influence of the age of the parent has been studied by several investigators, especially by Rosenfeld, Sadler and Bertillon. I have not been able to refer to the original work of Bertillon and shall therefore confine myself to citing, in its proper place, one of his conclusions bearing on the case. Dr. Rosenfeld gives the following table of the sexes of more than thirty thousand births in Vienna, arranged according to the age of the father. I add the percentage E_m for each age:

VIENNA STATISTICS OF BIRTHS.

Age of father	Male children	Female children	100 M : F	E_m
Under 25	873	767	113.7	6.5
25 to 30	6,090	5,717	106.5	3.3
30 35	11,987	11,291	106.2	3.1
35 40	3,605	3,559	101.3	0.6
40 50	622	502	128.9	10.7
Over 50				

The table shows a decided preponderance of male children in the case of young and old fathers as compared with those in middle life. The conclusion thence-drawn is that male unisexuality is at its maximum in young and old people.

Francke, from the statistics of Norway, reached the same conclusions as regards young fathers, but the opposite as regards old ones. His numbers for the ratio of male to female births, arranged according to the age of the father, are as follows:

INFLUENCE OF THE AGE OF THE PARENTS

NORWEGIAN STATISTICS OF BIRTHS.

Age of father	100 M : F	E_m
Below 20	117.0	7.9
20 to 25	101.5	0.7
25 30	109.0	4.4
30 35	105.9	2.9
35 40	102.6	1.3
40 45	104.6	2.2
45 50	103.8	1.8
50 55	98.4	—0.8
55 60	97.9	—1.0
Over 60	99.8	—0.1

Rosenfeld gives a similar classification, arranged according to the age of the mother, as follows:

VIENNA STATISTICS OF BIRTHS.

Age of mother	Male children	Female children	100 M : F	E_m
Below 17	19	7	271.4	46.1
17 to 20	366	341	107.3	3.6
20 25	4,444	4,161	106.8	3.3
25 30	7,287	6,759	107.8	3.7
30 40	9,907	9,356	105.9	2.8
Over 40	1,412	1,396	101.1	0.5

The enormous preponderance of male births in the case of mothers under 17 years of age is probably the result of accident and not expressive of a general law, the births, 26 in number, being too few to base a determinate conclusion upon. If we combine all the mothers under 20 years of age, the result will be:

$$100 \text{ M} : \text{F} = 110.6; \quad E_m = 5.0$$

The numbers now show a marked preponderance of male children borne by very young mothers, which drops to the normal at the age of 20 and falls below it at the age of 40.

It may be noted in this connection that the ratio of male and female children is, in the general average, somewhat above the normal, possibly indicating an imperfection of the record by not including all female children. This, however, will not alter the conclusion.

All these conclusions as regards the age of the parent seem to me to lack a solid foundation, from the fact that the ages of the two parents are not completely distinguished. I shall discuss this difficulty after setting forth the results of the genealogical statistics collected by myself.

The effect of difference of age between father and mother was investigated by Sadler, who laid down the general law that the older parent has a preponderating influence in the direction of determining children of his or her own sex. But Ahlfeldt reached the opposite conclusion, finding that when the father was more than 10 years older than the mother, there was a preponderance of female children instead of the normal excess of male children. His numbers are, however, too few to base any conclusions upon; and the same is probably true of the statistics used by Sadler.

I have not attempted to investigate this subject by age because the data are not at hand for the purpose. Instead of doing this, I have taken the order of progression of children in families as found in the genealogies already cited. In each family the sex of the several children was tabulated in the order, first born, second born, third born, etc. Then the total number of first-born children of each sex, the second born, and so on, was taken. The results are shown in the following table in which the first column gives the order of birth. This is followed by the respective number of male and female first-born children. In the same line the numbers are given for the second born, and so on. In families of more than 14 children, the fourteenth and those following are all tabulated together, as their separate numbers are too small to base a conclusion upon.

The fourth column gives the total number of children; the fifth the excess of males, and the sixth the percentage of this excess.

COMPARISON OF MALE AND FEMALE CHILDREN IN THE ORDER OF BIRTH IN AMERICAN FAMILIES.

Order of Birth	Males	Females	Sum	Excess of Males	E_r	Corrected E_m
1	3,906	3,265	7,171	641	8.9	6.7
2	3,261	2,987	6,248	274	4.4	2.2
3	2,605	2,532	5,137	73	1.4	—0.8
4	2,145	2,024	4,169	121	3.0	0.8
5	1,766	1,651	3,417	115	3.4	1.2
6	1,406	1,338	2,744	68	2.5	0.3
7	1,102	1,020	2,122	82	3.9	1.7
8	782	754	1,536	28	1.8	—0.4
9	592	513	1,105	79	7.3	5.1
10	407	356	763	51	6.7	4.5
11	246	221	467	25	5.3	3.1
12	142	108	250	34	13.6	11.4
13	71	52	123	19	15.4	13.2
14 to 17	51	42	93	9	9.8	7.6
Total	18,482	16,863	35,345	1,619	4.6	

It will be seen that the excess of male births in the general average markedly exceeds its normal value. We must regard this divergence as unreal and attribute it to the greater liability of a female child to be omitted from the record. As this omission would be probably about equal in the case of all the successive children, we may assume that the values of E'_m are all equally in error from this cause. The normal value from the statistics of birth being about 2.4, while the count gives 4.6, we subtract the excess, 2.2, from each separate value of E'_m and thus obtain corrected values of the percentage of excess, which are found in the last column.

It will be seen from the numbers of this column that the excess of males among first-born children exceeds 6 per cent. This shows that there are about eight males to seven females of this class. But, in the case of the second child, the percentage of excess drops to 2.2, which is slightly blow the normal and, in the case of the third child, it becomes negative, showing that, after we correct the supposed defect of the record, there is actually a slight excess of female births.

The rapidity of the drop from 6.7 in the case of the first birth to 2.2 in the case of the second and then to a negative quantity in the case of the third, seems to show quite conclusively that the excess of males in the number of the first-born children is not attributable to the age of the mother, but to the fact that it is a first child, irrespective of age. That the fall is too rapid to be the effect of age is shown in the following way: The difference of age at the birth of the first and third child is not likely to have been more, in the general average, than three years. Now a drop of 4 in the percentage in three years would imply a drop of twice this amount between the ages of 17 and 24, which we may take as the probable range in the case of a first child. The approach to uniformity in the percentage in these cases where the marriage must have been at such different ages, precludes the supposition that age is the main factor in the case.

Continuing our study of the table, we find a remarkable uniformity in the number of male and female births up to the eighth child. In the case of the second child the excess is still fairly well marked. Thus we may conclude that the tendency toward male excess, though greatly diminished, is probably not wholly obliterated in the case of the second child. But from the fourth to the eighth inclusive, the deviations are so small that we may regard them as the effects of accident. In the case of the six children from the third to the eighth, it would seem that the birth of the two sexes is equally probable. Then, from the ninth child onward we find an excess of males which generally exceeds the normal all through. But it is not at all certain that this arises from a unisexual tendency in the case of older parents. It is quite possible that it may be attributed to first-born children after remarriage, the table having been constructed without any reference to the mother and giving only children in families by the same father. It must also be noted that the total numbers beyond the tenth child become too small to predicate a very certain conclusion upon. A more complete investigation of the subject will therefore be necessary before it can be said with certainty whether the results derived by Rosenthal in the Vienna statistics in the case of old parents are correct, or whether we here have to do with the first-born of second or third wives.

It might appear, at first sight, that these statistics do not decide whether the variation in the proportion of male and female, as the family advances in number, are due to the male or female parent. But a consideration of the ratio between the number of acts on the part of the two parents who are concerned in the case, decides the probabilities in favor of the mother.

A more conclusive investigation than has as yet been made is necessary to absolutely decide whether, as has been suggested in this paper, the part of the father is completely asexual. To make this investigation, it is necessary to compare the statistics of births by mothers of one and the same class with fathers of different ages. Since the ratio of male to female is the same, at least from the third to the eighth birth, the preferable method is to confine the investigations to those births which may be grouped all together, so far as the mother is concerned. We then compare the sex of each child of this class with the age of its father and, by a sufficient accumulation of cases, ascertain whether the ratio varies with that age.

8. EXAMINATION OF CERTAIN OTHER CONDITIONS WHICH HAVE BEEN SUPPOSED TO INFLUENCE THE PRODUCTION OF SEX.

It has sometimes been supposed that the destruction of an important fraction of the male population of a country by war, such as has occasionally been known in history, has resulted in a greater preponderance of male offspring in the country so affected. A very slight analysis of the supposed cause will show that this proposition belongs to a class which require very strong proof. Granting the truth of the proposition: since those who were killed in war could not subsequently have taken part in the propagation of the race, it would follow that those who returned in safety showed a unisexual tendency in the male direction.

That a tendency of this sort could be produced in one man by the mere death of another is a notion that hardly needs to be refuted. If such an effect is real, it would therefore have to be the result of privations and other evils suffered in war, and not of the mere destruction of life, a process which Nature is carrying on all the time. The question would then be whether privations and sufferings generally produce a male unisexual tendency. This idea seems to be conclusively negatived by the fact that the male preponderance is not shown to be a function of the wealth of the country, or the condition of the great mass of the population.

Nevertheless, in order that none of my conclusions might be based on *a priori* reasoning, and in order to answer the objection that there may be something peculiar in the effect of privations suffered in war, which differentiates them from other privations, I have examined the population statistics found in the New York census for 1865, and the United States census for 1870, enumerating the sexes of children who, from their ages, must have been born about the close of the civil war. In the case of the United States census I confined the examination to the Southern States, because there it was that the suffering and privations were the greater. The result, comprising enumeration of the sex of more than 100,000 children, showed that the male preponderance was as nearly as possible the normal one, and that not the slightest influence of the war could be detected.

It has also been maintained that the practice of polygamy has been found productive of the unisexual tendency in the female direction. The data for deciding this question are insufficient; but I find that, in the only region of the United States where such an effect would be likely to be observable, there is the usual preponderance of male births. Analysis will show that this proposition also belongs to the most improbable class. The only polygamous practices which could reasonably be supposed to affect sex are so far from rare that any unisexual tendency arising from it would be brought to light by a very slight examination. The author believes that the preceding paper contains sufficient matter to disprove the supposition, without the necessity of further inquiry.

9. SUMMARY OF CONCLUSIONS.

I do not present the following summary of conclusions as being, in all cases, so well established as not to be worthy of farther investigation. Whether well or ill established, they are those indicated by the statistics, and I earnestly hope that other investigators, more especially concerned with the subject, will take it up with more extended data and test each conclusion separately. With this proviso we may say that the following propositions are indicated by the statistics with a greater or less degree of probability.

I. The preponderance of male over female births probably varies with the race. Although remarkably uniform in all branches of the Semitic race, it seems to be either non-existent or quite small in the Negro race.

II. There are no important differences as regards capacity for producing children of one sex rather than the other which are permanent in the individual. All fathers and all mothers are equally likely to have children of either sex, except for the slight variations that may be due to age. In view of the great variety of conditions on which this conclusion is based, it seems in the highest degree unlikely that there is any way by which a parent can affect the sex of his or her offspring.

III. The most natural inference from all the statistical data is that the functions of the father in generation are entirely asexual, the sex being determined wholly by the mother. If so, it cannot be said that one father is more likely than another to have children of either sex. This conclusion requires to be tested by making a classification of the sex of third born and following children according to the age of the father.

IV. The sex is not absolutely determined at any one moment or by any one act, but is the product of a series of accidental causes, some acting in one direction and some in another, until a preponderance in one direction finally determines it. The statistics of twins and triplets seem to show very strongly that these accidents occur after conception, but throw no light upon the question of the time which they occupy.

V. The first born child of any mother is more likely to be a male in the proportion of about 8 to 7. There is probably a smaller preponderance in the case of the second child. But there is no conclusive evidence that, after a mother has had two children, there is any change in her tendencies.

VI. The observed preponderance of male births in the Semitic race is due mainly to the unisexual tendency of the mother in the case of a first child.

APPENDIX.

MATHEMATICAL THEORY OF THE EFFECT OF A UNISEXUAL TENDENCY.

The statistical theory on which the preceding research is based, being presumably susceptible of other applications than that here made, will now be developed. So far as generality is concerned, nothing will be lost by taking the special problem, considered in section IV preceding, as a basis of investigation. The data of the problem will be as follows:

1. An indefinite number of pairs of parents, each pair of which may have an indefinite number of children of either sex. The treatment of this subject will include the general case of an indefinite number of causes, each of which may, on each trial, be productive of one or the other of two different effects.

2. Taking the general average of the whole mass of couples, there is a certain normal probability, p, that a child, taken at random, will be male, and the probability $1-p$ that it will be female.

3. It may be that this probability is the same for every individual couple of the whole mass. But it may also be that, for some of the couples, the probability is greater than p. In this case it will necessarily follow that for certain other couples the probability is less than p, the latter quantity being the average for the whole mass.

4. In order not to complicate the problem too greatly, we shall suppose that each of the individual couples belongs to one of three classes; a class for which the probability of having a male child has the normal value p, another for which it is greater than p by an unknown quantity a, and a third for which it is less than p by the same quantity. We designate these classes by A, B and C; A representing couples with probability $p+a$; B, those with probability p; and C, those with the probability $p-a$. The numbers of classes A and C are necessarily equal.

Let us put:

h, the fraction of the total belonging to the two equal classes A and C;
h', the fraction of the whole mass belonging to class B.

We shall then have

$$h + h' = 1.$$

Proceeding according to the method of probabilities, we suppose a parent couple taken at random from the mass. The respective probabilities that this couple will belong to the classes A, B and C are

$$\tfrac{1}{2}h, \; h' \text{ and } \tfrac{1}{2}h.$$

The probabilities of a male child are, in these several classes:

For class A, $p+a$.
  B, p. (1)
  C, $p-a$.

Then by the principles of the theories of probabilities, if a couple be taken at random from the whole mass, the respective combined probabilities that the couple will be of one of the classes, and the child a male, will be:

In class A, $\tfrac{1}{2}h(p+a)$.
B, $h' p$.
C, $\tfrac{1}{2}h(p-a)$. (2)

of which the sum is p, as it should be.

The problem before us is to find a criterion for deciding whether the quantity a, which we may consider as the unisexual factor, and which we shall call the *coefficient of unisexuality*, is or is not of appreciable magnitude. Such a criterion is afforded by a count of males and females in families of two or more children. The theory requires that, in a family of a given number of children, we express the probable respective numbers of males and females in terms of the factor a.

The problem now assumes the following form: A parent couple, taken at random from the whole mass, has n children; what is the probability that s of these children will be males and $n-s$ females?

Using the notation

$$\begin{bmatrix} n \\ s \end{bmatrix} = \frac{n(n-1)(n-2)\ldots(n-s+1)}{1.2.3\ldots s}$$

we have the well-known theorem that, if the probability of an event on a single trial is μ, the probability of its occurring s times on n trials is

$$P_s^{(n)} = \begin{bmatrix} n \\ s \end{bmatrix} \mu^s (1-\mu)^{n-s}$$

Putting for μ the three values of the probabilities given in (1) we find that the probabilities in question are:

For class A, $\begin{bmatrix} n \\ s \end{bmatrix}(p+a)^s (1-p-a)^{n-s}$

For class B, $\begin{bmatrix} n \\ s \end{bmatrix} p^s (1-p)^{n-s}$ (3)

For class C, $\begin{bmatrix} n \\ s \end{bmatrix}(p-a)^s (1-p+a)^{n-s}$

Multiplying these expressions, as in (2), by the respective factors $\tfrac{1}{2}h$, h' and $\tfrac{1}{2}h$, putting for brevity

$$n-s=r$$
$$1-p=q$$

and taking the sum of the products, we find the probability that a family of n children taken at random from the whole mass, will comprise s males and r females to be

$$P_{r,s}^{(n)} = \left\{ \tfrac{1}{2}h((p+a)^s(q-a)^r + (p-a)^s(q+a)^r) + h'p^s q^r \right\} \begin{bmatrix} n \\ s \end{bmatrix} \quad (4)$$

This expression may now be developed in even powers of a, the coefficients of the odd powers all vanishing. In the form

$$P_{r,s}^{(n)} = (A_0 + A_2 a^2 + A_4 a^4 + ..) \begin{bmatrix} n \\ s \end{bmatrix}$$

the values of the first two coefficients are

$$A_0 = (h + h') p^s q^r = p^s q^r$$

$$A_2 = h p^{s-2} q^{r-2} \left(p^2 \begin{bmatrix} r \\ 2 \end{bmatrix} + q^2 \begin{bmatrix} s \\ 2 \end{bmatrix} - rspq \right)$$

$$= \tfrac{1}{2} h \{ n(n-1) p^2 - 2(n-1) sp + s(s-1) \} p^{s-2} q^{r-2}$$

For our present purpose these terms suffice. To investigate unisexual deviations it will also lead to no appreciable error to suppose

$$p = q = \tfrac{1}{2}$$

The value of $P_{r,s}^{(n)}$, that is, the probability that a family of n children will consist of s males and r females now becomes

$$P_{r,s}^{(n)} = \begin{bmatrix} n \\ s \end{bmatrix} \left\{ \frac{1}{2^n} + \frac{(r-s)^2 - n}{2^{n-1}} h a^2 \right\} \quad (5)$$

We may use this formula to express the probability in question for the case of a family of any number of children, distributed in any way among the two sexes. We shall now form these expressions for families of various numbers of children. In doing this families in which the sexual distributions are the reverse of each other will be combined. For example, the equal probabilities that a family of five will be wholly male and wholly female will be added into one sum; as will the probabilities of 4 of one sex and 1 of the other, whichever sex it may be.

The pair of probabilities thus combined would be rigorously equal when, and only when, there is an equal probability of male and female children. But not only is the error involved in the assumption of inequality unimportant for the present purpose but, resulting as it does in giving too small a probability for a preponderance of male and too large for the preponderance of females, it is nearly self-compensatory when we combine families of inversely distributed sexes.

The computation of the formula (5) is shown in the following table. To enable the essential numbers of this table to be understood without the necessity of going through all the mathematical formulæ, I shall state their significance and application. On the left are found certain possible values of n, the number of children in a family from 2 to 12 inclusive. Each block of numbers connected with a single value of n relates solely to families of that number of children.

In the next column are given all possible distributions between the two sexes which the family can have. Complementary families as regards sex are combined. For example, a family of three children must consist either of three children of one sex, whether male or female, and none of the other; or it comprises two of one sex, whichever it may be, and one of the other. The two lines correspond to these cases.

The three following columns contain numbers employed in computing the probabilities as found in the expression on the right. The denominators of the fractions which enter into these probabilities are written after the sign $\div$ of division, and, in each set relating to one value of n, the fractions are reduced to the least common denominator, but not to their lowest terms. This form of expression is used for convenience in tracing the law of the numbers and continuing the table. The probability is expressed as the sum of two terms: one a pure number; the other a coefficient of the factor ha^2. The purely numerical term shows what the respective probabilities of the division of sexes found in the second column will be in case of no unisexual tendency. For example, in a family of four children there will then be one chance of all four being of one sex, three chances of one being of one sex and one of the other, and three chances of an equal division, making eight chances in all. Hence, in a great mass of such families, we shall have one-eighth all of the same sex, four-eighths, or one-half, with a preponderance 3 to 1, and three-eighths with an equal division.

The next term shows how this probability is modified in case of a unisexual tendency. The symbol h expresses the fraction of the whole number of parents which have such a tendency. The tendency in one-half of this fraction of cases will be in the male, in the other in the female direction. The symbol a is the unknown amount of this tendency.

These expressions for the probability are rigorous when n is 2 or 3. But, when n has a greater value, terms in the higher powers of a really exist, the highest power being n, or $n-1$, according as n is even or odd, but, as a must always be a rather small factor, these high powers may be neglected.

MATHEMATICAL THEORY

Construction of the Numerical Formulæ.

n	r, s	$\begin{bmatrix} n \\ s \end{bmatrix}$	2^{n-1}	$(s-r)^2 - n$	Probability			
2	2, 0	1	2	2	$1 +$	2	$+$	$2ha^2$
	1, 1	2		-2	$1 +$	2	$-$	$2ha^2$
3	3, 0	1	4	6	$1 +$	4	$+$	$3ha^2$
	2, 1	3		-2	$3 +$	4	$-$	$3ha^2$
4	4, 0	1	8	12	$1 +$	8	$+$	$3ha^2$
	3, 1	4		0	$4 +$	8		
	2, 2	6		-4	$3 +$	8	$-$	$3ha^2$
5	5, 0	1	16	20	$1 +$	16	$+$	$5ha^2 + 2$
	4, 1	5		4	$5 +$	16	$+$	$5ha^2 + 2$
	3, 2	10		-4	$10 +$	16	$-$	$10ha^2 + 2$
6	6, 0	1	32	30	$1 +$	32	$+$	$15ha^2 + 8$
	5, 1	6		10	$6 +$	32	$+$	$30ha^2 + 8$
	4, 2	15		-2	$15 +$	32	$-$	$15ha^2 + 8$
	3, 3	20		-6	$10 +$	32	$-$	$30ha^2 + 8$
7	7, 0	1	64	42	$1 +$	64	$+$	$21ha^2 + 16$
	6, 1	7		18	$7 +$	64	$+$	$63ha^2 + 16$
	5, 2	21		2	$21 +$	64	$+$	$21ha^2 + 16$
	4, 3	35		-6	$35 +$	64	$-$	$105ha^2 + 16$
8	8, 0	1	128	56	$1 +$	128	$+$	$7ha^2 + 8$
	7, 1	8		28	$8 +$	128	$+$	$28ha^2 + 8$
	6, 2	28		8	$28 +$	128	$+$	$28ha^2 + 8$
	5, 3	56		-4	$56 +$	128	$-$	$28ha^2 + 8$
	4, 4	70		-8	$35 +$	128	$-$	$35ha^2 + 8$
9	9, 0	1	256	72	$1 +$	256	$+$	$9ha^2 + 16$
	8, 1	9		40	$9 +$	256	$+$	$45ha^2 + 16$
	7, 2	36		16	$36 +$	256	$+$	$72ha^2 + 16$
	6, 3	84		0	$84 +$	256		
	5, 4	126		-8	$126 +$	256	$-$	$126ha^2 + 16$
10	10, 0	1	512	90	$1 +$	512	$+$	$45ha^2 + 128$
	9, 1	10		54	$10 +$	512	$+$	$270ha^2 + 128$
	8, 2	45		26	$45 +$	512	$+$	$585ha^2 + 128$
	7, 3	120		6	$120 +$	512	$+$	$360ha^2 + 128$
	6, 4	210		-6	$210 +$	512	$-$	$630ha^2 + 128$
	5, 5	252		-10	$126 +$	512	$-$	$630ha^2 + 128$
11	11, 0	1	1024	110	$1 +$	1024	$+$	$55ha^2 + 256$
	10, 1	11		70	$11 +$	1024	$+$	$385ha^2 + 256$
	9, 2	55		38	$55 +$	1024	$+$	$1045ha^2 + 256$
	8, 3	165		14	$165 +$	1024	$+$	$1155ha^2 + 256$
	7, 4	330		-2	$330 +$	1024	$-$	$330ha^2 + 256$
	6, 5	462		-10	$462 +$	1024	$-$	$2310ha^2 + 256$
12	12, 0	1	2048	132	$1 +$	2048	$+$	$33ha^2 + 256$
	11, 1	12		88	$12 +$	2048	$+$	$264ha^2 + 256$
	10, 2	66		52	$66 +$	2048	$+$	$858ha^2 + 256$
	9, 3	220		24	$220 +$	2048	$+$	$1320ha^2 + 256$
	8, 4	495		4	$495 +$	2048	$+$	$495ha^2 + 256$
	7, 5	792		-8	$792 +$	2048	$-$	$1584ha^2 + 256$
	6, 6	924		-12	$462 +$	2048	$-$	$1386ha^2 + 256$

The method of using the numbers is, from the statistics for each value of n, to form conditional equations having h and a as unknown quantities. These unknowns are to be determined by a solution of the equations. It will be seen that h and a cannot be determined separately, but only the combination ha^2. We may therefore suppose $h = 1$ without any loss of generality so far as these equations are concerned.

We now make a practical application of this theory by determining the numerical value of the unisexual tendency, a, in the respective cases of twins and triplets, as enumerated in section 5 preceding. The statistics of twins there cited show that, of such pairs, 0.646 are unisexual and 0.354 bisexual. Equating these percentages to the expressions for the probability we find

$$\tfrac{1}{2} + 2ha^2 = 0.646$$
$$\tfrac{1}{2} - 2ha^2 = 0.354$$

Subtracting these from each other we find $4ha^2 = 0.292$, and hence, supposing $h = 1$,

$$a^2 = 0.073$$
$$a = 0.27$$

We may now consider the case of triplets in two ways. Proceeding, as in the case of twins, by equating each probability to the fraction indicating the proportional number of the families to which it relates, we have the equations:

$$\tfrac{1}{4} + 3ha^2 = 0.499 \qquad \tfrac{3}{4} - 3ha^2 = 0.501$$

Solving these we derive, after putting $h = 1$,

$$a^2 = 0.0831 \qquad a = 0.29$$

We may also proceed in another way by substituting in the expressions for the respective probabilities of unisexual and bisexual triplets the value of ha^2 derived from the case of twins. This will give, as has already been stated, the percentage 46.9 for unisexual triplets instead of 49.9, as has been found from observation. It may be added that this relation is not changed by changing the value of h; it is therefore indifferent what value we assign to h.

ON SKEW PROBABILITY CURVES

Karl Pearson

On Skew Probability Curves.

IN a memoir, entitled "Contributions to the Mathematical Theory of Evolution. II. Skew Variation in Homogeneous Material" (*Phil. Trans.* 186, A, pp. 343–414), and noticed in your columns by Mr. Francis Galton (January 31, 1895), I have dealt with four types of skew frequency curves.

Last Tuesday, Prof. Edgworth drew my attention to the fact that a portion of my results has been anticipated by Mr. E. L. De Forest in vols. vi., ix., and x. of *The Analyst*, an excellent American mathematical journal, the acquaintance of which, I am ashamed to say, I have only to-day made for the first time.

So far as Mr. De Forest's priority is concerned, it covers the special class of curve I have in my memoir termed Type III. He has fully worked out the geometry of this type, and I consider his deduction of it, if somewhat more lengthy than mine, to have the advantage of greater generality. So far as my own memoir is concerned, a knowledge of Mr. De Forest's memoir would not have led me to rewrite pp. 373-6 of mine, which deal with this type, because my discussion there is only a branch of my general treatment of a series of skew frequency curves. I should, however, have referred to Mr. De Forest's priority and the excellency of his work. In particular I should have cited the whole of his numerical table iii. x. p. 69, which gives the values of the frequency in excess and defect of the mode, and the probable errors in excess and defect, for a considerable range of values. These results are only given by algebraic or empirical formulæ in my paper. The statisticians among your readers, who may be proposing to deal with skew frequency, would find a copy of Mr. De Forest's Table III. of considerable service should they come across a curve of Type III. KARL PEARSON.

University College, London, July 24.

CRITERION
FOR
THE REJECTION
OF
DOUBTFUL OBSERVATIONS

Benjamin Peirce

THE ASTRONOMICAL JOURNAL.
No. 45.

CRITERION FOR THE REJECTION OF DOUBTFUL OBSERVATIONS.
By BENJAMIN PEIRCE, LL. D.,
PERKINS PROFESSOR OF ASTRONOMY AND MATHEMATICS IN HARVARD UNIVERSITY.

1. In almost every true series of observations, some are found, which differ so much from the others as to indicate some abnormal source of error not contemplated in the theoretical discussions, and the introduction of which into the investigations can only serve, in the present state of science, to perplex and mislead the inquirer. Geometers have, therefore, been in the habit of rejecting those observations which appeared to them liable to unusual defects, although no exact criterion has been proposed to test and authorize such a procedure, and this delicate subject has been left to the arbitrary discretion of individual computers. The object of the present investigation is to produce an exact rule for the rejection of observations, which shall be legitimately derived from the fundamental principles of the Calculus of Probabilities.

2. It is proposed to determine in a series of N observations, the limit of error, beyond which all observations involving so great an error may be rejected, provided there are as many as n such observations.

The principle upon which it is proposed to solve this problem is, *that the proposed observations should be rejected when the probability of the system of errors obtained by retaining them is less than that of the system of errors obtained by their rejection multiplied by the probability of making so many, and no more, abnormal observations.*

In determining the probability of these two systems of errors, it must be carefully observed that, because observations are rejected in the second system, the corresponding observations of the first system must be regarded, not as being limited to their actual values, but only as surpassing the limit of rejection.

3. Let e be the base of the hyperbolic system of logarithms. Let m be the number of unknown quantities.

$$P = \varphi \varDelta \cdot \varphi \varDelta' \cdot \varphi \varDelta'' \ldots \left(\frac{\psi x}{\varphi(x \varepsilon)}\right)^n = \frac{1}{\varepsilon^{n'} (2\pi)^{\frac{1}{2}n'}} e^{-\frac{\Sigma \varDelta^2 - n x^2 \varepsilon^2}{2 \varepsilon^2}} (\psi x)^n$$

$$= \frac{1}{\varepsilon^{n'} (2\pi)^{\frac{1}{2}n'}} e^{\frac{1}{2}(-N + m + n x^2)} (\psi x)^n$$

The probability of the second system of errors is

$$P_1 = x'^{n'} x^n, \varphi \varDelta_1 \cdot \varphi \varDelta_1' \ldots = \frac{x'^{n'} x^n}{\varepsilon_1^{n'} (2\pi)^{\frac{1}{2}n'}} e^{-\frac{\Sigma \varDelta_1^2}{2 \varepsilon_1^2}}$$

$$= \frac{y'^{n'} y^n}{\varepsilon_1^{n'} 2 \pi^{\frac{1}{2}n'}} e^{\frac{1}{2}(-n' + m)}.$$

Let N be the whole number of observations.
Let n be the number of observations proposed to be rejected.
Let $n' = N - n$ be the number of observations to be retained.
Let $\varDelta, \varDelta', \varDelta''$, &c. be the system of errors when no observation is rejected.
Let $\varDelta_1, \varDelta_1', \varDelta_1''$, &c. be the system of errors when n observations are rejected.
Let ε denote the mean error when no observation is rejected.
Let ε_1 denote the mean error when n observations are rejected.

Let $\varphi \varDelta = \frac{1}{\varepsilon \sqrt{2\pi}} e^{-\frac{\varDelta^2}{2\varepsilon^2}}$ denote the probability of an error $\varDelta$, in the first system, with the corresponding notation for the second system.

Let x be the ratio of the required limit of error for the rejection of n observations to the mean error ε.

Let $\psi x = \frac{2}{\sqrt{2\pi}} \int_x^\infty e^{-\frac{1}{2} x^2}$ be the probability of an error in the first system, which exceeds the required limit.

Let y be the probability, supposed to be unknown, of such an abnormal observation that it is rejected upon account of its magnitude.

Let $y' = 1 - y$ be the probability that an observation is not of the abnormal character which involves its rejection.

Let $\Sigma \cdot \varDelta^2 = (N - m) \varepsilon^2$ denote the sum of the squares of all the errors $\varDelta, \varDelta'$, &c.

Let $\Sigma \cdot \varDelta_1^2 = (n' - m) \varepsilon_1^2$ denote the sum of the squares of all the errors $\varDelta_1, \varDelta_1'$, &c.

4. The probability of the first system of errors embodying the condition that n observations exceed the limit $x \varepsilon$, is

5. To authorize the proposed rejection of n observations we must have

$$P < P_1$$

which gives at once

$$\left(\frac{\varepsilon_1}{\varepsilon}\right)^{n'} e^{\frac{1}{2}n(x^2-1)} (\psi x)^n < y'^{n'} y^n.$$

6. The value of y must be determined by the condition that P_1 is a maximum, and therefore that $y'^{n'} y^n$ is a maximum. The differential of the logarithm of this equation gives then for the maximum

$$\frac{y'}{n'} = \frac{y}{n}$$

$$y = \frac{n}{N}, \quad y' = \frac{n'}{N}$$

$$y'^{n'} y^n = \frac{n^m n^{m'}}{N^n}$$

which may be denoted by Q^N; and *the required condition of rejection becomes*, by extracting the Nth root,

$$\left(\frac{\varepsilon_1}{\varepsilon}\right)^{y'} \left(e^{\frac{1}{2}(x^2-1)} \psi x\right)^y < Q \qquad (A.)$$

in which

$$Q = y'^{y'} y^y = y^y (1-y)^{1-y}. \qquad (B.)$$

7. The relation of ε' to ε must depend upon the nature of the equations which correspond to the rejected observations. But in the form in which the problem is here presented, the attention is directed to no special case; and it is sufficient for a first approximation to suppose that the excess of $\Sigma \mathcal{A}^2$ over $\Sigma \mathcal{A}_1^2$ is only equal to the sum of the squares of the rejected observations, which gives the equation

$$(1-m') \varepsilon^2 - y x^2 \varepsilon^2 = (y' - m') \varepsilon_1^2$$

in which

$$m' = \frac{m}{N}.$$

Hence,

$$\frac{\varepsilon_1}{\varepsilon} = \left(\frac{1-m'-y x^2}{y'-m'}\right)^{\frac{1}{2}}. \qquad (C.)$$

The combination of (A), (B), and (C) for various values of m' and y' gives a set of values of x, which will answer for most cases. Table I. contains such values for different values of m' and y.

8. When the attention is directed to the rejection of special observations, the values of ε' and ε can be determined by direct investigation, for which the usual forms of computation in the method of least squares afford various facilities; and the application of the criterion involved in (A) is simplified by means of Tables II. and III. Table II. contains Brigg. log. Q for different values of y, and Table III. contains Brigg. log. R for values of x, in which

$$R = e^{\frac{1}{2}(x^2-1)} \psi x. \qquad (D.)$$

Previously to the publication of these tables, computers will easily determine ψx from BESSEL's table of $\int e^{-t^2}$, published in the *Fundamenta Astronomiæ*, or more conveniently from the tables given by ENCKE in his memoir, *Ueber die Methode der Kleinsten Quadrate*, in the Berlin Jahrbuch for 1834.

The slow rate at which R changes will be found greatly to facilitate the determination of the criterion, as will be seen in the following examples :—

Example 1. To determine the limit of rejection of a single observation in the case of thirty observations of the vertical semidiameter of *Venus*, made by Professor COFFIN and Lieutenant PAGE with the mural-circle, at Washington, in the year 1846.

In the reduction of these observations, I have assumed two unknown quantities, of which one is a constant and the other is proportional to the horizontal parallax. I have then found for the residual errors

$-0''.63$	$+0''.27$	$+0''.60$	$+0''.30$	$-0''.10$	$-0''.19$
$+0.45$	-0.43	-0.17	-0.05	$+0.16$	-0.52
$+0.34$	$+0.08$	-0.23	-0.45	-0.47	-0.51
-0.26	-0.11	$+0.83$	$+0.11$	$+0.04$	-0.49
-0.28	$+0.13$	$+0.69$	$+0.45$	-0.12	$+0.50$

and I also find $\quad$ log. $\varepsilon^2 = 9.2368.$

We have then, in this case,

$$N = 30, \quad n = 1, \quad n' = 29, \quad m = 2$$

whence

$$\frac{\varepsilon'}{\varepsilon} = \left(\frac{28-x^2}{27}\right)^{\frac{1}{2}}, \quad Q^N = \frac{29^{29}}{30^{30}}$$

$-N$ log. $Q = 1.9040.$

For a first approximation let log. $R = 9.3$ which gives log. $R - N$ log. $Q = 1.2 < \left(\frac{\varepsilon}{\varepsilon'}\right)^{29}$

$$\text{log. } \varepsilon^2 - \text{log. } \varepsilon'^2 > .08$$
$$\text{log. } (28-x^2) < 1.35$$
$$28-x^2 < 22.4$$
$$x^2 > 5.6$$
$$x > 2.4.$$

If this value of the limit is substituted in the next approximation, we have $\quad$ log. $R = 9.248$

whence $\quad$ log. $\varepsilon^2 -$ log. $\varepsilon'^2 > .0795$

$$x^2 > 5.51$$
$$x \varepsilon > 0''.97$$

which is so much greater than each of the above errors of observation, that neither of them is to be rejected.

Example 2. To determine the limit of rejection of one or two observations in the case of fifteen observations of the vertical semidiameter of *Venus*, made by Lieutenant HENNDON, with the meridian-circle at Washington, in the year 1846.

The same method of reduction used in this example as in the preceding gave for the residual errors

$-0''.30$	$+0''.48$	$+0''.63$	$-0''.22$	$+0''.18$
-0.44	-0.24	-0.13	-0.05	$+0.39$
$+1.01$	$+0.06$	-1.40	$+0.20$	$+0.10$

log. $\varepsilon^2 = 9.53050.$

For the criterion of rejecting one observation, we have

$$N = 15, \quad n = 1, \quad n' = 14, \quad m = 2$$

whence

$$\frac{\varepsilon'}{\varepsilon} = \left(\frac{13-x^2}{12}\right)^{\frac{1}{2}}, \quad Q^N = \frac{14^{14}}{15^{15}}$$

$-N$ log. $Q = 1.5955.$

For a first approximation let log. $R = 9.3$
which gives log. $R - N$ log. $Q = 0.90 < \left(\dfrac{\varepsilon}{\varepsilon'}\right)^{14}$

$$\log. \varepsilon^2 - \log. \varepsilon'^2 > 0.13$$
$$\log. (13 - x^2) < 0.95$$
$$13 - x^2 < 8.9$$
$$x^2 > 4.1$$
$$x > 2.0.$$

If this value of the limit is substituted in the next approximation, we have
$$\log. R = 9.309$$
whence
$$\log. \varepsilon^2 - \log. \varepsilon'^2 > 0.1292$$
$$x^2 > 4.083$$
$$x \varepsilon > 1''.18$$

which shows that the error $1''.40$ should be rejected.

For the criterion for rejecting two observations, we have
$$N = 15, \quad m = 2, \quad n' = 13, \quad m = 2$$

whence $\dfrac{\varepsilon}{\varepsilon'} = \left(\dfrac{13 - 2x^2}{11}\right)^{\frac{1}{2}}, \quad Q^N = \dfrac{2^2 \cdot 13^{13}}{15^{15}}$

$-\dfrac{N}{n}$ log. $Q = 1.2790$.

For a first approximation let log. $R = 9.3$
which gives log. $R - \dfrac{N}{n}$ log. $Q = 0.58 < \left(\dfrac{\varepsilon}{\varepsilon'}\right)^{6\frac{1}{2}}$

$$\log. \varepsilon^2 - \log. \varepsilon'^2 > 0.18$$
$$\log. (13 - 2x^2) < 0.86$$
$$13 - 2x^2 < 7.24$$
$$x^2 > 2.88$$
$$x > 1.7.$$

If this value of the limit is adopted as a second approximation, we have $\log. R = 9.360$
whence $\log. \varepsilon^2 - \log. \varepsilon'^2 > 0.1882$
$$x^2 > 2.930$$
$$x \varepsilon > 0''.997$$

which shows that the two observations corresponding to errors $+1''.01$ and $-1''.40$ should be rejected.

FIRST PART OF TABLE III. LOG. R.

Argument x	.00	.01	.02	.03	.04	.05	.06	.07	.08	.09
1.0	9.5015	9.4992	9.4970	9.4947	9.4924	9.4902	9.4880	9.4857	9.4835	9.4813
1	.4791	.4769	.4747	.4725	.4704	.4682	.4661	.4639	.4618	.4597
2	.4575	.4554	.4533	.4512	.4491	.4471	.4450	.4429	.4409	.4388
3	.4367	.4347	.4327	.4307	.4286	.4266	.4246	.4226	.4206	.4187
4	.4167	.4147	.4128	.4108	.4088	.4069	.4050	.4030	.4011	.3992
5	.3973	.3954	.3935	.3916	.3897	.3878	.3860	.3841	.3822	.3804
6	.3786	.3767	.3749	.3731	.3713	.3694	.3676	.3658	.3640	.3622
7	.3604	.3587	.3569	.3551	.3533	.3516	.3498	.3481	.3464	.3446
8	.3429	.3412	.3395	.3378	.3360	.3343	.3327	.3310	.3293	.3276
9	.3259	.3243	.3226	.3209	.3193	.3176	.3160	.3144	.3127	.3111
2.0	.3095	.3079	.3063	.3046	.3030	.3014	.2999	.2983	.2967	.2951
1	.2935	.2920	.2904	.2888	.2873	.2857	.2842	.2827	.2811	.2796
2	.2780	.2765	.2750	.2735	.2720	.2705	.2690	.2675	.2660	.2645
3	.2630	.2615	.2600	.2586	.2571	.2556	.2542	.2527	.2512	.2498
4	.2483	.2469	.2455	.2440	.2426	.2412	.2398	.2383	.2369	.2355
5	.2341	.2327	.2313	.2299	.2285	.2272	.2258	.2244	.2230	.2217
6	.2203	.2189	.2176	.2162	.2149	.2135	.2122	.2108	.2105	.2082
7	.2068	.2055	.2042	.2029	.2016	.2002	.1989	.1976	.1963	.1950
8	.1937	.1924	.1911	.1898	.1886	.1873	.1860	.1847	.1835	.1822
9	.1810	9.1797	9.1784	9.1772	9.1759	9.1747	9.1735	9.1722	9.1710	9.1697
3.0	9.1685									

ON PEIRCE'S CRITERION

C.S. Peirce
and
Chas. Schott

XXIII.

ON PEIRCE'S CRITERION.

By Benjamin Peirce.

My dear Sir, — I perceive that the theory of my criterion has been frequently misunderstood. I presume this to be due in a great degree to the conciseness of the argument with which it was published; and I propose to remedy this defect.

The problem which I undertook to solve was the following. There being given certain observations, of which the greater portion is to be regarded as normal and subject to the ordinary law of error adopted in the method of least squares, while a smaller unknown portion is abnormal and subject to some obscure source of error, to ascertain the most probable hypothesis as to the partition of the observations into normal and abnormal. The principle adopted in my solution of the problem is the universally recognized doctrine that the measure of the probability of an hypothesis compared with other hypotheses equally probable in other respects is the probability that the event will occur under the hypothesis, and that the most probable hypothesis is that under which the event is most probable. This is the literal expression of the mathematical analysis published in Gould's Astronomical Journal for 1852. The Criterion has been used otherwise than in the Coast Survey, and especially by my friend Dr. Gould himself. Dr. Gould's tables have greatly facilitated its use, and his sound judgment given in favor of its validity is at least as valuable as that of any living geometer. It has also been much used by that excellent authority Mr. Schott, as in a letter hereto appended.

The evidence, by which certain observations are placed in the doubtful list and subjected to scrutiny, whether they should be rejected, must be exclusively the magnitude of the errors which they involve, when these errors are computed as if they were normal observations. This would not seem to be obnoxious to the charge of inconsistency, any more than is the ordinary *Reductio ad Absurdum*, in using a method as correct in an observation where it was finally rejected. An hypothesis which leads to improbable results is itself improbable to just that extent.

It is not likely that a series of observations of any great extent was ever made, in which some were not rejected on account of the magnitude of the deviations from the arithmetic mean. The object of the Criterion is to systematize this species of rejection, and reduce it to a form of exact computation. Wherever it has been applied, I believe that it has been found to accord with the best judgment of observers and computers. This fact has been to me an agreeable surprise, which has more than balanced the unfavorable criticism, having its foundation in misconception. It has been a surprise, because the theory was altogether *a priori*, and independent of comparison with observation.

I might add concerning the charge of inconsistency, that it would seem to be almost equally applicable to the case where two sets of observations made by different observers, and the arithmetical mean of each set differing materially from the arithmetical mean of the whole, the difference is explained by personal equation. The argument for the hypothesis of personal equation must rigidly assume the same form with that by which the Criterion is established.

There might be a doubt as to the reality of such abnormal observations with their obscure sources of error. I am frank to admit that in most cases science will detect the system of the sources of error, and free the observations from them. But even such a case as the familiar one of writing down a wrong figure must be included among these sources, and is evidently insusceptible of correction; whereas, if it be sufficiently large, it will be eliminated by the Criterion. Another case, which must often have occurred in transit observations by eye and ear, depends upon the erroneous mental sub-division of the record, of which I have given the analysis in another place. I can point out many observations which were manifestly erroneous on this account, and which it would be too vast a labor to undertake to rectify. Before the judgment of the Criterion all such errors disappear, if they are large enough to be of serious injury.

That some of the observations which are not rejected may be abnormal, notwithstanding the smallness of their errors, must be admitted. This possibility was fully recognized in the geometrical development which was given in the Astronomical Journal; and I am not aware that there has been any criticism adverse to the mathematics of that article.

To Hon. C. P. PATTERSON,
    Superintendent U. S. Coast Survey.

NOVEMBER 22, 1877.

Respecting Prof. Peirce's Criterion, I venture a few remarks: —

Large errors may arise either from an *accumulation* of a number of small ones, having their origin in different sources, as is recognized in the theory of combination of errors, and as such they may be regarded as *normal* (even if quite large), provided they belong to a series of an indefinitely large number of observations. Practically, we have but a few observations (most frequently less than one hundred), and while certain sources of error may combine to the production of large ones, their *actual* appearance in a *short* series of observations must injuriously affect the most probable result (say the mean) deducible from the series. Here we need the means of separation, and the Criterion logically performs this.

Or large errors may arise from bad observations (due to inattention of observer, without being aware of it), from the presence in this particular case of an unsuspected constant error, or even from an accidental slip (necessarily not altogether outside the possibility of its being due to other causes admissible): all such large errors having no recognized place in the adopted law of the occurrence of error must be subject to rejection, for which we need the Criterion.

Supposing the probable error of a result before rejection $= \varepsilon$ and after rejection $= \varepsilon_1$, the latter value has some times been adopted, which I think is generally not admissible, the value ε would better be retained as evidence that rejections have been made. If ε_1 is retained, we may fall into the error, of again trying on the Criterion[*] based upon ε_1.

Having used the Criterion for the last twenty years in various investigations, I found it uniformly gave excellent discrimination, and do not remember a single case where it came in conflict with proper judgment based upon experience. Of course in those instances where we have antecedently a knowledge of ε, we would employ this value in preference to a value deduced from a *single* series of observations: in such cases, observations are frequently retained by the Criterion which otherwise would have been rejected. If it errs at all, it may sometimes appear to cut too deep; but our general experience is that

---

[*] I remember a rather curious case of this kind, where an observer rejected successively three times, each time deducing and starting from a new probable error, when he became alarmed for the safety of the rest of his observations and stopped further testing them.

probable errors deduced by the method of least squares generally prove too small, whenever we can bring them to the test of different methods: this, of course, is due to the presence of unknown constant errors peculiar to each method.

Yours very respectfully,
CHAS. A. SCHOTT.

CARLILE P. PATTERSON, Esq.,
  Superintendent U. S. Coast Survey.

ON THE THEORY
OF
ERRORS OF OBSERVATIONS

C.S. Peirce

APPENDIX No. 21.

ON THE THEORY OF ERRORS OF OBSERVATIONS, BY ASSISTANT C. S. PEIRCE.

The object of this paper is to give a general account of the theory of errors of observations, with the design of showing what the limitations to the applicability of the method of least squares are, and what course is to be pursued when that method fails. We shall begin with an elementary account of the general principles of the subject, in order to state them with a little more accuracy than is commonly done.

The notation employed is one which has been suggested by the study of the logic of relations. Small Roman letters will denote objects partly indeterminate. Thus m may denote a man, without saying what man. Small Italics will be used for relative terms; thus l may denote a lover. The correlates of such relative terms will be written after them on the same line; thus t m or l m may denote a tooth of a man or a lover of a man, if m denotes man, t a tooth of and l a lover of something undetermined. Then, l w will denote a lover of a woman, it being indeterminate what lover and what woman. $t\, l$ w will denote a tooth of a lover of a woman. If we wish to denote that which is a lover of all women, we must have a symbol to denote all women. As [x] is commonly used in the method of least squares to denote the sum of all the quantities x, so we may write [w] to denote all women, and then l [w] will denote something which is a lover of all women, or we may write the same thing thus, l^w. A relative term has a double indeterminacy, being indeterminate in reference to the relate and also in reference to the correlate. A lover may be this lover, that lover, or the other lover, and each of these may be lover of this, or that, or the other. Corresponding, therefore, to the [l], which denotes all lovers, we may write $\left\{ l \right\}$ to denote the lover to whomever he is a lover. Thus, $\left\{ l \right\}^w$ will denote a lover of nothing but women, or we may write the same thing thus, $'$w. We may denote "loved by" by K^l.

Corresponding to any absolute term, as man, there is a relative term, "man that is," as in the expression "a man that is rich." I shall denote a relative of this sort by the symbol for the absolute term, with an inverted comma after it, as $m_,$. Thus, if b denotes anything black, $m_,$ b will denote a man that is black.

Let V be the relative "a general name which is applicable to." Thus, Vm will denote a general term which is applicable to some man. V^m will denote a general term which is applicable to every man. $V^m_, V^w$ will be a general term which is applicable to every man and to every woman. $K V [V^m_, V^w]$ will denote that to which every general term is applicable which is applicable to every man and to every woman; in other words, this denotes either a man or a woman. I shall write this for short m $+\!\!,$ w.

Zero is defined by the general equation $x +\!\!, 0 = x$, whatever x may be. Then, zero generally denotes nothing.

Unity is defined by the general equation $x_, 1 = x$, whatever x may be, and then 1 generally will denote anything.

But 1 and 0 have sometimes to be interpreted as relative terms. Now, it can be proved by the principles of the logic of relatives that so considered $0^x = 0$, unless $x = 0$, when $0^0 = 1$; and that $1x = 1$, unless $x = 0$, when $10 = 0$. It follows that 0^x is such a logical function* of x that it signifies "the case of the non-existence of," while $1x$ is such a logical function of x that it signifies "the case of the existence of."

---

* A mathematical function of x, such as ϕx, is something whose value is obtained by mathematical processes when x is given. A logical function of x, of which we have, as common examples, letters with a subscript x, as P_x, is something whose signification is logically deducible when the signification of x is known.

Since [m] denotes all men, we may naturally write $\frac{[m]}{m}$ to denote what all men become when that factor is removed which makes [m] refer to *men* rather than to anything else; that is to say, to denote the *number* of men. We may write this for short $[\mathbf{m}]$ with heavy brackets. Then t being a relative term, ("a tooth of,") by $[\mathbf{t1}]$ will be denoted the total number of teeth in the universe. But $[\mathbf{t}]$ will be used as equivalent to $\frac{[t\,1]}{[1]}$, or the average number of teeth that anything has. But "anything" is not to be taken here in an absolute sense. We always limit our considerations entirely to a certain class. As De Morgan expresses it, we always have a limited universe. When we reckon up the number of all things to find the average number of teeth *per* thing, it would be absurd to count among things the virtues, shades of color, days or seconds of time. Anything which belongs to the limited universe under consideration is called, in the theory of probabilities, an *event*. An expression like $[\mathbf{{}^s_t}]$, where t is a relative term, is termed a relative number, average number, or probable number. If the relative term to which the average number refers is one of those relatives which are formed by adding a comma to the symbol for an absolute term, as m, then the relative number is called a *probability*. For example, $[\mathbf{m,}]$ is the average number of men that anything is, but it is usually called the probability that anything is a man.

The importance of average numbers arises from the fact that all our knowledge really consists of nothing but average numbers; for all our knowledge is derived from induction, and its analogue, hypothesis. Now, the scientific conduct of this kind of reasoning is highly complex, because all sorts of precautions have to be attended to, and it has to be accompanied by a great deal of deduction. But the general nature of induction is everywhere the same, and is completely typified in the following example. From a bag of mixed black and white beans I take out a handful, and count the number of black and the number of white beans, and I assume that the black and white are nearly in the same ratio throughout the bag. If I am in error in this conclusion, it is an error which a repetition of the same process must tend to rectify. It is, therefore, a valid inference. But it clearly teaches me nothing in reference to the color of any particular bean. Of that I am as ignorant as before. The case is not in the least altered if I find all the beans of my handful to be black, or all to be white. I can still infer only the approximate general ratio, and it is only this I express when I say the observation makes the probability of any one bean being white or being black very great; for a probability is itself only an average number. At first thought it is hard to admit this; but the difficulty will be in great measure removed if we consider how it is that the knowledge of average numbers becomes useful in particular cases. Suppose we know the relative number of black beans in a certain bag; then, if we draw a large number of beans out of it, we know that the total number of black beans we shall draw will be equal to the number of drawings multiplied by the average number of beans in the bag. Suppose we know the relative numbers of black beans in a large number of bags, containing different proportions; then, if the beans are well mixed up in each, we may only draw a single bean from each, and yet we can predict nearly the total number of black beans which would be drawn by simply taking the sum of the relative numbers. If the black beans had a value while the white ones were worthless, then the total number of black beans which would be drawn would be the important thing to know. But as knowledge derives its practical importance from its influence upon our conduct, let us suppose that at every drawing we have our choice between two bags to draw from; then the man who knew the relative number of black beans in every bag would act in every case as though the bean he would draw from the bag which contained the larger proportion of black ones were known to be black, and as though the bean he would draw from the other were certainly white. Strictly speaking, he would know nothing about the beans that would be drawn in the particular case, but he would have a knowledge which would be so far equivalent to that that it would influence his conduct in the particular case. This is the only knowledge we ever have, a knowledge of what assumption to make in the particular case in order to do the best in the long run. Whenever, then, we have to do with a *value*, the sum-total of which in the long run is the only thing which concerns us, the average amount of it is important to be known; but in all other cases the average numbers are of no consequence.

It is evident that in the example just given it would be a valuable increase of knowledge to know, for instance, what the difference in the relative number of black beans at the top and bottom of a bag was, and any limitation of the "universe" used which should separate a relative number into two different ones would be advantageous.

There are many problems in probabilities, which, being solved, give a relative number composed of two terms, one known and the other unknown. Such an indeterminate result shows that a wider "universe" must be adopted for one of the terms of the relative number.

The fundamental arithmetical formulæ relating to relative numbers are as follows:

We have seen that the relative number of things that are men, or the probability that a thing is a man, is equal to $\frac{[m_t\ 1]}{[1]}$. By "thing" here is meant any object of our limited universe, as, for example, an animal. But we may wish to consider the relative number of animals that are men when our limited universe is a wider class than animal. In this case, a denoting an animal, we write this probability $\frac{[m_t\ a]}{[a]}$. Let us suppose that, our universe being "day," we wish to know the probability that if it thunders on any day it will rain on that day. To say that if it thunders it will rain on the same day is the same as to say that every day on which it thunders is a day on which it rains. Then let t be a day on which it thunders and r a day on which it rains, and the probability in question is $\frac{[t_t\ r]}{[t]}$. In general, the probability that if one event happens another will happen is equal to the probability that both will happen, divided by the probability that the first will happen.

Let us now see how to express the probability that a certain quantity will have a certain value. It is clearly implied that the quantity is defined in some other way than by its value. It might be, for instance, the length of time a bird will be on wing. Let x be this quantity, and let n be any definite value. Then $\frac{[x_t\ n]}{[x]}$, or the number of cases in which the time a bird is flying has that value, divided by the whole number of cases of a bird's flying, is the probability that a bird will fly for that length of time. But since time is continuous, the length of time a bird may be up may have an infinite number of different values. Consequently, the probability of any one is *zero*. We, therefore, seek the probability that the time lies between n and $n + \triangle n$; and if $\triangle n$ is infinitesimal, (say d n,) then the probability is proportional to d n. We may, therefore, write this probability thus, $[\dot{n}_x]$ d n. Here n_x denotes the case of the value of x being about n.

The probability that if one quantity, x, has a value lying between m and $m + d\ m$, then another quantity, y, has a value lying between n and $n + d\ n$, will, according to what has been said, be equal to the probability that both x and y will have the supposed values, divided by the probability that x will have the supposed value, or will be $\frac{[m_{x_t}\ n_y]\ d\ m.\ d\ n}{[m_x]\ d\ m}$, or, since the d m disappeared by cancellation, $\frac{[m_{x_t}\ n_y]}{[m_x]}$ d n.

Given, the probability of A, the probability of B, and the probability that if A happens, B happens. Required, the probability that if B happens, A happens.

The probability of A is [A].

The probability of B is [B].

The probability that if A then B is $\frac{[A_t\ B]}{[A]}$.

The probability that if B then A is $\frac{[A_t B]}{[B]}$.

Then we have—

$$\frac{[A_t B]}{[B]} = \frac{[A_t B]}{[A]} \times [A] \div [B]$$

or the probability, if B happens that A happens, is equal to the probability if A happens that B happens, multiplied by the probability of A and divided by the probability of B.

We now pass to the theory of observations. An observation gives us the value of a certain quantity which is connected with an unknown quantity in such a way as to be partly dependent on the latter value, and partly on accidental circumstances, not capable of being separately taken account of.

These accidental variations are, however, in all cases subject to a statistical law, so that (observations of a certain kind forming the limited universe, X being the unknown quantity, $\varXi$ the quantity observed) the quantity,

$$\frac{[\xi \varXi_t \, x_X]}{[x_X]} d\xi$$

or the probability that if the unknown quantity X has a certain value x, then the observed quantity $\varXi$ will have a value between ξ and $\xi + d\xi$, is a certain arithmetical function of the values ξ and x. If we write ϵ for $\xi - x$, or the error of observation, then we may put φ for such a function that—

$$\frac{[\xi \varXi_t \, x_X]}{[x_X]} = \varphi(\epsilon, x)$$

The special form of the function φ is called the law of the facility of the errors. Except so far as this law is known, an observation can afford us no information whatever. The following conditions are invariably fulfilled by this function. (It must be understood that only *real* quantities are considered.)

1. It has but one value for each set of values of its variables.
2. Its value is always positive and less than unity.
3. It vanishes when ϵ is infinite.
4. Its integral relatively to ϵ from $-\infty$ to $+\infty$ is always unity.

Beyond this the form of the function is determined by the peculiarities of the kind of observations.

The probability that if the observed quantity $\varXi$ has the value ξ, then the unknown quantity X has the value x is—

$$\frac{[x_{X_t} \, \xi \varXi]}{[\xi \varXi]} dx$$

The value of this probability is for any particular kind of observations an arithmetical function of ϵ and ξ, which we may write $\psi(\epsilon, \xi) d\epsilon$.

The probability that the unknown quantity X has the value x without reference to observation is $[x_X] dx$. This is in any case a function of x, which may be written $\varPsi x \cdot dx$.

The probability that the quantity given by observation $\varXi$ has the value ξ, without reference to the value of the unknown quantity, is $[\xi \varXi] d\xi$. This an arithmetical function of ξ, which may be $\varPhi \xi \cdot d\xi$.

If $\varPhi \xi$, $\varPsi x$, and $\varphi(\epsilon, x)$ are given, then we can obtain $\psi(\epsilon, \xi)$ thus:

$$\psi(\epsilon, \xi) = \frac{\varphi(\epsilon, x)}{\varPhi \xi} \varPsi x$$

Suppose a number of independent observations to be made. Then we shall have a series of functions—

$$\varphi_1(\epsilon_1, x) \qquad \varPhi_1 \xi_1$$
$$\varphi_2(\epsilon_2, x) \qquad \varPhi_2 \xi_2$$
$$\varphi_3(\epsilon_3, x) \qquad \varPhi_3 \xi_3$$
$$\text{\&c.} \qquad \text{\&c.}$$

then the probability that if the quantities observed have the values ξ_1, ξ_2, ξ_3, &c., the unknown quantity X has the value x will be—

$$\Psi x \cdot \frac{\varphi_1(\varepsilon_1, X)}{\Phi_1 \xi_1} \cdot \frac{\varphi_2(\varepsilon_2, X)}{\Phi_2 \xi_2} \cdot \frac{\varphi_3(\varepsilon_3, X)}{\Phi_3 \xi_3} \cdot \&c.$$

or—

$$\Psi x \cdot \prod_1^n \frac{\varphi_1(\varepsilon_1, X)}{\Phi_1 \xi_1}$$

The probability $\Psi x \cdot dx$ antecedent to *all* observations will be simply dx, and, therefore, the factor Ψx may be omitted in the above expression.

It will be perceived that observation never gives us to know a *number* expressing the value of the unknown quantity, but only a *function* expressing the probability of each value. It happens, however, in a very comprehensive case, that this function assumes a form which involves but two constants, so that in this case observation may be said to give us two numbers, a value for the unknown quantity, and the probable error of that value.

Mr. Crofton's method of considering this case seems to me to make it more comprehensible than any other. Suppose that the unknown quantity X has been observed twice, the values given by observation being ξ_1 and ξ_2. Put $[\xi]$ for $\xi_1 + \xi_2$. What then is the probability that if x is the value of X, the sum of the values given by the two observations will be $[\xi]$. It is clearly—

$$\int_{-\infty}^{+\infty} \varphi_1\bigl(\varepsilon_1, x\bigr) \cdot \varphi_2\bigl([\varepsilon] - \varepsilon_1, x\bigr) \cdot d\varepsilon_1$$

Developing $\varphi_2\bigl([\varepsilon] - \varepsilon_1, x\bigr)$ by powers of ε_1, this integral becomes—

$$\varphi_2\bigl([\varepsilon], x\bigr) - \varphi_2'\bigl([\varepsilon], x\bigr) \int_{-\infty}^{+\infty} \varepsilon_1 \varphi_1\bigl(\varepsilon_1, x\bigr) \cdot d\varepsilon_1$$

$$+ \tfrac{1}{2} \varphi_2''\bigl([\varepsilon], x\bigr) \int_{-\infty}^{+\infty} \varepsilon_1^2 \varphi_1\bigl(\varepsilon_1, x\bigr) \cdot d\varepsilon_1$$

$$+ \tfrac{1}{6} \varphi_2'''\bigl([\varepsilon], x\bigr) \int_{-\infty}^{+\infty} \varepsilon_1^3 \varphi_1\bigl(\varepsilon_1, x\bigr) \cdot d\varepsilon_1 - \&c.$$

If in place of two we have n observations, the probability that the sum of all will be $[\xi]$ is—

$$\prod_1^{n-1}\left(1 - \int_{-\infty}^{+\infty} \varepsilon_i \varphi_i(\varepsilon_i, x) \cdot D_i + \tfrac{1}{2} \int_{-\infty}^{+\infty} \varepsilon_i^2 \varphi_i(\varepsilon_i, x) \cdot D^2 \varepsilon_i - \&c.\right) \varphi\, n\bigl([\varepsilon], x\bigr)$$

Of these co-efficients—

$$\int_{-\infty}^{+\infty} \varepsilon_i \varphi_i\bigl(\varepsilon_i, x\bigr)$$

is the probable or mean value of the error of the observed quantity—

$$\tfrac{1}{2} \int_{-\infty}^{+\infty} \varepsilon_i^2 \varphi_i'\bigl(\varepsilon_i, x\bigr)$$

is half the probable value of the square of this error, and so on. The probable value of the error is often less than the probable value of its square, owing to positive and negative errors balancing. But the co-efficients which involve the cube and higher powers of the error may often become insignificant. This, for example, will be the case if n is very great; for then, in comparison with the sum of all the errors, the value of any one will be very small. In fact, in this case we may neglect a quantity a certain number of times, say M, larger than what we could neglect before, and may take a unit of measurement M times larger. Then if—

$$[\epsilon], [\epsilon^2], [\epsilon^3], \&c.,$$

be the probable value of the error, the error square, the error cube, &c., on the old scale of measures, their numerical values on the new scale will be—

$$\left[\frac{\epsilon}{M}\right], \left[\frac{\epsilon^2}{M^2}\right], \left[\frac{\epsilon^3}{M^3}\right], \&c.$$

Consequently if M is sufficiently large, the higher co-efficients may be neglected. It also frequently happens that the error of each observation is due to the combined effect of a great number of independent or nearly independent causes, each one of which alone would produce but an insignificant effect. In this case, by the same reasoning the higher co-efficients will disappear.

The manner in which sensation and volition are propagated through the nerves is unknown, but it must be by some very complicated process, because it is very slow compared with the rate of propagation of sound. It is, therefore, probable that there may be variations of the rate of passage, and that the velocity through each small portion of the whole length of the nerve is to some extent independent of the velocity through the other parts. If this is so, the whole time of propagation would be subject to a variation, the probable values of whose cube and higher powers would be insignificant. However this may be, it appears to be a fact that in all carefully-made observations, the error of which is due to the inevitable inaccuracy of the action of the human nerves, the probable values of the cube, &c., of the errors are very small.

Whenever these quantities disappear it can be proved by an analytical process which need not be reproduced here that—

$$\varphi(\epsilon, x) = \frac{h}{\sqrt{\partial}} \ominus^{-h^2(\epsilon - \dot{E})^2}$$

where h and E are quantities which depend upon x. We thus reach the fundamental formula of the method of least squares.

It is not the object of this paper to explain that method itself. Only it may be remarked that in the deduction of it which is usually given, it is assumed that what we wish to obtain from observations in any case (whether the method of least squares is applicable or not) is the most probable value of the unknown quantity. This is not the case, for there is but an infinitesimal probability in favor of any one value. Suppose that the cause of error in observing the place of a star were a nearly simple oscillation of the image about its mean point. Then the most probable errors would be the extreme ones. But we should much prefer to assume as the value the probable or mean value than the most probable value, which would be removed the furthest possible from that value. The conception of the matter is this. What observation has to teach us is a *function*, $\psi(\epsilon, \xi)$, not a mere number. But in cases to which the method of least squares is applicable, this function is completely determined by two numbers, which are the value of the unknown quantity derived from the application of that method and the value of its probable error.*

It is assumed in treatises on least squares that $\Phi\xi . d\xi$, or the probability (without regard to the value of the unknown quantity) that the quantity observed will have a value between ξ and $\xi + d\xi$, is equal to $d\xi$. When this is not the case it is only necessary to weight each observation by dividing by $\Phi\xi$.

---

* The term probable error is here used in its usual but unanalogical sense, and not for the probable value of the error, which is always assumed to be zero in least squares, or else determined by some special research.

If the probability of error does not follow that law which the method of least squares supposes, that is to say, if the probability of the error x in the mean of a large number of observations is not equal to $\frac{h}{\sqrt{\pi}} e^{-h^2 x^2}$, where h is a constant independent of x, then it must at least be of this form if h be considered to be a function of x. Now, h^2 is the weight which has to be assigned to an observation in the application of the method of least squares; and therefore when this method does not apply in its unmodified form, it is only necessary to find what function of x, h must be in order to give the required law of facility of errors, and then proceed according to the method of least squares, after having weighted each observation by h^2. Let the equation which represents the facility of error be plotted, the error being taken as abcissæ, and the probability of that error as ordinate; then plot on the same diagram the curve $y = \frac{1}{\sqrt{\pi}} e^{-x^2}$. Let us suppose, then, that a certain value of x, y, is A times as great for the actual curve of errors as it is for the normal least-squares curve. Now, if h be increased in the ratio A, the ordinates will be increased in this same ratio, and the abcissæ will be diminished in the inverse ratio so that the area of the curve is preserved. But if the ordinates at any one point are to be increased in the ratio A, then the abcissæ at that point must be contracted in the inverse ratio, so as to preserve the area; so that if we had a function f x such that $D_x f x = \frac{1}{a}$, then the probability of this function of the error would follow the law $y = \frac{1}{\sqrt{\pi}} e^{-(fx)^2}$, and consequently $\frac{1}{(D_x fx)^2}$ is the weight which has to be assigned to any such observation in applying the method of least squares. It will be observed that since the weight depends on the value of the error, it will be necessary first to make an approximate solution of the problem in order to get an approximate value of the error from which to determine the weight, so that when the method of least squares is not applicable in its unmodified form, an approximate method must necessarily be resorted to.

Let us now proceed to consider the method of ascertaining the law of facility of error. In any case, if the error is compounded of an infinite number of infinitely small errors, or approximates to being so, then the law of facility of errors is of that general form which the method of least squares prescribes, and nothing remains indeterminate excepting the value of h. Observation has sufficiently shown that in transit-observations the law of error is of this sort. I copy, for example, from Chauvenet's Astronomy the following table, taken from Bessel's "Fundamenta Astronomiæ," showing the errors made by Bradley in observing Sirius and Procyon:

Between—	No. of errors by the theory.	No. of errors by experience.
0.0 and 0.1	95	94
0.1 and 0.2	89	88
0.2 and 0.3	78	78
0.3 and 0.4	64	58
0.4 and 0.5	50	51
0.5 and 0.6	36	36
0.6 and 0.7	24	26
0.7 and 0.8	15	14
0.8 and 0.9	9	10
0.9 and 1.0	5	7
Over 1.0	5	8

Fechner, in his "Elemente der Psychophysik," has, in connection with his psychophysical law, discussed the applicability of the method to cases of sensation generally such as the estimate of the relative weights of two masses, and he finds that if v be the energy of the force which produces the excitation of any nerve, then if log v be considered as the observed quantity, the errors of the observed quantity will follow the law of least squares. Strictly speaking, the law of least squares recognizes the possibility of any error positive or negative, however great, although the probability of indefinitely great errors will be indefinitely small. It occurred to me that in the case of the emersion of a star from an occultation, since it was impossible to strike the chronograph-key too early, while it might be struck indefinitely too late, the law of least squares could hardly apply, and I have made some experiments upon this point, which I will narrate in detail at the end of this paper, merely remarking in this place that the divergence from the theoretical law proved to be insignificant. On the other hand, there are many cases in which we have no reason to expect that the errors will vary according to the least-squares curve. Let us consider, for example, a chemical analysis. Here the error is generally not due to the combined action of a very great number of very small causes, but, on the contrary, there are generally two or three leading causes of error, depending upon some defect in the theory of the analysis or on some error in the manipulation, which is likely to result from a single one, such as cannot be regarded as in any way compounded of a large number of independent parts. Thus, a drop may be spilled or a portion thrown out of the crucible too small to be detected, but the whole drop is spilled at once, or the whole portion goes at once. In very exact analysis, in which such causes are almost altogether eliminated, the remaining and chief cause of error will be an error of weighing, due to a want of delicacy of the balance, and will be of the same nature as an error of computation, due to the fact that the number of decimal places used has been limited. The method of considering such errors is treated by Dr. Bremiker, in the preface to his "Tabula Logarithmorum Sex Decimalium," a work to which the attention of chemists ought to be drawn. When the law of facility of errors cannot be deduced *a priori* from the consideration of the causes to which it is due, a large number of experimental observations must be made upon a known quantity in order to find in what manner the errors vary, or the same series of observations may be used both to determine what the value of the unknown quantity is and also what the law of the variation of errors is. Thus, in the method of least squares, h is to be determined empirically, and the common way of doing it is to use the actual observations by which the unknown quantity is determined to determine also the value of h. In doing this, it should be remembered that a precise and trustworthy determination can only be obtained from a large number of observations. This procedure amounts, in fact, to adding an additional unknown quantity, a very obvious fact, and yet one which is habitually overlooked. Encke, in his "Astronomisches Jahrbuch" for 1834, has given the most complete formulæ that I have anywhere found for determining the value of h, as well as its uncertainty. These formulæ require correction on account of the circumstance just mentioned. They should be as follows:

When it is necessary to combine, by least squares, observations of different orders of precision, they are weighted proportionately to h^2. If we have two series of observations, one of which is as accurate as you please, and the other as inaccurate as you please, a better result than that which the most accurate series of measures gives can always be got by combining with it the least accurate series, provided the proper weights are given to the two series. This proposition seems paradoxical, and is not admitted by many very competent heads, but I cannot see how the conclusion can possibly be evaded. It does not depend at all upon any of the peculiar principles induced by the method of least squares, but rests on the fundamental axioms of probabilities. Indeed, it may conveniently be based directly upon the principles of logic itself. The least accurate series of measures offers certain facts, which may be used as premises, and it cannot be that if those facts are properly used they leave us in greater ignorance than we were before. Additional facts must increase our knowledge, if the proper inferences are made from them, and, therefore, an additional series of observations, if it have any weight at all, must, if its proper weight be assigned to it, improve the value of the unknown quantity. On the other hand, when it is considered that there is an uncertainty in the value of h, it may be that if the two series of observations differ greatly in accuracy, and the value of h is not determined with much precision, it may be better at once to take the result of the best series of observations than to combine the two series with the best weights that we are able to give.

Let—
  x_1 be the value from any set of observations;
  ϵ_1 the mean error of this set;
  $g_1 \epsilon_1$ the mean error of ϵ_1;
  w_1 the true weight;
  E the mean error of weight *one*; and
  x the truly weighted mean.

$$x = \frac{\Sigma_1(w_1 x_1)}{\Sigma_1 w_1}$$

The best approximation we can get to x will be subject to two kinds of error: first, those arising from errors of x_1; and, secondly, those arising from errors in our assumed values of ϵ_1, from which we derive w_1 by the formula,

$$w_1 = \frac{E^2}{\epsilon_1^2}$$

The mean error of w_i will be $w_i^2 g_i^2$

$$D_{x_2} x = \frac{w_2}{\Sigma_1 w_1}$$

$$D_{w_1} x = \frac{x_1 \Sigma_1 w_1 - \Sigma_1 w_1 x}{(\Sigma_1 w_1)^2} = \frac{x_1 - x}{\Sigma_1 w_1}$$

Then we have—

$$\epsilon^2 = \Sigma_1 \left(\frac{x_1 - x}{\Sigma_1 w_1}\right)^2 w_i^2 g_i^2 + \Sigma_1 \left(\frac{w_1}{\Sigma_1 w_1}\right)^2 \epsilon_i^2 = \frac{\Sigma_1(x_1-x)^2 w_1 g_1^2 + E^2 \Sigma_1 w_1}{(\Sigma_1 w_1)^2}$$

These are the two common rules by which ϵ may be calculated. Call their results ϵ' and ϵ'', so that—

$$\epsilon' = \frac{E}{\sqrt{\Sigma_1 w_1}}$$

$$\epsilon'' = \sqrt{\frac{\Sigma_1 w_1 (x_1-x)^2}{(m-1)\Sigma_1 w_1}}$$

where m is the number of determinations.
The first gives—

$$\epsilon^2 = \epsilon'^2 + \frac{\Sigma_1(x_1-x)^2 w_1^2 g_1^2}{(\Sigma_1 w_1)^2}$$

If m = 2,

$$(x_1 - x) = x_1 - \frac{w_1 x_1 + w_2 x_2}{w_1 + w_2} = \frac{w_2}{w_1 + w_2} x_1 - x_2$$

$$x_2 - x = -\frac{w_1}{w_1 + w_2}(x_1 - x_2)$$

$$\epsilon^2 = \epsilon'^2 + \frac{w_1^2 w_2^2}{(w_1 + w_2)^4}(g_1^2 + g_2^2)(x_1 - x_2)^2$$

Put $\frac{w_2}{w_1} = r$,

$$\epsilon^2 = \epsilon'^2 + \frac{r^2}{(1+r)^4}(g_1^2 + g_2^2)(x_1 - x_2)^2$$

$$\epsilon'' = \sqrt{\frac{w_1 w_2^2 + w_1^2 w_2}{(w_1 + w_2)^3}(x_1 - x_2)^2} = \sqrt{\frac{w_1 w_2}{(w_1 + w_2)^2}(x_1 - x_2)^2}$$

$$\epsilon''^2 = \frac{w_1 w_2}{(w_1 + w_2)^2}(x_1 - x_2)^2 = \frac{r}{(1+r)^2}(x_1 - x_2)^2$$

$$\epsilon^2 = \epsilon'^2 + \frac{r}{(1+r)^2}(g_1^2 + g_2^2)\epsilon''^2$$

Now, suppose $\varepsilon^2 < \varepsilon_1^2 < \varepsilon_2^2$; then $r < 1$; then—

$$\varepsilon_1^2 > \varepsilon'^2 + \frac{r}{(1+r)^2}(g_1^2 + g_2^2)\varepsilon''^2$$

But—

$$\varepsilon' = \varepsilon_1 \sqrt{\frac{w_1}{w_1 + w_2}} = \frac{\varepsilon_1}{\sqrt{1+r}} = \varepsilon_2 \sqrt{\frac{r}{1+r}}$$

$$\varepsilon'^2 = \varepsilon_1^2 \frac{1}{1+r} = \varepsilon_2^2 \frac{r}{1+r}$$

$$\frac{\varepsilon_1^2}{\varepsilon_2^2} = \frac{w_2}{w_1} = r$$

$$\varepsilon_1^2 = r\,\varepsilon_2^2$$

$$\varepsilon_1^2 - \varepsilon'^2 = \varepsilon_1^2 \frac{r}{1+r}$$

$$\varepsilon_1^2 \frac{r}{1+r} > \frac{r}{1+r^2}(g_1^2 + g_2^2)\varepsilon''^2$$

$$\frac{\varepsilon_1^2}{\varepsilon''^2}(1+r) > (g_1^2 + g_2^2)$$

$$\varepsilon_2^2 r = \varepsilon_1^2$$

$$r\frac{\varepsilon_2^2}{\varepsilon''^2}(1+r) > (g_1^2 + g_2^2)$$

$$r\frac{\varepsilon_1^2 + \varepsilon_2^2}{\varepsilon''^2} > g'_2 + g_2^2$$

Unless this condition is satisfied, the combination is worse than the best determination.

It is generally admitted that m_1 being the number of observations from which ε_1 has been determined,

$$g_i = \frac{1}{\sqrt{2m_i}}$$

Then the condition is—

$$2r\frac{\varepsilon_1^2 + \varepsilon_2^2}{\varepsilon''^2} > \frac{1}{m_1} + \frac{1}{m_2}$$

or we may write—

$$\frac{\varepsilon_1^2 + \varepsilon_2^2}{(x_1 - x_2^2)} > \frac{g_1^2 + g_2^2}{(1+r)^2}$$

or—

$$\frac{\varepsilon_1^2 + \varepsilon_2^3}{(x_1 - x_2)^2}(1+r)^2 > \frac{\frac{1}{m_1} + \frac{1}{m_2}}{2}$$

or we may write—

$$\frac{\frac{(x_1 - x)^2}{\varepsilon_1^2} + \frac{(x_2 - x)^2}{\varepsilon_2^2}}{\frac{1}{\varepsilon^2} + \frac{1}{\varepsilon_2^2}} < \frac{-4\,\varepsilon_1^2}{\frac{1}{m_1} + \frac{1}{m_2}}$$

Some writers upon the subject have wished to assign a smaller weight to those observations which differ largely from the mean than to those which come close to it. They have reasoned as if h, or the precision of an observation, were something which belonged to a single observation; whereas, in fact, it is a statistical quantity altogether, and belongs only to an observation as a member of a certain series. We may have a large series of observations for which h has a certain value, and those observations may perhaps be separated into two series on some principle or other for which h shall have two different values, and if this can be done there is an advantage in doing it. It is, in fact, limiting our universe. In probabilities generally h is a mean or probable quantity for a series of observations, and if we can divide our universe into two parts, getting different values of h, it will be an increase of knowledge to do so. For example, suppose that some of the observations were taken under one set of circumstances and the rest under another set of circumstances. That would afford a principle upon which the observations could be distinguished, and if the value of h for the two sets turned out different, it would be an advantage to separate them and to give them different weights. Now, the degree of discordance of observations from the most probable value of the unknown quantity may be taken as a means of estimating the relative degrees of care, &c., used in making them, and so to discriminate between them. But it would certainly be very absurd to allow no weight to the fact that we have endeavored to make them all with equal care. It must never be forgotten that h is a statistical quantity; not one which belongs to a single observation, but one which belongs to an infinite series of observations.

It is entirely in accordance with the method of least squares to reject discordant observations, and this has always been done, even by those who object to an exact criterion for determining what observations should be rejected. For example, Mr. Glaisher says that no observation should be rejected excepting obvious mistakes, thereby admitting that it is proper sometimes to reject observations, and nobody is more opposed to the rejection of observations than he. But no line of demarkation can be drawn between mistakes which are obvious and mistakes which are not obvious. In some cases it may be obvious that 53 has been written by the recorder instead of 35. In other cases it may be doubtful whether it ought to be called an obvious mistake or whether there may be some doubt hanging over it, and there is every grade of probability, from the greatest to the least; and when we examine into the facts of observation, and do not attempt to make our way through a vacuous space of pure theory, it will be found that the occasional rejection of observations is justified from every point of view; and if observations are to be rejected, exact criteria are necessary to determine upon principles of probabilities in what cases they should be rejected. The criterion of Professor Peirce is the only one which conforms rigidly to those principles, and, indeed, I am not aware that it has been attacked upon the ground of not conforming to the principles of probabilities, although it has been attacked on the ground that no such criterion should be used, and that no observation should be rejected except upon principles of guesswork, for that is what it amounts to to say that none but obvious mistakes should be rejected. Experience has shown that the errors which this criterion rejects are almost precisely those which a person of sound judgment would pronounce to be obvious mistakes, but some other criteria have been proposed, which are confessedly inexact, but which have the advantage of involving less calculation, but these are no better than the unaided judgment of an experienced person, and in some cases not so good.

Account of the experiments.

These experiments were made in order to study the distribution of errors in the observation of a phenomenon not seen coming on, as in the case of a transit, but sudden, as in the case of the emersion of a star from behind the moon. The time was noted upon a Hipp chronoscope, which is a modification of an invention of Wheatstone's. The train of clock-work moved by weight is regulated by the vibration of a little spring or reed striking against a toothed wheel a thousand times a second. There are two hands, one of which marks tenths of a second, and the other thousandths of a second. These hands are thrown into gear when the first event occurs, and out of gear when the second event occurs, so that the amount that they have moved measures the interval. The manner in which they are thrown in and out of gear is this: The axis of one of the wheels of the main train is hung, and the axis of one of the wheels of the hand-gearing passes completely through it and comes out behind, where it rests upon a spring, which spring is influenced by an electro-magnet. There are two crown-wheels, one upon the hollow axis belonging to the main train already mentioned, the other facing it at a very small distance from it, and fixed in position and upon the axis of the wheel belonging to the hand-gearing, which moves backward and forward inside, and the other axis as described. There is a little arm, which will catch in the teeth of one or other of these crown-wheels. Before the first event it is in the teeth of the fixed crown-wheel, which thus prevents the hands from turning round. When the first event occurs this arm is thrown forward into the teeth of the rapidly rotating crown-wheel, and thus the hands begin to turn round. When the second event occurs the arm is thrown into the teeth of the first crown-wheel, and so the hands are suddenly stopped.

It will be observed that although the instrument only registers to thousandths of a second, yet if an event can be repeated many times with a variation of time much smaller than that, the instrument ought, theoretically, in the mean of a large number of observations, to give a much closer result than 0.001 second; for when the first event occurs, and the arm is thrown into the moving crown-wheel, it probably will not strike exactly in any catch, but will strike on the inclined side of a tooth. If it strikes on the forward side of the tooth, the hands will be carried forward by the fraction of a thousandth of a second more as the arm glides down this side to the bottom of the catch. But if it strikes on the back side of the tooth, the hands will be carried relatively back the fraction of the thousandth part of a second as the arm glides back to the bottom of the catch. Now, if the top of the tooth is midway between the bottom of two catches, it is equally likely to be carried forward or back. The same thing occurs when the second event happens and the arm strikes upon the fixed crown-wheel. An error in the marked interval will thus result, which error may amount to -0.001 second in the extreme, or to $+0.001$ second, and any one error between these limits is as likely as any other; consequently, these errors, being entirely incidental and independent of one another, they will balance one another in the mean of a large number of observations, and thus a much higher degree of accuracy may be reached. This, however, is a matter which has no influence on the experiments which I have made, inasmuch as the interval measured by me was a variable one, and in point of fact I have never been able to make the instrument work with such nicety as to measure much closer than 0.001 second. In the descriptions of this instrument which I have seen, only two instrumental corrections have been mentioned; one is owing to the rate with which the instrument goes, and the other is with reference to the time occupied by the arm in passing from one crown-wheel to the other. To determine these two constants, a little apparatus accompanies the instrument, by which the time of the fall of a ball from different heights may be registered, and by registering the time from two different heights these two corrections, one of which is proportional to the time and the other is a constant, may be determined. The ball is held in a pair of jaws; when these jaws separate, the contact is broken, the hands begin to move, and the ball begins to fall. Care should be taken that the ball is so small that the jaws cannot be separated for any appreciable time before the ball is free to fall, but if the spring with which they open is sufficiently strong the ball may fall freely from the very first. At the bottom the ball strikes upon a platform made of wood and covered with green cloth, and it throws this platform down upon two metallic springs below it, through which contact is made again, so that the hand stops, and then the platform is held down by a catch.

As the instrument came from the makers it was found that when the ball struck upon the platform it threw one of the springs down, so that the contact was made, and then immediately interrupted before there was time for the hand to stop, so that a slight error of about 0.001 second arose in this way. This was usually corrected by putting little wooden wedges beneath these springs, so that they could not be thrown down in this way. In order further to test the instrument, I made use of a break-circuit chronometer, and measured the interval of two seconds upon the instrument for this purpose. It was necessary to employ two telegraph-repeaters. There are two ways in which this can be arranged, so as to correct a break-circuit chronometer with a Hipp chronoscope. It is sufficient to describe one of them. The arrangement is shown in the accompanying figure:

Bat. is the battery; Ch., the chronoscope; Chr., the chronometer; R., a resistance-coil; and I and II, two telegraph-repeaters. I is a common telegraph-repeater, with the non-conducting screw so far raised that when the armature once flies up the magnet cannot bring it down again. F is the magnet end of this repeater; B, the end at the second circuit. When the first circuit is broken the armature flies up and instantly breaks the second circuit. H is arranged differently from common repeaters. As long as there is a current through the first circuit and the armature is held down, there is no connection in the second circuit. When the first circuit is broken, the armature, under the influence of a very strong spring, flies up for a distance of a tenth of a millimeter, and there makes the connection in the second circuit.

This repeater can be extemporized out of a common relay. The resistance-coil should always be used in connection with the chronoscope in such a way that when the circuit is broken in the first place the current shall be so weak as just to be able to hold the hands still, while when it is made again the current shall be so strong as to make the circuit as quickly as possible.

The rate as given by the break-circuit chronometer did not agree with that found by the fall-apparatus, and indeed there was a slight discrepancy in the rate given by the latter for different heights of fall. Professor J. E. Oliver suggested to me that this discrepancy was due to a retardation of the instrument when the hands were geared in, which took place somewhat gradually, and I found that this was the case, and that the ear could detect that when the hands were geared in, during the space of three-fourths of a second, the note produced by the vibration of the reed was lowered about the sixth of a tone. The supposition that this took place uniformly sufficiently accounted for all the discrepancy, and this gave me two more instrumental constants, viz: the amount of retardation on gearing in the hands, and the time during which that retardation was brought about. With this instrument as well as with the other chronometer I made a large number of experiments upon the time occupied in answering signals of various kinds, such as the emission of points upon paper from behind a screen, the appearance of induction-sparks from a Ruhmkorff coil, flashes of light thrown upon a screen, sudden changes from one magic-lantern figure to another, &c., the general result of which was to confirm the facts already in our possession, and which are due to the researches of Hirsch, Daumbusch, and others. But there was one series of experiments which deserves particular description. I employed a young man about eighteen years of age, who had had no previous experience whatever in observations, to answer a signal consisting of a sharp sound like a rap, the answer being made upon a telegraph-operator's key nicely adjusted. Five hundred observations were made on every week-day during a month, twenty-four days' observations in all. The results are given in the accompanying table, and are also shown upon plate No. 27. On this plate the abscissæ represent the interval of time between the signal and the answer as indicated on the Hipp chronoscope, the ordinates measure the number of observations, which were subject to a large amount of error. The curve has, however, not been plotted directly from the observations, but after they have been smoothed off by the addition of adjacent numbers in the table eight times over, so as to diminish the irregularities of the curve. The smoother curve on the figures is a mean curve for every day drawn by eye so as to eliminate the irregularities entirely. It was found that after the first two or three days the curve differed very little from that derived from the theory of least squares. It will be noticed that on the first day, when the observer was entirely inexperienced, the observations scattered to such an extent that I have been obliged to draw the curve upon a different scale from that adopted for the other days. It will also be seen that the personal equation from the mean amount by which the answer came too late rapidly decreased for the first five days, until it was about one seventh of a second, and that it then gradually increased until the twelfth day, when it amounted to about 0.22 seconds. But while the personal equation was thus first diminishing and afterward increasing, the probable error or range of errors was constantly decreasing after the twelfth day. There was some variation in the personal equation, but not much, but the range of errors continually decreased as long as the observations lasted, and so remarkably that for the twenty-fourth day the probable error does not exceed one-eightieth of a second. I think that this clearly demonstrates the value of such practice in training the nerves for observation, for it can hardly be supposed that the best observer has so small a range of error as this, and I would therefore recommend that transit-observers be kept in constant training by means of some observations of an artificial event which can be repeated with rapidity, so that several hundred can be taken daily without great labor, and I do not think that it is essential that these observations should very closely imitate the transit of a star over wires, inasmuch as it is the general condition of the nerves which it is important to keep in training more than anything peculiar to this or that kind of observation.

THE UNITED STATES COAST SURVEY.

Details of the experiments.

FIRST DAY, JULY 1, 1872.

Thousandths of a second.	Number of observations.	Thousandths of a second.	Number of observations.	Thousandths of a second.	Number of observations.	Thousandths of a second.	Number of observations.	Thousandths of a second.	Number of observations.	Thousandths of a second.	Number of observations.	Thousandths of a second.	Number of observations.	Thousandths of a second.	Number of observations.	Thousandths of a second.	Number of observations.
158	1	348	0	389	1	430	0	471	1	512	1	553	0	593	1	633	2
240	1	9	1	390	3	1	2	2	4	3	2	4	1	4	1	4	1
261	1	350	0	1	2	2	0	3	4	4	2	5	1	5	0	5	1
277	1	1	0	2	3	3	3	4	2	5	3	6	2	6	0	6	3
296	1	2	0	3	4	4	2	5	1	6	2	7	2	7	1	7	1
312	.1	3	1	4	2	5	1	6	2	7	1	8	0	8	0	8	1
3	0	4	2	5	1	6	5	7	1	8	0	9	1	9	1	9	0
4	0	5	2	6	4	7	1	8	3	9	2	560	3	600	1	640	0
5	1	6	0	7	2	8	2	9	1	520	1	1	1	1	0	1	0
6	1	7	1	8	0	9	0	480	5	1	0	2	1	2	2	2	0
7	0	8	2	9	3	440	2	1	2	2	0	3	0	3	0	3	0
8	0	9	1	400	1	1	3	2	1	3	1	4	0	4	0	4	0
9	0	360	0	1	3	2	1	3	0	4	0	5	2	5	0	5	0
320	0	1	3	2	0	3	1	4	4	5	3	6	0	6	1	6	1
1	2	2	3	3	0	4	1	5	2	6	1	7	2	7	0	7	0
2	0	3	2	4	3	5	3	6	2	7	1	8	0	8	1	8	0
3	1	4	1	5	2	6	3	7	0	8	2	9	1	9	1	9	0
4	1	5	0	6	2	7	5	8	1	9	1	570	1	610	0	650	0
5	2	6	2	7	2	8	2	9	6	530	2	1	1	1	1	1	0
6	0	7	3	8	3	9	1	490	2	1	2	2	2	2	0	2	2
7	0	8	1	9	3	450	1	1	0	2	1	3	0	3	0	3	0
8	0	9	1	410	2	1	1	2	6	3	2	4	1	4	0	4	0
9	0	370	1	1	3	2	1	3	2	4	0	5	0	5	0	5	0
330	0	1	1	2	3	3	2	4	0	5	3	6	1	6	1	6	0
1	0	2	1	3	1	4	2	5	1	6	0	7	1	7	0	7	0
2	1	3	0	4	4	5	3	6	2	7	1	8	1	8	0	8	0
3	0	4	1	5	3	6	3	7	3	8	1	9	2	9	0	9	1
4	0	5	1	6	6	7	2	8	2	9	2	580	1	620	1	660	1
5	0	6	0	7	2	8	2	9	2	540	3	1	2	1	1	1	0
6	0	7	2	8	2	9	1	500	0	1	3	2	0	2	1	2	0
7	2	8	1	9	2	460	3	1	3	2	1	3	3	3	2	3	0
8	0	9	1	420	2	1	6	2	3	3	1	4	2	4	0	4	1
9	1	380	1	1	1	2	1	3	5	4	1	5	0	5	1	682	1
340	0	1	1	2	0	3	3	4	0	5	0	6	2	6	1	692	1
1	0	2	0	3	3	4	1	5	2	6	1	7	1	7	0	732	1
2	1	3	0	4	2	5	1	6	4	7	0	8	0	8	2	748	1
3	1	4	1	5	4	6	3	7	6	8	2	9	1	9	0	757	1
4	2	5	0	6	3	7	6	8	4	9	2	590	1	630	0	900	1
5	2	6	4	7	0	8	4	9	1	550	1	1	2	1	1	919	1
6	0	7	2	8	4	9	4	510	2	1	1	592	2	632	2	978	1
347	3	388	1	429	2	470	4	511	0	552	2						

Details of the experiments.—Continued.

SECOND DAY, JULY 5, 1872.

Thousandths of a second.	Number of observations.	Thousandths of a second.	Number of observations.	Thousandths of a second.	Number of observations.	Thousandths of a second.	Number of observations.	Thousandths of a second.	Number of observations.	Thousandths of a second.	Number of observations.	Thousandths of a second.	Number of observations.	Thousandths of a second.	Number of observations.
97	1	176	3	203	3	230	8	257	5	283	1	309	1	335	0	361	1
106	1	7	1	4	4	1	10	8	7	4	5	310	0	6	1	2	0
119	1	8	1	5	3	2	6	9	2	5	0	1	0	7	0	3	1
124	1	9	2	6	1	3	5	260	3	6	1	2	0	8	0	4	0
127	1	180	2	7	4	4	1	1	2	7	0	3	2	9	1	5	0
129	1	1	2	8	4	5	5	2	3	8	4	4	2	340	0	6	2
134	1	2	1	9	3	6	3	3	5	9	0	5	0	1	0	7	0
136	1	3	1	210	3	7	7	4	5	290	1	6	2	2	0	8	0
139	1	4	0	1	6	8	7	5	1	1	2	7	1	3	0	9	1
140	1	5	2	2	4	9	7	6	3	2	1	8	3	4	1	370	0
144	1	6	0	3	5	240	6	7	2	3	1	9	1	5	0	1	0
153	1	7	1	4	4	1	1	8	8	4	1	320	0	6	2	2	0
154	1	8	2	5	4	2	2	9	1	5	0	1	0	7	0	3	0
155	1	9	0	6	7	3	3	270	3	6	4	2	1	8	0	4	0
159	1	190	1	7	4	4	6	1	3	7	4	3	0	9	0	5	0
164	1	1	2	8	8	5	4	2	3	8	1	4	0	350	1	6	0
5	1	2	4	9	2	6	7	3	1	9	3	5	0	1	0	7	0
6	0	3	2	220	7	7	3	4	1	300	5	6	1	2	3	8	0
7	1	4	1	1	7	8	4	5	0	1	1	7	1	3	0	9	1
8	1	5	4	2	3	9	2	6	2	2	3	8	0	4	0	380	0
9	0	6	1	3	7	250	5	7	2	3	0	9	0	5	1	1	1
170	1	7	4	4	6	1	1	8	2	4	1	330	0	6	0	2	0
1	0	8	3	5	3	2	7	9	5	5	2	1	0	7	0	3	1
2	2	9	3	6	4	3	5	280	3	6	1	2	0	8	0	4	1
3	0	200	2	7	2	4	5	1	1	7	2	3	0	9	1	5	1
4	0	1	6	8	2	5	6	282	1	308	3	334	0	360	0	386	1
175	6	202	7	229	9	256	3										

THIRD DAY, JULY 6, 1872.

Thousandths of a second.	Number of observations.	Thousandths of a second.	Number of observations.	Thousandths of a second.	Number of observations.	Thousandths of a second.	Number of observations.	Thousandths of a second.	Number of observations.	Thousandths of a second.	Number of observations.	Thousandths of a second.	Number of observations.	Thousandths of a second.	Number of observations.
90	2	142	1	164	2	186	4	208	5	229	3	250	0	271	1	292	1
106	1	3	1	5	3	7	4	9	4	230	2	1	1	2	2	3	1
111	1	4	3	6	3	8	2	210	3	1	6	2	2	3	0	4	0
113	1	5	1	7	5	9	2	1	4	2	4	3	5	4	2	5	1
124	2	6	3	8	2	190	2	2	3	3	2	4	1	5	1	6	0
5	2	7	5	9	5	1	6	3	1	4	3	5	2	6	1	7	0
6	2	8	5	170	4	2	4	4	9	5	2	6	2	7	0	8	0
7	2	9	2	1	8	3	5	5	9	6	2	7	2	8	1	9	0
8	2	150	4	2	1	4	10	6	5	7	1	8	2	9	0	300	0
9	0	1	7	3	3	5	10	7	1	8	4	9	4	290	1	1	0
130	1	2	2	4	4	6	1	8	2	9	3	260	1	1	1	2	0
1	0	3	0	5	6	7	4	9	7	240	2	1	2	2	3	3	1
2	2	4	3	6	2	8	5	220	3	1	1	2	1	3	1	318	1
3	0	5	1	7	3	9	10	1	4	2	3	3	2	4	0	323	1
4	1	6	1	8	6	200	8	2	7	3	1	4	1	5	1	347	1
5	0	7	5	9	2	1	3	3	5	4	0	5	3	6	0	356	1
6	2	8	1	180	6	2	4	4	3	5	3	6	0	7	1	364	1
7	0	9	3	1	5	3	4	5	2	6	4	7	0	8	0	365	1
8	1	160	4	2	2	4	4	6	5	7	2	8	1	9	1	381	1
9	1	1	6	3	2	5	8	7	8	8	1	9	1	280	0	445	1
140	3	2	1	4	8	6	4	228	3	249	1	270	0	291	1	452	1
141	1	163	3	185	2	207	5										

THE UNITED STATES COAST SURVEY.

Details of the experiments—Continued.

FOURTH DAY, JULY 8, 1872.

Thousandths of a second.	Number of observations.	Thousandths of a second.	Number of observations.	Thousandths of a second.	Number of observations.	Thousandths of a second.	Number of observations.	Thousandths of a second.	Number of observations.	Thousandths of a second.	Number of observations.	Thousandths of a second.	Number of observations.	Thousandths of a second.	Number of observations.	Thousandths of a second.	Number of observations.	Thousandths of a second.	Number of observations.
65	1	148	1	169	2	190	6	211	10	231	3	251	1	271	2	291	0		
97	1	9	1	170	4	1	9	2	6	2	5	2	2	2	1	2	0		
111	2	150	0	1	7	2	9	3	9	3	3	3	2	3	2	3	0		
126	1	1	1	2	5	3	4	4	8	4	3	4	1	4	0	4	1		
131	1	2	2	3	4	4	7	5	6	5	3	5	1	5	0	5	0		
2	0	3	2	4	5	5	9	6	1	6	3	6	2	6	0	6	0		
3	0	4	1	5	3	6	5	7	6	7	2	7	0	7	1	7	2		
4	1	5	2	6	6	7	6	8	1	8	3	8	0	8	0	8	0		
5	1	6	0	7	10	8	5	9	5	9	2	9	2	9	1	9	2		
6	0	7	3	8	4	9	5	220	5	240	3	260	1	280	0	314	1		
7	0	8	0	9	4	200	10	1	5	1	1	1	2	1	0	319	1		
8	1	9	5	180	2	1	12	2	5	2	1	2	0	2	0	323	1		
9	0	160	2	1	1	2	8	3	4	3	1	3	4	3	1	327	1		
140	0	1	3	2	8	3	6	4	8	4	1	4	3	4	0	343	1		
1	1	2	2	3	13	4	10	5	3	5	0	5	0	5	0	347	1		
2	0	3	1	4	10	5	6	6	4	6	1	6	0	6	0	352	1		
3	1	4	3	5	5	6	6	7	3	7	3	7	1	7	1	355	1		
4	1	5	2	6	4	7	3	8	3	8	0	8	1	8	0	365	1		
5	0	6	3	7	3	8	6	9	5	9	1	9	0	9	0	381	1		
6	1	7	3	8	6	9	6	230	4	250	0	270	0	290	0	521	1		
147	1	168	2	189	6	210	8												

FIFTH DAY, JULY 9, 1872.

Thousandths of a second.	Number of observations.	Thousandths of a second.	Number of observations.	Thousandths of a second.	Number of observations.	Thousandths of a second.	Number of observations.	Thousandths of a second.	Number of observations.	Thousandths of a second.	Number of observations.	Thousandths of a second.	Number of observations.	Thousandths of a second.	Number of observations.	Thousandths of a second.	Number of observations.	Thousandths of a second.	Number of observations.
62	1	82	0	101	7	120	3	139	8	158	2	177	6	196	1	215	0		
3	0	3	1	2	5	1	3	140	7	9	8	8	3	7	0	6	1		
4	0	4	0	3	2	2	4	1	5	160	6	9	6	8	0	7	0		
5	2	5	0	4	3	3	5	2	8	1	9	180	5	9	0	8	1		
6	0	6	2	5	4	4	1	3	8	2	7	1	3	200	1	9	0		
7	1	7	0	6	0	5	4	4	11	3	3	2	5	1	0	220	1		
8	1	8	0	7	1	6	2	5	4	4	4	3	3	2	1	1	0		
9	0	9	0	8	3	7	4	6	8	5	3	4	5	3	2	2	0		
70	0	90	2	9	2	8	4	7	6	6	3	5	3	4	2	3	0		
1	0	1	1	110	3	9	1	8	4	7	3	6	4	5	0	4	0		
2	1	2	3	1	2	130	6	9	3	8	4	7	4	6	0	5	1		
3	1	3	3	2	2	1	6	150	5	9	2	8	3	7	1	237	1		
4	1	4	0	3	0	2	5	1	4	170	7	9	2	8	2	247	1		
5	0	5	3	4	3	3	4	2	5	1	6	190	2	9	2	249	1		
6	1	6	2	5	4	4	7	3	8	2	5	1	1	210	2	256	1		
7	2	7	2	6	6	5	7	4	4	3	8	2	4	1	2	259	1		
8	2	8	0	7	4	6	8	5	7	4	6	3	3	2	3	270	1		
9	1	9	0	8	4	7	6	6	10	5	3	4	1	3	0	285	1		
80	2	100	3	119	1	138	9	157	5	176	3	195	1	214	0	310	1		
81	0																		

Details of the experiments—Continued.

SIXTH DAY, JULY 10, 1872.

Thousandths of a second.	Number of observations.	Thousandths of a second.	Number of observations.	Thousandths of a second.	Number of observations.	Thousandths of a second.	Number of observations.	Thousandths of a second.	Number of observations.	Thousandths of a second.	Number of observations.	Thousandths of a second.	Number of observations.	Thousandths of a second.	Number of observations.
66	1	117	0	137	2	157	5	177	4	197	3	217	1	237	1	257	0
72	1	8	1	8	0	8	6	8	3	8	3	8	3	8	1	8	0
75	1	9	1	9	5	9	7	9	7	9	1	9	2	9	2	9	1
87	2	120	1	140	5	160	7	180	3	200	5	220	3	240	0	260	1
88	1	1	1	1	3	1	7	1	4	1	1	1	1	1	0	1	0
101	2	2	3	2	6	2	3	2	11	2	8	2	1	2	1	2	0
2	0	3	2	3	3	3	10	3	9	3	2	3	1	3	0	3	0
3	0	4	2	4	4	4	6	4	7	4	4	4	1	4	3	4	1
4	1	5	1	5	1	5	12	5	6	5	0	5	2	5	1	272	1
5	1	6	0	6	6	6	2	6	8	6	2	6	1	6	1	277	1
6	1	7	0	7	8	7	4	7	9	7	1	7	0	7	0	280	1
7	1	8	1	8	3	8	5	8	2	8	2	8	1	8	0	285	1
8	1	9	2	9	4	9	6	9	7	9	1	9	3	9	0	287	2
9	2	130	1	150	5	170	9	190	7	210	4	230	1	250	0	290	1
110	0	1	4	1	4	1	1	1	6	1	3	1	0	1	0	316	1
1	1	2	2	2	7	2	9	2	7	2	3	2	0	2	0	327	1
2	2	3	0	3	4	3	5	3	5	3	1	3	1	3	0	367	1
3	2	4	5	4	7	4	5	4	6	4	4	4	0	4	0	376	1
4	0	5	4	5	4	5	5	5	2	5	3	5	0	5	0	392	1
5	1	136	1	156	5	176	7	196	7	216	3	236	0	256	0	411	1
116	3																

SEVENTH DAY, JULY 15, 1872.

Thousandths of a second.	Number of observations.	Thousandths of a second.	Number of observations.	Thousandths of a second.	Number of observations.	Thousandths of a second.	Number of observations.	Thousandths of a second.	Number of observations.	Thousandths of a second.	Number of observations.	Thousandths of a second.	Number of observations.	Thousandths of a second.	Number of observations.
65	1	107	0	129	2	151	5	173	6	195	4	217	3	238	1	269	0
67	1	8	1	130	6	2	4	4	3	6	4	8	5	9	0	260	0
76	1	9	2	1	1	3	6	5	0	7	7	9	4	240	1	1	1
82	1	110	1	2	2	4	5	6	5	8	4	220	1	1	1	2	0
89	1	1	0	3	2	5	2	7	4	9	6	1	4	2	0	3	1
90	1	2	1	4	1	6	5	8	2	200	2	2	2	3	2	4	0
1	0	3	0	5	0	7	4	9	6	1	4	3	4	4	1	5	1
2	0	4	4	6	1	8	5	180	6	2	1	4	3	5	2	6	0
3	0	5	3	7	1	9	2	1	9	3	5	5	1	6	3	7	0
4	0	6	2	8	2	160	3	2	13	4	5	6	5	7	2	8	1
5	0	7	3	9	2	1	4	3	6	5	2	7	3	8	2	278	1
6	1	8	0	140	1	2	5	4	8	6	4	8	4	9	0	282	1
7	0	9	0	1	1	3	4	5	7	7	7	9	4	250	2	301	2
8	0	120	0	2	2	4	2	6	6	8	4	230	1	1	1	314	1
9	0	1	1	3	3	5	7	7	7	9	6	1	5	2	1	336	1
100	1	2	0	4	2	6	4	8	4	210	4	2	1	3	0	357	1
1	0	3	0	5	3	7	14	9	5	1	5	3	3	4	0	273	1
2	2	4	0	6	2	8	4	190	4	2	4	4	2	5	1	376	1
3	0	5	0	7	7	9	2	1	7	3	8	5	2	6	0	386	1
4	1	6	1	8	4	170	7	2	1	4	2	6	1	7	1	459	1
5	0	7	2	9	2	1	3	3	7	5	5	237	3	258	0	687	1
106	1	128	1	150	1	172	5	194	5	216	0						

THE UNITED STATES COAST SURVEY.

Details of the experiments—Continued.

EIGHTH DAY, JULY 16, 1872.

Thousandths of a second.	Number of observations.	Thousandths of a second.	Number of observations.	Thousandths of a second.	Number of observations.	Thousandths of a second.	Number of observations.	Thousandths of a second.	Number of observations.	Thousandths of a second.	Number of observations.	Thousandths of a second.	Number of observations.	Thousandths of a second.	Number of observations.	Thousandths of a second.	Number of observations.	Thousandths of a second.	Number of observations.	Thousandths of a second.	Number of observations.	Thousandths of a second.	Number of observations.	Thousandths of a second.	Number of observations.	Thousandths of a second.	Number of observations.
61	1	133	0	151	2	169	5	187	9	204	9	221	5	238	1	255	1										
85	1	4	0	2	2	170	7	8	10	5	10	2	7	9	1	6	0										
105	1	5	0	3	1	1	3	9	3	6	3	3	2	240	0	7	1										
111	1	6	0	4	4	2	3	190	11	7	8	4	1	1	2	8	0										
119	1	7	2	5	5	3	5	1	6	8	5	5	4	2	1	9	0										
120	0	8	2	6	3	4	7	2	8	9	5	6	1	3	1	260	1										
1	0	9	1	7	2	5	4	3	11	210	4	7	2	4	1	1	0										
2	1	140	0	8	4	6	4	4	8	1	3	8	3	5	0	2	1										
3	1	1	0	9	1	7	8	5	8	2	10	9	5	6	0	3	0										
4	0	2	0	160	3	8	3	6	8	3	7	230	4	7	1	4	1										
5	0	3	3	1	2	9	5	7	7	4	9	1	2	8	0	5	1										
6	1	4	1	2	2	180	6	8	8	5	3	2	3	9	0	274	1										
7	0	5	0	3	4	1	9	9	8	6	3	3	1	250	0	288	2										
8	0	6	4	4	0	2	5	200	6	7	6	4	1	1	1	312	1										
9	0	7	2	5	3	3	6	1	4	8	5	5	3	2	0	318	1										
130	0	8	2	6	4	4	10	2	13	9	2	6	0	3	2	335	1										
1	1	9	2	7	4	5	4	203	7	220	10	237	1	254	0	360	1										
132	1	150	1	168	3	186	11																				

NINTH DAY, JULY 17, 1872.

Thousandths of a second.	Number of observations.	Thousandths of a second.	Number of observations.	Thousandths of a second.	Number of observations.	Thousandths of a second.	Number of observations.	Thousandths of a second.	Number of observations.	Thousandths of a second.	Number of observations.	Thousandths of a second.	Number of observations.	Thousandths of a second.	Number of observations.
71	1	136	1	154	4	172	2	189	6	206	9	223	2	240	1	257	0
77	1	7	0	5	3	3	6	190	9	7	3	4	4	1	0	8	1
88	1	8	1	6	1	4	5	1	6	8	6	5	5	2	0	9	0
106	1	9	2	7	3	5	8	2	5	9	8	6	0	3	0	260	0
114	1	140	1	8	3	6	6	3	8	210	5	7	5	4	2	1	1
115	1	1	1	9	2	7	6	4	6	1	8	8	1	5	0	2	0
124	1	2	1	160	1	8	4	5	10	2	6	9	4	6	4	3	0
5	0	3	3	1	2	9	7	6	7	3	7	230	2	7	1	4	1
6	0	4	0	2	2	180	2	7	6	4	5	1	0	8	0	276	1
7	1	5	1	3	1	1	8	8	14	5	6	2	3	9	1	281	2
8	0	6	3	4	5	2	7	9	7	6	6	3	4	250	2	286	1
9	2	7	2	5	4	3	5	200	8	7	4	4	2	1	0	301	2
130	0	8	0	6	5	4	8	1	10	8	7	5	1	2	0	302	1
1	0	9	2	7	2	5	5	2	6	9	3	6	2	3	1	307	1
2	2	150	2	8	2	6	8	3	8	220	8	7	2	4	1	347	1
3	0	1	2	9	5	7	6	4	8	1	3	8	3	5	1	368	1
4	2	2	1	170	8	188	11	205	6	222	5	239	4	256	0	505	1
135	1	153	2	171	2												

H. Ex. 112——28

Details of the experiments—Continued.

TENTH DAY, JULY 18, 1872.

Thousandths of a second.	Number of observations.	Thousandths of a second.	Number of observations.	Thousandths of a second.	Number of observations.	Thousandths of a second.	Number of observations.	Thousandths of a second.	Number of observations.	Thousandths of a second.	Number of observations.	Thousandths of a second.	Number of observations.	Thousandths of a second.	Number of observations.
81	1	165	0	183	5	200	5	217	7	234	4	251	6	268	0	285	1
122	1	6	2	4	1	1	4	8	10	5	2	2	2	9	0	6	0
130	1	7	2	5	6	2	5	9	9	6	9	3	4	270	0	7	0
150	1	8	0	6	4	3	6	220	10	7	6	4	2	1	0	8	1
1	1	9	1	7	5	4	5	1	9	8	4	5	5	2	2	9	0
2	0	170	2	8	4	5	7	2	5	9	9	6	2	3	0	290	0
3	0	1	3	9	5	6	7	3	4	240	4	7	1	4	1	1	0
4	0	2	3	190	3	7	10	4	11	1	4	8	2	5	1	2	0
5	0	3	1	1	4	8	13	5	4	2	1	9	1	6	0	3	0
6	1	4	1	2	3	9	6	6	9	3	6	260	2	7	0	4	1
7	1	5	2	3	7	210	11	7	8	4	7	1	2	8	0	5	0
8	0	6	1	4	1	1	8	8	10	5	0	2	2	9	0	6	0
9	1	7	1	5	9	2	11	9	5	6	2	3	0	280	0	7	0
160	0	8	5	6	7	3	5	230	8	7	3	4	1	1	1	8	0
1	1	9	2	7	6	4	8	1	9	8	2	5	3	2	0	9	1
2	2	180	2	8	2	5	10	2	7	9	2	6	0	3	0	302	1
3	1	1	4	199	8	216	4	233	1	250	3	267	1	284	0	446	1
164	3	182	5														

ELEVENTH DAY, JULY 19, 1872.

Thousandths of a second.	Number of observations.	Thousandths of a second.	Number of observations.	Thousandths of a second.	Number of observations.	Thousandths of a second.	Number of observations.	Thousandths of a second.	Number of observations.	Thousandths of a second.	Number of observations.	Thousandths of a second.	Number of observations.	Thousandths of a second.	Number of observations.
68	1	158	1	178	1	198	7	218	11	238	8	258	1	278	2	298	0
114	1	9	3	9	0	9	8	9	5	9	5	9	1	9	0	9	0
135	1	160	1	180	3	200	7	220	10	240	2	260	3	280	0	300	1
141	2	1	1	1	1	1	9	1	7	1	7	1	1	1	1	1	0
2	2	2	3	2	4	2	5	2	3	2	5	2	2	2	0	2	0
3	0	3	1	3	2	3	4	3	3	3	3	3	0	3	0	3	1
4	1	4	0	4	2	4	7	4	9	4	1	4	1	4	0	4	0
5	1	5	1	5	2	5	4	5	11	5	6	5	1	5	0	5	1
6	1	6	3	6	3	6	5	6	2	6	2	6	2	6	0	6	0
7	0	7	0	7	6	7	9	7	8	7	3	7	0	7	1	7	1
8	0	8	2	8	3	8	6	8	4	8	3	8	1	8	0	8	0
9	1	9	3	9	5	9	7	9	2	9	6	9	0	9	0	9	0
150	0	170	0	190	3	210	12	230	5	250	4	270	2	290	0	310	0
1	1	1	2	1	5	1	12	1	3	1	5	1	1	1	0	1	0
2	0	2	0	2	12	2	5	2	7	2	2	2	0	2	3	2	3
3	0	3	0	3	9	3	5	3	4	3	2	3	1	3	1	360	1
4	2	4	2	4	10	4	7	4	5	4	2	4	3	4	0	379	1
5	2	5	2	5	10	5	5	5	4	5	2	5	1	5	0	400	1
6	2	6	3	6	5	6	7	6	4	6	0	6	0	6	0	437	1
157	0	177	3	197	4	217	5	237	3	257	4	277	1	297	0	491	1

THE UNITED STATES COAST SURVEY.

Details of the experiments—Continued.

TWELFTH DAY, JULY 20, 1872.

Thousandths of a second.	Number of observations.	Thousandths of a second.	Number of observations.	Thousandths of a second.	Number of observations.	Thousandths of a second.	Number of observations.	Thousandths of a second.	Number of observations.	Thousandths of a second.	Number of observations.	Thousandths of a second.	Number of observations.	Thousandths of a second.	Number of observations.	Thousandths of a second.	Number of observations.
95	1	175	1	193	2	211	1	229	6	247	3	265	5	282	0	299	0
109	1	6	0	4	2	2	2	230	8	8	1	6	3	3	2	300	1
126	1	7	0	5	0	3	6	1	4	9	2	7	5	4	3	1	0
141	1	8	0	6	5	4	5	2	3	250	2	8	1	5	0	2	0
157	1	9	1	7	0	5	5	3	9	1	7	9	4	6	1	3	0
162	1	180	0	8	4	6	7	4	6	2	3	270	4	7	0	4	1
3	0	1	1	9	1	7	5	5	6	3	3	1	2	8	0	5	1
4	1	2	1	200	3	8	8	6	5	4	3	2	1	9	2	314	1
5	0	3	0	1	2	9	6	7	4	5	5	3	1	290	0	316	1
6	0	4	0	2	2	220	12	8	4	6	6	4	3	1	0	318	1
7	0	5	0	3	1	1	8	9	6	7	5	5	1	2	1	322	1
8	0	6	0	4	1	2	11	240	11	8	1	6	1	3	0	329	1
9	0	7	1	5	2	3	4	1	4	9	3	7	1	4	0	354	1
170	1	8	0	6	4	4	4	2	7	260	2	8	0	5	0	365	1
1	0	9	0	7	5	5	6	3	6	1	4	9	1	6	1	374	1
2	2	190	0	8	3	6	5	4	5	2	3	280	0	7	1	390	1
3	1	1	3	9	4	7	5	5	5	3	1	281	3	298	1	391	1
174	0	192	0	210	2	228	13	246	6	264	1						

THIRTEENTH DAY, JULY 22, 1872.

Thousandths of a second.	Number of observations.	Thousandths of a second.	Number of observations.	Thousandths of a second.	Number of observations.	Thousandths of a second.	Number of observations.	Thousandths of a second.	Number of observations.	Thousandths of a second.	Number of observations.	Thousandths of a second.	Number of observations.	Thousandths of a second.	Number of observations.	Thousandths of a second.	Number of observations.
132	1	166	0	185	0	204	0	223	8	242	8	261	14	280	3	299	0
140	1	7	0	6	0	5	1	4	5	3	11	2	8	1	0	300	0
148	1	8	0	7	0	6	3	5	9	4	8	3	4	2	3	1	0
9	1	9	0	8	1	7	3	6	5	5	9	4	7	3	6	2	0
150	0	170	0	9	0	8	4	7	5	6	7	5	2	4	2	3	0
1	0	1	0	190	0	9	2	8	9	7	9	6	6	5	1	4	1
2	0	2	0	1	0	210	2	9	4	8	7	7	4	6	0	5	0
3	0	3	0	2	0	1	6	230	8	9	9	8	5	7	4	6	1
4	0	4	0	3	0	2	4	1	13	250	5	9	6	8	0	7	0
5	0	5	0	4	1	3	1	2	7	1	9	270	5	9	0	8	0
6	0	6	0	5	2	4	3	3	6	2	8	1	6	290	1	9	0
7	1	7	1	6	0	5	2	4	13	3	8	2	3	1	1	310	1
8	1	8	0	7	0	6	3	5	8	4	5	3	2	2	0	1	1
9	0	9	0	8	2	7	2	6	6	5	13	4	5	3	1	2	1
160	0	180	0	9	0	8	4	7	3	6	2	5	5	4	1	3	0
1	0	1	0	200	1	9	4	8	10	7	7	6	3	5	2	4	0
2	0	2	0	1	0	220	5	9	5	8	4	7	2	6	3	5	0
3	0	3	0	2	0	1	10	240	5	9	7	8	2	7	1	6	1
4	1	184	1	203	2	222	5	241	7	260	5	279	3	298	2	373	1
165	0																

Details of the experiments—Continued.

FOURTEENTH DAY, JULY 23, 1872.

Thousandths of a second.	Number of observations.	Thousandths of a second.	Number of observations.	Thousandths of a second.	Number of observations.	Thousandths of a second.	Number of observations.	Thousandths of a second.	Number of observations.	Thousandths of a second.	Number of observations.	Thousandths of a second.	Number of observations.	Thousandths of a second.	Number of observations.	Thousandths of a second.	Number of observations.
63	1	185	0	203	4	221	6	239	13	256	12	273	2	290	2	307	0
105	1	6	0	4	2	2	9	240	6	7	1	4	0	1	0	8	
119	1	7	4	5	3	3	5	1	10	8	6	5	3	2	0	9	
132	1	8	2	6	5	4	10	2	9	9	2	6	0	3	0	310	0
157	1	9	2	7	2	5	9	3	11	260	6	7	0	4	0	1	
158	1	190	0	8	1	6	6	4	8	1	6	8	2	5	0	2	
166	1	1	3	9	8	7	6	5	9	2	2	9	1	6	0	3	
174	1	2	1	210	5	8	11	6	11	3	5	280	1	7	1	4	0
5	0	3	1	1	8	9	14	7	5	4	3	1	1	8	0	5	0
6	0	4	0	2	1	230	10	8	8	5	1	2	1	9	0	6	0
7	0	5	2	3	5	1	11	9	3	6	4	3	1	300	0	7	2
8	0	6	4	4	4	2	5	250	5	7	2	4	0	1	0	8	0
9	1	7	3	5	3	3	8	1	7	8	1	5	0	2	0	9	1
160	1	8	2	6	2	4	6	2	9	9	0	6	1	3	0	359	1
1	1	9	1	7	4	5	9	3	9	270	1	7	0	4	1	470	1
2	0	200	4	8	7	6	8	4	8	1	3	8	0	5	1	693	1
3	0	1	4	9	3	7	11	255	7	272	4	289	1	306	2	705	1
184	3	202	1	220	4	238	12										

FIFTEENTH DAY, JULY 24, 1872.

Thousandths of a second.	Number of observations.	Thousandths of a second.	Number of observations.	Thousandths of a second.	Number of observations.	Thousandths of a second.	Number of observations.	Thousandths of a second.	Number of observations.	Thousandths of a second.	Number of observations.	Thousandths of a second.	Number of observations.	Thousandths of a second.	Number of observations.	Thousandths of a second.	Number of observations.
73	1	177	0	195	2	213	4	231	5	249	4	267	1	285	1	302	1
105	1	8	0	6	1	4	9	2	11	250	7	8	7	6	0	3	0
110	1	9	1	7	1	5	8	3	8	1	4	9	1	7	1	4	0
112	1	180	0	8	3	6	4	4	8	2	6	270	4	8	2	5	0
140	1	1	0	9	1	7	8	5	9	3	3	1	3	9	1	6	0
148	1	2	0	200	0	8	7	6	15	4	4	2	4	290	1	7	0
158	1	3	0	1	1	9	9	7	7	5	7	3	1	1	0	8	0
166	1	4	1	2	4	220	5	8	12	6	8	4	1	2	1	9	2
7	0	5	0	3	1	1	5	9	10	7	6	5	2	3	1	310	1
8	1	6	1	4	2	2	10	240	15	8	3	6	1	4	0	1	1
9	1	7	3	5	2	3	12	1	9	9	1	7	1	5	2	2	0
170	0	8	0	6	1	4	7	2	8	260	2	8	1	6	1	3	0
1	0	9	0	7	0	5	6	3	14	1	3	9	3	7	0	4	0
2	0	190	0	8	5	6	7	4	4	2	8	280	1	8	0	5	1
3	0	1	0	9	4	7	8	5	4	3	3	1	0	9	0	406	1
4	2	2	2	210	6	8	11	6	7	4	1	2	1	300	0	443	1
5	1	3	4	1	5	9	8	7	7	5	1	3	0	301	0	467	2
176	0	194	1	212	5	230	12	248	8	266	2	284	1				

THE UNITED STATES COAST SURVEY.   221

Details of the experiments—Continued.

SIXTEENTH DAY, JULY 25, 1872.

Thousandths of a second.	Number of observations.	Thousandths of a second.	Number of observations.	Thousandths of a second.	Number of observations.	Thousandths of a second.	Number of observations.	Thousandths of a second.	Number of observations.	Thousandths of a second.	Number of observations.	Thousandths of a second.	Number of observations.	Thousandths of a second.	Number of observations.	Thousandths of a second.	Number of observations.
67	1	170	2	186	0	202	2	218	6	234	8	250	6	266	4	282	2
86	1	1	0	7	3	3	4	9	8	5	10	1	10	7	5	3	0
103	1	2	0	8	3	4	1	220	8	6	8	2	6	8	6	4	2
114	1	3	0	9	0	5	10	1	5	7	5	3	3	9	3	5	0
157	1	4	0	190	1	6	3	2	3	8	7	4	5	270	4	6	0
8	0	5	0	1	0	7	6	3	3	9	7	5	4	1	4	7	1
9	0	6	3	2	2	8	4	4	6	240	6	6	10	2	1	8	0
160	0	7	0	3	4	9	3	5	10	1	11	7	7	3	3	9	0
1	1	8	1	4	1	210	1	6	7	2	11	8	4	4	2	290	0
2	1	9	1	5	0	1	7	7	8	3	12	9	3	5	0	1	1
3	0	180	2	6	2	2	5	8	9	4	9	260	0	6	0	303	1
4	0	1	1	7	1	3	3	9	12	5	7	1	2	7	0	312	1
5	0	2	2	8	3	4	3	230	15	6	6	2	3	8	1	313	1
6	0	3	2	9	4	5	2	1	13	7	7	3	2	9	1	333	1
7	0	4	1	200	5	6	4	2	6	8	7	4	0	280	4	357	1
8	0	185	3	201	6	217	5	233	8	249	8	265	2	281	2	784	1
169	0																

SEVENTEENTH DAY, JULY 26, 1872.

Thousandths of a second.	Number of observations.	Thousandths of a second.	Number of observations.	Thousandths of a second.	Number of observations.	Thousandths of a second.	Number of observations.	Thousandths of a second.	Number of observations.	Thousandths of a second.	Number of observations.	Thousandths of a second.	Number of observations.	Thousandths of a second.	Number of observations.	Thousandths of a second.	Number of observations.
76	1	195	0	216	1	237	5	258	9	279	10	300	4	321	1	342	1
104	1	6	2	7	0	8	2	9	7	280	4	1	0	2	0	3	0
128	1	7	0	8	1	9	5	260	11	1	7	2	1	3	1	4	1
147	1	8	0	9	2	240	7	1	9	2	6	3	1	4	1	5	3
162	1	9	0	220	3	1	2	2	8	3	7	4	1	5	1	6	0
166	1	200	1	1	3	2	4	3	5	4	7	5	1	6	1	7	1
180	1	1	1	2	0	3	5	4	7	5	3	6	2	7	0	8	2
1	0	2	0	3	4	4	4	5	8	6	6	7	0	8	0	9	0
2	0	3	1	4	3	5	9	6	5	7	7	8	2	9	0	350	0
3	0	4	0	5	2	6	2	7	3	8	3	9	5	330	0	1	0
4	0	5	0	6	0	7	5	8	7	9	5	310	6	1	0	2	0
5	0	6	1	7	2	8	4	9	2	290	5	1	1	2	1	3	1
6	3	7	2	8	1	9	8	270	6	1	6	2	4	3	0	366	1
7	0	8	5	9	3	250	5	1	5	2	7	3	0	4	0	372	1
8	0	9	4	230	6	1	7	2	6	3	5	4	1	5	0	380	1
9	0	210	2	1	2	2	4	3	7	4	5	5	4	6	0	390	1
190	1	1	2	2	4	3	5	4	5	5	3	6	1	7	1	392	1
1	0	2	2	3	3	4	7	5	6	6	5	7	2	8	1	394	1
2	0	3	2	4	3	5	7	6	11	7	1	8	0	9	1	448	1
3	1	4	2	5	4	6	5	7	2	8	3	9	0	340	0	467	1
194	0	215	2	236	3	257	7	278	5	299	3	320	2	341	1		

Details of the experiments—Continued.

EIGHTEENTH DAY, JULY 27, 1872.

Thousandths of a second.	Number of observations.	Thousandths of a second.	Number of observations.	Thousandths of a second.	Number of observations.	Thousandths of a second.	Number of observations.	Thousandths of a second.	Number of observations.	Thousandths of a second.	Number of observations.	Thousandths of a second.	Number of observations.	Thousandths of a second.	Number of observations.	Thousandths of a second.	Number of observations.
184	2	201	3	218	0	235	4	252	16	269	5	286	6	302	2	318	0
5	0	2	1	9	3	6	10	3	6	270	11	7	4	3	1	9	0
6	0	3	2	220	2	7	5	4	10	1	10	8	1	4	0	320	0
7	0	4	2	1	3	8	2	5	11	2	6	9	1	5	1	1	0
8	1	5	0	2	0	9	5	6	8	3	6	290	1	6	1	2	1
9	0	6	6	3	4	240	6	7	7	4	5	1	3	7	0	3	0
190	0	7	1	4	6	1	5	8	7	5	7	2	1	8	0	4	0
1	0	8	1	5	2	2	3	9	8	6	5	3	1	9	0	5	0
2	1	9	3	6	0	3	8	260	13	7	5	4	2	310	0	6	0
3	0	210	0	7	5	4	6	1	9	8	3	5	0	1	0	7	0
4	0	1	0	8	3	5	5	2	5	9	6	6	0	2	0	8	0
5	0	2	3	9	8	6	7	3	16	280	3	7	0	3	0	9	0
6	0	3	3	230	7	7	14	4	11	1	1	8	0	4	0	330	1
7	1	4	1	1	2	8	8	5	4	2	4	9	2	5	1	341	1
8	0	5	3	2	10	9	11	6	6	3	5	300	0	6	0	366	1
9	1	6	4	3	3	250	12	7	8	4	5	301	0	317	0	367	1
200	0	217	4	234	6	251	3	268	5	285	4						

NINETEENTH DAY, JULY 29, 1872.

Thousandths of a second.	Number of observations.	Thousandths of a second.	Number of observations.	Thousandths of a second.	Number of observations.	Thousandths of a second.	Number of observations.	Thousandths of a second.	Number of observations.	Thousandths of a second.	Number of observations.	Thousandths of a second.	Number of observations.	Thousandths of a second.	Number of observations.	Thousandths of a second.	Number of observations.
119	1	200	0	217	0	234	5	251	11	267	8	283	2	299	1	315	0
148	1	1	0	8	1	5	2	2	10	8	6	4	1	300	1	6	0
152	1	2	0	9	2	6	7	3	13	9	11	5	3	1	2	7	0
157	1	3	1	220	3	7	7	4	15	270	6	6	5	2	1	8	1
187	1	4	0	1	3	8	9	5	11	1	10	7	4	3	0	9	0
8	0	5	1	2	3	9	4	6	12	2	12	8	4	4	0	320	0
9	0	6	0	3	0	240	11	7	7	3	9	9	5	5	0	1	1
190	1	7	2	4	1	1	8	8	11	4	10	290	1	6	0	2	0
1	1	8	0	5	0	2	7	9	6	5	4	1	0	7	0	3	0
2	0	9	0	6	2	3	8	260	11	6	5	2	1	8	2	4	1
3	0	210	2	7	7	4	7	1	8	7	5	3	2	9	1	334	1
4	0	1	0	8	0	5	10	2	11	8	4	4	1	310	1	359	1
5	0	2	1	9	3	6	11	3	4	9	2	5	1	1	1	306	1
3	0	3	0	230	3	7	8	4	4	280	7	6	1	2	0	591	1
7	0	4	1	1	4	8	11	5	4	1	5	7	0	3	0	779	1
8	1	5	1	2	3	9	13	266	8	282	4	298	1	314	0	831	1
199	1	216	1	233	6	250	7										

THE UNITED STATES COAST SURVEY.

Details of the experiments—Continued.

TWENTIETH DAY, JULY 30, 1871.

Thousandths of a second.	Number of observations.	Thousandths of a second.	Number of observations.	Thousandths of a second.	Number of observations.	Thousandths of a second.	Number of observations.	Thousandths of a second.	Number of observations.	Thousandths of a second.	Number of observations.	Thousandths of a second.	Number of observations.
130	1	195	0	214	2	233	10	252	12	271	1	290	1	309	1	328	0
159	1	6	0	5	3	4	4	3	9	2	8	1	1	310	0	9	1
177	0	7	0	6	1	5	8	4	4	3	7	2	3	1	0	330	1
8	0	8	1	7	5	6	6	5	7	4	5	3	1	2	1	1	0
9	0	9	2	8	2	7	9	6	9	5	3	4	0	3	0	2	0
180	0	200	0	9	2	8	5	7	8	6	8	5	1	4	0	3	0
1	0	1	0	220	9	9	6	8	8	7	6	6	2	5	0	4	0
2	0	2	1	1	6	240	13	9	6	8	2	7	3	6	0	5	0
3	1	3	2	2	2	1	12	260	8	9	4	8	0	7	0	6	0
4	0	4	2	3	2	2	9	1	9	280	5	9	1	8	0	7	1
5	0	5	0	4	4	3	4	2	8	1	2	300	0	9	0	8	0
6	1	6	1	5	3	4	6	3	14	2	7	1	2	320	1	9	0
7	1	7	0	6	4	5	7	4	7	3	2	2	0	1	0	340	0
8	0	8	2	7	4	6	8	5	11	4	0	3	1	2	0	1	0
9	0	9	0	8	3	7	7	6	6	5	1	4	0	3	0	2	1
190	0	210	2	9	2	8	9	7	8	6	1	5	0	4	1	360	0
1	0	1	2	230	4	9	3	8	7	7	3	6	1	5	0	486	1
3	0	2	1	1	4	250	12	9	6	8	2	7	0	6	0	754	1
2	0	213	0	232	4	251	9	270	12	289	3	308	1	327	1	817	1
194	0																

TWENTY-FIRST DAY, JULY 31, 1872.

Thousandths of a second.	Number of observations.	Thousandths of a second.	Number of observations.	Thousandths of a second.	Number of observations.	Thousandths of a second.	Number of observations.	Thousandths of a second.	Number of observations.	Thousandths of a second.	Number of observations.	Thousandths of a second.	Number of observations.
75	1	193	0	209	0	225	7	241	10	257	8	273	1	288	0	303	0
137	1	4	0	210	1	6	7	2	7	8	7	4	4	9	3	4	0
160	1	5	0	1	3	7	7	3	10	9	10	5	2	290	1	5	0
161	2	6	1	2	1	8	6	4	8	260	5	6	3	1	2	6	0
181	1	7	1	3	3	9	9	5	3	1	11	7	1	2	2	7	0
2	0	8	2	4	0	230	8	6	10	2	6	8	3	3	2	8	0
3	1	9	0	5	3	1	10	7	11	3	8	9	4	4	0	9	0
4	0	200	1	6	5	2	6	8	11	4	6	280	4	5	0	310	0
5	1	1	3	7	5	3	10	9	9	5	7	1	1	6	0	1	0
6	0	2	2	8	2	4	4	250	7	6	3	2	2	7	1	2	1
7	0	3	0	9	2	5	8	1	8	7	7	3	1	8	1	3	1
8	0	4	2	220	9	6	8	2	8	8	5	4	0	9	1	331	1
9	2	5	2	1	4	7	11	3	9	9	3	5	4	300	0	348	1
190	0	6	3	2	6	8	9	4	7	270	4	6	1	1	0	349	1
1	2	7	0	3	5	9	10	5	10	1	1	287	5	302	1	438	1
192	1	208	2	224	3	240	10	256	10	272	3						

TWENTY-SECOND DAY, AUGUST 1, 1872.

Thousandths of a second.	Number of observations.	Thousandths of a second.	Number of observations.	Thousandths of a second.	Number of observations.	Thousandths of a second.	Number of observations.	Thousandths of a second.	Number of observations.	Thousandths of a second.	Number of observations.	Thousandths of a second.	Number of observations.
139	1	204	0	219	2	233	2	247	7	261	14	275	12	289	6	303	1
157	1	5	2	220	0	4	7	8	10	2	8	6	3	290	0	4	0
171	1	6	1	1	1	5	6	9	3	3	12	7	4	1	1	5	0
192	1	7	2	2	6	6	6	250	8	4	7	8	8	2	1	6	0
3	1	8	0	3	1	7	6	1	13	5	8	9	4	3	0	7	1
4	1	9	1	4	4	8	12	2	8	6	17	280	2	4	2	8	0
5	0	210	1	5	4	9	10	3	14	7	8	1	4	5	0	9	1
6	1	1	2	6	4	240	4	4	8	8	6	2	7	6	2	325	1
7	1	2	2	7	4	1	7	5	6	9	5	3	3	7	1	339	1
8	0	3	2	8	1	2	10	6	8	270	5	4	1	8	2	342	1
9	0	4	1	9	6	3	7	7	6	1	9	5	4	9	1	382	1
200	0	5	1	230	5	4	3	8	8	2	7	6	5	300	1	406	1
1	0	6	3	1	10	5	6	9	10	3	9	7	3	1	0	410	1
2	0	7	0	232	6	216	10	260	9	274	2	288	1	302	0	689	1
203	0	218	0														

REPORT OF THE SUPERINTENDENT OF

Details of the experiments—Continued.

TWENTY-THIRD DAY, AUGUST 2, 1872.

Thousandths of a second.	Number of observations.	Thousandths of a second.	Number of observations.	Thousandths of a second.	Number of observations.	Thousandths of a second.	Number of observations.	Thousandths of a second.	Number of observations.	Thousandths of a second.	Number of observations.	Thousandths of a second.	Number of observations.	Thousandths of a second.	Number of observations.	Thousandths of a second.	Number of observations.
142	1	184	0	202	1	220	1	238	5	256	10	274	7	292	0	310	1
166	1	5	1	3	0	1	6	9	9	7	10	5	3	3	4	1	0
7	2	6	0	4	1	2	5	240	8	8	11	6	1	4	1	2	0
8	0	7	0	5	0	3	6	1	8	9	9	7	3	5	2	3	0
9	0	8	0	6	1	4	5	2	7	260	8	8	4	6	0	4	1
170	0	9	1	7	1	5	6	3	12	1	9	9	7	7	0	5	0
1	0	190	0	8	0	6	4	4	9	2	13	280	3	8	2	6	1
2	2	1	0	9	0	7	3	5	7	3	6	1	2	9	0	7	0
3	0	2	0	210	1	8	4	6	7	4	11	2	3	300	0	8	0
4	0	3	1	1	0	9	5	7	6	5	7	3	4	1	0	9	0
5	1	4	0	2	3	230	4	8	6	6	6	4	1	2	1	320	0
6	0	5	0	3	3	1	9	9	12	7	8	5	0	3	0	1	0
7	0	6	3	4	2	2	8	250	9	8	8	6	3	4	0	2	1
8	0	7	0	5	4	3	6	1	8	9	6	7	1	5	0	3	1
9	1	8	0	6	1	4	6	2	10	270	14	8	1	6	1	340	1
180	1	9	0	7	0	5	6	3	12	1	3	9	1	7	0	371	1
1	0	200	0	8	5	6	4	4	9	2	7	290	2	8	0	406	1
2	0	201	0	219	2	237	2	255	5	273	4	291	1	309	1	661	1
183	1																

TWENTY-FOURTH DAY, AUGUST 3, 1872.

Thousandths of a second.	Number of observations.	Thousandths of a second.	Number of observations.	Thousandths of a second.	Number of observations.	Thousandths of a second.	Number of observations.	Thousandths of a second.	Number of observations.	Thousandths of a second.	Number of observations.	Thousandths of a second.	Number of observations.	Thousandths of a second.	Number of observations.	Thousandths of a second.	Number of observations.
119	1	198	2	213	3	228	9	243	12	257	12	271	3	285	1	299	0
134	1	9	1	4	4	9	4	4	10	8	5	2	1	6	4	300	0
156	1	200	2	5	1	230	7	5	16	9	8	3	4	7	0	1	0
172	1	1	1	6	1	1	11	6	9	260	6	4	1	8	1	2	1
175	1	2	0	7	4	2	12	7	7	1	9	5	5	9	0	3	0
188	2	3	0	8	7	3	18	8	11	2	2	6	0	290	2	4	0
9	0	4	1	9	5	4	5	9	13	3	6	7	2	1	0	5	0
190	1	5	1	220	3	5	7	250	13	4	4	8	0	2	4	6	0
1	0	6	1	1	6	6	11	1	14	5	1	9	0	3	0	7	2
2	1	7	1	2	3	7	10	2	14	6	1	280	3	4	3	8	0
3	0	8	2	3	1	8	8	3	12	7	2	1	2	5	0	9	1
4	0	9	6	4	4	9	8	4	12	8	2	2	1	6	0	334	1
5	0	210	3	5	8	240	6	5	6	9	3	3	1	7	1	340	1
6	0	1	3	6	6	1	9	256	8	270	3	284	0	298	0	374	1
197	0	212	3	227	5	242	8										

No. 27. COAST SURVEY REPORT FOR 1870.

DIAGRAMS ILLUSTRATING APPENDIX No. 21.
ON THE THEORY OF ERRORS OF OBSERVATION
BY C. S. PEIRCE.

I July 1st
II July 5th
III July 6th
IV July 8th
V July 9th
VI July 10th
VII July 15th
VIII July 16th
IX July 17th
X July 18th
XI July 19th
XII July 20th
XIII July 22nd
XIV July 23rd
XV July 24th
XVI July 25th
XVII July 26th
XVIII July 27th
XIX July 29th
XX July 30th
XXI July 31st
XXII Aug 1st
XXIII Aug 2nd
XXIV Aug 3rd

NOTE ON THE THEORY
OF
THE ECONONY OF RESEARCH

C.S. Peirce

APPENDIX No. 14.

NOTE ON THE THEORY OF THE ECONOMY OF RESEARCH, BY ASSISTANT C. S. PEIRCE.

When a research is of a quantitative nature, the progress of it is marked by the diminution of the probable error. The results of non-quantitative researches also have an inexactitude or indeterminacy which is analogous to the probable error of quantitative determinations. To this inexactitude, although it be not numerically expressed, the term "probable error" may be conveniently extended.

The doctrine of economy, in general, treats of the relations between utility and cost. That branch of it which relates to research considers the relations between the utility and the cost of diminishing the probable error of our knowledge. Its main problem is, how, with a given expenditure of money, time, and energy, to obtain the most valuable addition to our knowledge.

Let r denote the probable error of any result, and write $s = \dfrac{1}{r}$. Let $Ur \cdot dr$ denote the infinitesimal utility of any infinitesimal diminution, dr, of r. Let $Vs \cdot ds$ denote the infinitesimal cost of any infinitesimal increase, ds, of s. The letters U and V are here used as functional symbols. Let subscript letters be attached to r, s, U, and V, to distinguish the different problems into which investigations are made. Then, the total cost of any series of researches will be

$$\Sigma_i \int V_i \, s_i \cdot ds_i;$$

and their total utility will be

$$\Sigma_i \int U_i \, r_i \cdot dr_i.$$

The problem will be to make the second expression a maximum by varying the inferior limits of its integrations, on the condition that the first expression remains of constant value.

The functions U and V will be different for different researches. Let us consider their general and usual properties. And, first, as to the relation between the exactitude of knowledge and its utility. The utility of knowledge consists in its capability of being combined with other knowledge so as to enable us to calculate how we should act. If the knowledge is uncertain, we are obliged to do more than is really necessary, in order to cover this uncertainty. And, thus, the utility of any increase of knowledge is measured by the amount of wasted effort it saves us, multiplied by the specific cost of that species of effort. Now, we know, from the theory of errors, that the uncertainty in the calculated amount of effort necessary to be put forth may be represented by an expression of the form

$$c\sqrt{a + r^2},$$

where a and c are constants. And, therefore, the differential coefficient of this, multiplied by the specific cost of the effort in question, say $\dfrac{h}{c}$, gives

$$Ur = h \frac{r}{\sqrt{a + r^2}}.$$

When a is very small compared with r this becomes nearly constant, and in the reverse case it is nearly proportional to r. An analogous proposition must hold for non-quantitative research.

Let us next consider the relation between the exactitude of a result and the cost of attaining it. When we increase our exactitude by multiplying observations, the different observations being independent of one another as to their cost, we know from the theory of errors that $\int V s \cdot ds$ is proportional to s^2, and that consequently Vs is proportional to s. If the costs of the different observations are not independent (which usually happens), the cost will not increase so fast relatively to the accuracy; but if the errors of the observations are not independent (which also usually happens), the cost will increase faster relatively to the accuracy; and these two perturbing influences may be supposed, in the long run, to balance one another. We may, therefore, take $Vs = ks$, where k represents the specific cost of the investigation.

We thus see that when an investigation is commenced, after the initial expenses are once paid, at little cost we improve our knowledge, and improvement then is especially valuable; but as the investigation goes on, additions to our knowledge cost more and more, and, at the same time, are of less and less worth. Thus, when chemistry sprang into being, Dr. Wollaston, with a few test tubes and phials on a tea-tray, was able to make new discoveries of the greatest moment. In our day, a thousand chemists, with the most elaborate appliances, are not able to reach results which are comparable in interest with those early ones. All the sciences exhibit the same phenomenon, and so does the course of life. At first we learn very easily, and the interest of experience is very great; but it becomes harder and harder, and less and less worth while, until we are glad to sleep in death.

Let us now apply the expressions obtained for Ur and Vs to the economic problem of research. The question is, having certain means at our disposal, to which of two studies they should be applied. The general answer is that we should study that problem for which the economic urgency, or the ratio of the utility to the cost

$$\frac{Ur \cdot dr}{Vs \cdot ds} = r^2 \frac{Ur}{Vs} = \frac{h}{k} \frac{r^4}{\sqrt{a + r^5}}$$

is a maximum. When the investigation has been carried to a certain point this fraction will be reduced to the same value which it has for another research, and the two must then be carried on together, until finally, we shall be carrying on, at once, researches into a great number of questions, with such relative energies as to keep the urgency-fraction of equal values for all of them. When new and promising problems arise they should receive our attention to the exclusion of the old ones, until their urgency becomes no greater than that of others. It will be remarked that our ignorance of a question is a consideration which has between three and four times the economic importance of either the specific value of the solution or the specific cost of the investigation in deciding upon its urgency.

In order to solve an economical problem, we may use as variables

$$x = \int Vs \cdot ds,$$

or the total cost of an inquiry, and

$$y = \frac{Ur \cdot dr}{Vs \cdot ds},$$

or the economic urgency. Then, C being the total amount we have to spend in certain researches, our equations will be

$$C = x_1 + x_2 + x_3 + \text{etc.}$$
$$y_1 = y_2 = y_3 = \text{etc.}$$

Then, expressing each y in terms of x, we shall have as many equations as unknown quantities.

When we have to choose between two researches only, the solution may be represented graphically, as follows:

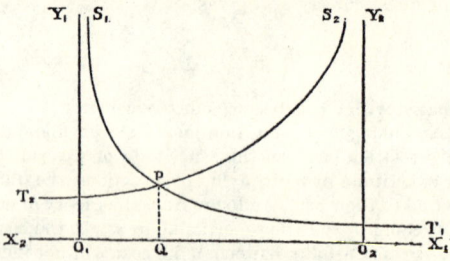

From any point O_1 taken as an origin, draw the axis of abscissas $O_1\ X_1$, along which x_1, the total cost of the first investigation, is to be measured. Draw also the axis of ordinates $O_1\ Y_1$, along which y_1, the economic urgency of the first investigation, is to be measured. Draw the curve $S_1\ T_1$ to represent the relations of x_1 and y_1. Take, on the axis $O_1\ X_1$, a point O_2 such that $O_1\ O_2$ shall measure the total cost of the two investigations. Let x_2, the total cost of the second investigation, be measured on the same axis as x_1, but in the opposite direction. From O_2 draw the axis of ordinates $O_2\ Y_2$ parallel to $O_1\ Y_1$, and measure y_2, the economic urgency of the second investigation, along this axis. Draw the curve $S_2\ T_2$ to represent the relations of x_2 and y_2. Then, the two curves $S_1\ T_1$ and $S_2\ T_2$ will generally cut one another at one point, and only one, between the axes $O_1\ Y_1$ and $O_2\ Y_2$. From this point, say P, draw the ordinate P Q, and the abscissas O_1 Q and O_2 Q will measure the amounts which ought to be expended on the two inquiries.

According to the usual values of U and V, we shall have

$$y = \frac{1}{4} \frac{h\,k}{x\sqrt{a\,x^2 + \frac{1}{2}k\,x}}.$$

In this case, when there are two inquiries, the equation to determine x_1 will be a biquadratic. Two of its roots will be imaginary, one will give a negative value of either x_1 or x_2, and the fourth, which is the significant one, will give positive values of both.

Let us now consider the economic relations of different researches to one another. 1st, as alternative methods of reaching the same result, and 2d, as contributing different premises to the same argument.

Suppose we have two different methods of determining the same quantity. Each of these methods is supposed to have an accidental probable error and a constant probable error, so that the probable errors, as derived from n observations in the two ways, are:

$$r_1 = \sqrt{R_1^2 + \frac{\rho_1^2}{n}} \quad \text{and} \quad r_2 = \sqrt{R_2^2 + \frac{\rho_2^2}{n}}.$$

The probable error of their weighted mean is

$$\frac{1}{\sqrt{\frac{1}{r_1^2} + \frac{1}{r_2^2}}},$$

if their constant probable errors are known. The sole utility of any observation of either is to reduce the error of the weighted mean; hence,

$$U\,r_1 = D_{r_1}(r_1^{-2} + r_2^{-2})^{-\frac{1}{2}} = (r_1^{-2} + r_2^{-2})^{-\frac{3}{2}} r_1^{-3}.$$

And as the cost is proportional to the number of observations

$$V\,s_1 = k_1 \frac{1}{D_{n_1} s_1} = \frac{k_1}{D_{n_1}(R_1^2 + \rho_1^2 n_1^{-1})^{-\frac{1}{2}}} = \frac{2\,k_1\,r_1^3\,n_1^2}{\rho_1^2}.$$

Hence, the urgency is (omitting a factor common to the values for the two methods)

$$r_1^2 \frac{U\,r_1}{V\,s_1} = \frac{1}{k_1 \rho_1^2 \left(1 + n_1 \dfrac{R_1^2}{\rho_1^2}\right)^2}$$

And, as the urgency of the two methods ought to be the same at the conclusion of the work, we should have

$$\sqrt{k_1} \cdot \rho_1 \left(1 + n_1 \frac{R_1^2}{\rho_1^2}\right) = \sqrt{k_2} \cdot \rho_2 \left(1 + n_2 \frac{R_2^2}{\rho_2^2}\right),$$

which equation serves to determine the relative values of n_1 and n. We again perceive that the cost is the smallest consideration. The method which has the smallest accidental probable error is the one which is to be oftenest used in case only a small number of observations are made; but if a large number are taken the method with the larger accidental probable error is to be oftenest used, unless it has so much greater a probable constant error as to countervail this consideration. If one of the two methods has only $\frac{1}{p}$ th the accidental probable error of the other, but costs p^2 times as much, the rule should be to make the total cost of the two methods inversely proportional to the squares of their constant errors.

Let us now consider the case in which two quantities x_1 and x_2 are observed, the knowledge of which serves only to determine a certain function of them, y. In this case the probable error of y is and we shall have
$$\sqrt{D_{x_1}y \cdot r_1^2 + D_{x_2}y \cdot r_2^2},$$
$$Ur_1 = 2r_1 \frac{dy}{dx_1}$$

Vs_1 will have the same value as before; but neglecting now the constant error, we may write
$$Vs_1 = 2k_1 \rho_1 n_1^{\frac{1}{2}}.$$

Then the urgency (with omission of the common factor) is
$$\frac{\rho_1^2}{k_1 n_1^2} \cdot \frac{dy}{dx_1},$$
and, as the two urgencies must be equal, we have
$$\frac{n_1}{n_2} = \frac{\rho_1}{\rho_2} \sqrt{\frac{k_2}{k_1}} \sqrt{\frac{\frac{dy}{dx_1}}{\frac{dy}{dx_2}}}.$$

The following is an example of the practical application of the theory of economy in research: Given a certain amount of time, which is to be expended in swinging a reversible pendulum, how much should be devoted to experiments with the heavy end up, and how much to those with the heavy end down?

Let T_d be the period of oscillation with heavy end down, T_u the same with heavy end up. Let h_d and h_u be the distances of the center of mass from the points of support of the pendulum in the two positions. Then the object of the experiments is to ascertain a quantity proportional to
$$h_d T_d - h_u T_u.$$

Accordingly, if dT_d and dT_u are the probable errors of T_d and T_u, that of the quantity sought will be
$$\sqrt{h_d^2 (dT_d)^2 + h_u^2 (dT_u^2)}.$$

We will suppose that it has been ascertained, by experiment, that the whole duration of the swinging being C, and the excess of the duration of the swinging with heavy end down over that with heavy end up being x, the probable errors of the results are
$$dT_d = \sqrt{a + \left(b + \frac{c}{h_d^2}\right) \frac{1}{C+x}}$$
$$dT_u = \sqrt{a + \left(b + \frac{c}{h_u^2}\right) \frac{1}{C-x}}$$
where a, b, and c are constants. Then, the square of the probable error of the quantity sought will be
$$a(h_d^2 + h_u^2) + (b\,h_d^2 + c) \frac{1}{C+x} + (b\,h_u^2 + c) \frac{1}{C-x}.$$

The differential coefficient of this relatively to x is
$$-(b\,h_d^2 + c) \frac{1}{(C+x)^2} + (b\,h_u^2 + c) \frac{1}{(C-x)^2}.$$

Putting this equal to zero and solving, we find for the only significant root,

$$\frac{x}{C} = \frac{b(h_d^2 + h_u^2) + 2c}{b(h_d^2 - h_u^2)} - \sqrt{\left(\frac{b(h_d^2 + h_u^2) + 2c}{b(h_d^2 - h_u^2)}\right)^2 - 1}$$

when b vanishes, x reduces to zero, and the pendulum should be swung equally long in the two positions. When c vanishes, as it would if the pendulum experiment were made absolutely free from certain disturbing influences, we have

$$\frac{x}{C} = \frac{h_d - h_u}{h_d + h_u},$$

so that the duration of an experiment ought to be proportional to the distance of the center of mass from the point of support. This would be effected by beginning and ending the experiments in the two positions with the same amplitudes of oscillation.

It is to be remarked that the theory here given rests on the supposition that the object of the investigation is the ascertainment of truth. When an investigation is made for the purpose of attaining personal distinction, the economics of the problem are entirely different. But that seems to be well enough understood by those engaged in that sort of investigation.

S. Ex. 37——26

THE NUMERICAL MEASURE
OF
THE SUCCESS OF PREDICTIONS

C.S. Peirce

The numerical measure of the success of predictions.

Suppose we have a method by which questions of a certain kind, presenting two alternatives, can in every case be answered, though not always rightly. Suppose, further, that a large number of such answers have been tabulated in comparison with the events, so that we have given the following four numbers: —

(aa), the number of questions for which the answers were the first way and the events the first way;
(ab), the number of questions for which the answers were the first way and the events the second way;
(ba), the number of questions for which the answers were the second way and the events the first way;
(bb), the number of questions for which the answers were the second way and the events the second way.

Then the problem is, from these data to assign a numerical measure to the success or science of the method by which the answers have been produced. Mr. G. K. Gilbert (*Amer. meteorological journal*, September, 1884) has recently proposed a formula for this purpose; and I desire to offer another.

I make use of two principles. The first is, that any two methods are to be regarded as equal approximations to complete knowledge, which, in the long-run, would give the same values for (aa), (ab), (ba), and (bb). The second principle is, that if the answers had been obtained by selecting a determinate proportion of the questions by chance, to be answered by an infallible witness, while the rest were answered by an utterly ignorant person at random (using *yes* and *no* with determinate relative frequencies), then the approximation to knowledge in the answers so obtained would be measured by the fraction expressing the proportion of questions put to the infallible witness. The second witness may know *how often* he ought to answer 'yes;' but I give him no credit for that, because he is ignorant *when* he ought to answer 'yes.'

Let i be the proportion of questions put to the infallible witness, and let j be the proportion of questions which the ignorant witness answers in the first way. Then we have the following simple equations: —

$$(aa) = i\{(aa) + (ba)\} + (1-i)j\{(aa) + (ba)\},$$
$$(ab) = (1-i)j\{(ab) + (bb)\},$$
$$(ba) = (1-i)(1-j)\{(aa) + (ba)\},$$
$$(bb) = i\{(ab) + (bb)\} + (1-i)(1-j)\{(ab) + (bb)\}.$$

Now, whatever the method of predicting, these equations can always be satisfied by possible values of i and j, unless the answers are worse than if they had been taken at random. Consequently, in virtue of the two principles just enunciated, the value of i obtained by solving these equations is the measure of the science of the method. This value is,

$$i = \frac{(aa)}{(aa)+(ba)} - \frac{(ab)}{(ab)+(bb)},$$
$$= \frac{(aa)}{(aa)+(ba)} + \frac{(bb)}{(ab)+(bb)} - 1,$$
$$= \frac{(aa)(bb) - (ab)(ba)}{\{(aa)+(ba)\}\{(ab)+(bb)\}}.$$

Mr. Gilbert's formula has the same numerator, but a different denominator. It is, in the present notation,

$$i = \frac{(aa)(bb) - (ab)(ba)}{\{(aa)+(ab)+(ba)+(bb)\}^2 - (aa)^2 + (ab)^2 + (ba)^2 - (bb)^2}.$$

For Sergeant Finley's tornado-predictions, $(aa) = 28$, $(ab) = 72$, $(ba) = 23$, $(bb) = 2,680$. From these data, Mr. Gilbert finds $i = 0.216$, while my formula gives $i = 0.523$.

If the questions should present more than two alternatives, it would be necessary to assign relative values or measures to the different kinds of mistakes that might be made. I have a solution for this case.

Another problem is to measure the utility of the method of prediction. For this purpose, let p be the profit, or saving, from predicting a tornado, and let l be the loss from every unfulfilled prediction of a tornado (outlay in preparing for it, etc.); then the average profit per prediction would be,

$$\frac{p \cdot (aa) - l(ab)}{(aa) + (ab) + (ba) + (bb)}.$$

C. S. PEIRCE.

ON SMALL DIFFERENCES OF SENSATION

C.S. Peirce
and
J. Jastrow

ON SMALL DIFFERENCES OF SENSATION.

READ OCTOBER 17, 1884.

By C. S. PEIRCE and J. JASTROW.

The physiological psychologists assume that two nerve excitations alike in quality will only produce distinguishable sensations provided they differ in intensity by an amount greater than a fixed ratio. The least perceptible difference of the excitations divided by half their sum is what they call the *Unterschiedsschwelle*. Fechner* gives an experiment to prove the fact assumed, namely: He finds that two very dim lights placed nearly in line with the edge of an opaque body show but one shadow of the edge. It will be found, however, that this phenomenon is not a clearly marked one, unless the lights are nearly in range. If the experiment is performed with lateral shifting of one of the lights, and with a knowledge of the effects of a telescope upon the appearance of terrestrial objects at night, it will be found very far from conclusive.

The conception of the psychologists is certainly a difficult one to seize. According to their own doctrine, in which the observed facts seem fully to bear them out, the intensity of the sensation increases continuously with the excitation, so that the least increase of the latter must produce a corresponding increase of the former. And, indeed, the hypothesis that a continuous increase of the excitation would be accompanied by successive discrete increments of the sensation, gratuitous as it would be, would not be sufficient to account for a constant *Unterschiedsschwelle*. We are therefore forced to conclude that if there be such a phenomenon it has its origin, not in the faculty of sensation, but in that of comparing sensations. In short, if the phenomenon were established, we should be forced to say that there was a least perceptible difference of sensation— a difference which, though existing in sensation, could not be brought into consciousness by any effort of attention. But the errors of our judgments in comparing our sensations seem sufficiently accounted for by the slow and doubtless complicated process by which the impression is conveyed from the periphery to the brain; for this must be liable to more or less accidental derangement at every step of its progress. Accordingly we find that the frequencies of errors of different magnitudes follow the probability curve, which is the law of an effect brought about by the sum of an infinite number of infinitesimal causes. This theory, however, does not admit of an *Unterschiedsschwelle*. On the contrary, it leads to the method of least squares, according to which the multiplication of observations will indefinitely reduce the error of their mean, so that if of two excitations one were ever so little the more intense, in the long run it would be judged to be the more intense the majority of times. It is true that the astronomers themselves have not usually supposed that this would be the case, because (apart from constant errors, which have no relevancy to the present question) they have supposed this extreme result to be contrary to common sense. But it has seemed to us that the most satisfactory course would be to subject the question to the test of direct experiment. If there be a least perceptible difference, then when two excitations differing by less than this are presented to us, and we are asked to judge which is the greater, we ought to answer wrong as often as right in the long run. Whereas, if the theory of least squares is correct, we not

---

* Elemente der Psychophysik, I, p. 242.

only ought to answer right oftener than wrong, but we ought to do so in a predictible ratio of cases.*

We have experimented with the pressure sense, observing the proportion of errors among judgments as to which is the greater of two pressures, when it is known that the two are two stated pressures, and the question presented for the decision of the observer is, which is which? From the probability, thus ascertained, of committing an error of a given magnitude, the probable error of a judgment can be calculated according to the mathematical theory of errors. If, now, we find that when the ratio of the two pressures is smaller than a certain ratio, the erroneous judgments number one-half of the whole, while the mathematical theory requires them to be sensibly fewer, then this theory is plainly disproved, and the maximum ratio at which this phenomenon is observed the so-called *Unterschiedsschwelle*. If, on the other hand, the values obtained for the probable error are the same for errors varying from three times to one-fourth of the probable error (the smallest for which it is easy to collect sufficient observations), then the theory of the method of least squares is shown to hold good within those limits, the presumption will be that it extends still further, and it is possible that it holds for the smallest differences of excitation. But, further, if this law is shown to hold good for difference so slight that the observer is not conscious of being able to discriminate between them at all, all reason for believing in an *Unterschiedsschwelle* is destroyed. The mathematical theory has the advantage of yielding conceptions of greater definiteness than that of the physiologists, and will thus tend to improve methods of observation. Moreover, it affords a ready method for determining the sensibility or fineness of perception and allows of a comparison with the results of others; for, knowing the number of errors in a certain number of experiments, and accepting the conclusions of this paper, the calculated ratio to the total excitation of that variation of excitation, in judging which we should err one time out of four, measures the sensibility. Incidentally our experiments will afford additional information upon the value of the normal average sensibility for the pressure sense, which they seem to make a finer sense than it has hitherto been believed to be. But in this regard two things have to be noted: (1) Our value relates to the probable error or the value for the point at which an error is committed half the time; (2) in our experiments there were two opportunities for judging, for the initial weight was either first increased and then diminished, or *vice versa*, the subject having to say which of these two double changes was made. It would seem at first blush that the value thus obtained ought to be multiplied by $\sqrt{2}$ (1.414) to get the error of a single judgment. Yet this would hardly be correct, because the judgment, in point of fact, depended almost exclusively on the sensation of increase of pressure, the decrease being felt very much less. The ratio $\sqrt{2}$ (1.414) would therefore be too great, and 1.2 would perhaps be about correct. The advantage of having two changes in one experiment consists in this: If only one change were employed, then some of the experiments would have an increase of excitation only and the others a decrease only; and since the former would yield a far greater amount of sensation than the latter, the nature of the results would be greatly complicated; but when each experiment embraces a

---

* The rule for finding this ratio is as follows: Divide the logarithm of the ratio of excitations by the probable error and multiply the quotient by 0.477. Call this product t. Enter it in the table of the integral θt, given in most works on probabilities; θt is the proportion of cases in which the error will be less than the difference between the given excitations. In all these cases, of course, we shall answer correctly, and also by chance in one-half of the remaining cases. The proportion of erroneous answers is therefore (1−θt)÷2. In the following table the first column gives the quotient of the logarithm of the ratio of excitation, divided by the probable error, and the second column shows the proportion of erroneous judgments:

0.0	0.50
0.05	0.49
0.1	0.47
0.25	0.43
0.5	0.37
1.0	0.25

To guess the correct card out of a pack of fifty-two once in eleven times it would be necessary to have a sensation amounting to 0.37 of the probable error. This would be a sensation of which we should probably never become aware, as will appear below.

double change this difference in the amount of sensation caused by an increase and decrease of pressure affects every experiment alike, and the liability to error is constant.*

Throughout our observations we noted the degree of confidence with which the observer gave his judgment upon a scale of four degrees, as follows:

- 0 denoted absence of any preference for one answer over its opposite, so that it seemed nonsensical to answer at all.
- 1 denoted a distinct leaning to one alternative.
- 2 denoted some little confidence of being right.
- 3 denoted as strong a confidence as one would have about such sensations.

We do not mean to say that when zero was the recorded confidence, there was absolutely no sensation of preference for the answer given. We only mean that there was no sensation that the observer noticed when attending to his feelings of this sort as closely as he conveniently could, namely, closely enough to mark them on this scale. The scale of confidence fluctuated considerably. Thus, when Mr. Jastrow passed from experiments upon differences of weight of 60, 30, and 15 on the thousand to differences of 20, 10, and 5 on the thousand, although the accuracy of his judgments was decidedly improved, his confidence fell off very greatly, owing to his no longer having the sensation produced by a difference of 60 present to his memory. The estimations of confidence were also rough, and might be improved in future work. The average marks seem to conform to the formula—

$$m = c \log \frac{p}{1-p}$$

where m denotes the degree of confidence on the scale, p denotes the probability of the answer being right, and c is a constant which may be called the index of confidence.

To show that this formula approximates to the truth, we compare it with the average marks assigned to estimates of differences for which more than a hundred experiments were made. Mr. Jastrow's experiments are separated into groups, which will be explained below.

First group.

Ratio of pressures.	Peirce, observer. $c=1.25$. Mean confidence.		Jastrow, observer. $c=1.5$. Mean confidence.		Jastrow, observer. $c=0.0$. Mean confidence.	
	Observed.	Calculated.	Observed.	Calculated.	Observed.	Calculated.
1.015	0.14	0.10	0.30	0.2	0.34	0.27
1.030	0.30	0.35	0.40	0.42	0.55	0.56
1.060	0.70	0.70	0.85	0.87	1.02	1.12

Ratio of pressures.	Jastrow, observer. $c=0.25$. Mean confidence.		$c=0.4$. Mean confidence.	
	Observed.	Calculated.	Observed.	Calculated.
1.005	0.00	0.03	0.00	0.06
1.010	0.07	0.06	0.05	0.12
1.020	0.12	0.12	0.50	0.39

* The number of errors, when an increase of weight was followed by a decrease, was slightly less than when the first change was a decrease of pressure.

The judgments enunciated with any given degree of confidence were more likely to be right with greater differences than with smaller differences. To show this, we give the frequency of the different marks in Mr. Jastrow's second, third, and fourth groups.*

The apparatus used was an adaptation of a "Fairbanks" post-office scale; upon the end of the beam of which was fixed a square enlargement (about one-half inch square), with a flat top, which served to convey the pressure to the finger in a manner to be presently described. This was tightly covered with an India-rubber cap, to prevent sensations of cold, &c., from contact with the metal. A kilogram placed in the pan of the balance brought a pressure of one-fourth

---

* The result of our observations on the confidence connected with the judgments is as follows:

[Subject, Mr Peirce.]

Variations.	Average confidence.	Number of sets of 50.
Grams.		
60	.67	7
30	.28	6
15	.15	5

[Subject, Mr. Jastrow.]

60	.90	13
30	.51	12
15	.30	12
20	.11	12
10	.06	12
5	.00	10

In 1,125 experiments (subject, Mr. Peirce)—variations 15, 30, and 60 grams—there occurred confidence of 3, 35 times (3 per cent.); of 2, 102 times (9 per cent.); of 1, 282 times (25 per cent.); of 0, 706 times (63 per cent.). In these experiments there were 332 (29 per cent.) errors committed, of which 1 (0.3 per cent.) was made in connection with a confidence 3; 10 (3 per cent.) with a confidence 2; 51 (15 per cent.) with a confidence 1; 270 (81 per cent.) with a confidence 0. From which we find that in connection with a confidence of 3 there occurred 1 error in 35 cases (3 per cent.); with a confidence of 2, 10 errors in 102 cases (10 per cent.); with a confidence of 1, 51 errors in 282 cases (18 per cent.); with a confidence of 0, 270 errors in 706 cases (38 per cent.).

In 1,975 experiments (subject, Mr. Jastrow)—variations 15, 30, and 60 grams—there occurred confidence of 3, 62 times (3 per cent.); of 2, 196 times (10 per cent.); of 1, 594 times (30 per cent.); of 0, 1,123 times (57 per cent.). In these experiments there were 451 (23 per cent.) errors committed, of which 2 (0.4 per cent.) were made in connection with a confidence of 3; 12 (3 per cent.) with a confidence of 2; 97 (22 per cent.) with a confidence of 1, 340 (75 per cent.) with a confidence of 0. Again, in connection with a confidence of 3, errors occurred twice in 62 cases (3 per cent.); with a confidence of 2, 12 times in 196 cases (6 per cent.); with a confidence of 1, 97 times in 504 cases (16 per cent.); with a confidence of 0, 340 times in 1,123 cases (30 per cent.).

In 1,675 experiments (subject, Mr. Jastrow)—variations 5, 10, and 20 grams—there occurred confidences of 3, none; of 2, none; of 1, 115 times (7 per cent.); of 0, 1,560 times (93 per cent.). In these experiments there were 538 (32 per cent.) errors committed, of which 16 (3 per cent.) occurred in connection with a confidence of 1; 522 (97 per cent.) with a confidence of 0. Again, in connection with a confidence of 1, errors occurred 16 times in 115 cases (14 per cent.); with a confidence of 0, 522 times in 1,560 cases (34 per cent.).

Second group.

Ratio of weights.	Mark 0.	Mark 1.	Mark 2.	Mark 3.
1.015	110 right 66 wrong	51 right 17 wrong	3 right 2 wrong	1 right 0 wrong
1.030	106 right 35 wrong	72 right 11 wrong	23 right 1 wrong	2 right 0 wrong
1.060	86 right 8 wrong	75 right 1 wrong	54 right 2 wrong	24 right 0 wrong

of its weight upon the finger. The differential pressure was produced by lowering upon the pan of the balance a smaller pan into which the proper weights could be firmly fixed; this little pan had its bottom of cork, and was placed upon a piece of flannel which constantly remained in the pan of the balance. It was lifted off and on by means of a fine India-rubber thread, which was so much stretched by the weight as certainly to avoid any noise or jar from the momentum of the descending pan. A sufficient weight could also be hung on the beam of the balance, so as to take off the entire pressure from the finger at the end of each experiment. This weight could be applied or removed by means of a cam acting upon a lever; and its bearings upon the beam were guarded by India-rubber. It was found that the use of this arrangement, which removed all annoying irregularities of sensation connected with the removal and replacement of the greater (initial) pressure, rendered the results more uniform and diminished the probable error. It also shortened the time necessary for performing the experiments, so that a series of 25 experiments was concluded before the effects of fatigue were noticeable. It may be mentioned that certain causes tended to the constant decrease of the probable error as the experiments went on, these mainly being an increased skill on the part of the *operator* and an education of the sensibility of the *subject*. The finger was supported in such a way as to be lightly but firmly held in position, all the muscles of the arm being relaxed; and the India-rubber top of the brass enlargement at the end of the beam of the balance was never actually separated from the finger. The projecting arm of a filter-stand (the height of which could be adjusted) with some attachments not necessary to detail, gently prevented the finger from moving upwards under the pressure exerted by the weight in the pan. In the case of Mr. Peirce as subject (it may be noted that Mr. Peirce is left-handed, while Mr. Jastrow is strongly right-handed) the tip of forefinger, and in the case of Mr. Jastrow of the middle finger, of the left hand were used. In addition, a screen served to prevent the subject from having any indications whatever of the movements of the operator. It is hardly necessary to say that we were fully on our guard against unconsciously received indications.

The observations were conducted in the following manner: At each sitting three differential weights were employed. At first we always began and ended with the heaviest, but at a later period the plan was to begin on alternate days with the lightest and heaviest. When we began with the heaviest 25 observations* were made with that; then 25 with the middle one, and then 25 with the lightest; this constituted one-half of the sitting. It was completed by three more sets of 25, the order of the weights being reversed. When we began with the lightest the heaviest was used for the third and fourth sets. In this way 150 experiments on each of us were taken at one sitting of two hours.

A pack of 25 cards were taken, 12 red and 13 black, or *vice versa*, so that in the 50 experiments made at one sitting with a given differential weight, 25 red and 25 black cards should be used. These cards were cut exactly square and their corners were distinguished by holes punched in them so as to indicate the scale of numbers (0, 1, 2, 3) used to designate the degree of confidence of the judgment. The backs of these cards were distinguished from their faces. They were, in fact, made of ordinary playing-cards. At the beginning of a set of 25, the pack was well shuffled, and, the operator and subject having taken their places, the operator was governed by the color

Third and fourth groups.

[Marks 2 and 3 do not occur.]

Ratio of weights.	Mark 0.	Mark 1.
1.005	294 right 203 wrong	2 right 1 wrong
1.010	366 right 192 wrong	32 right 30 wrong
1.020	395 right 131 wrong	68 right 6 wrong

* At first a short pause was made in the set of 25, at the option of the subject; later this was dispensed with.

of the successive cards in choosing whether he should first diminish the weight and then increase it, or *rice versa*. If the weight was to be first increased and then diminished the operator brought the pressure exerted by the kilogram alone upon the finger of the subject by means of the lever and cam mentioned above, and when the subject said "change" he gently lowered the differential weight, resting in the small pan, upon the pan of the balance. The subject, having appreciated the sensation, again said "change," whereupon the operator removed the differential weight. If, on the other hand, the color of the card directed the weight to be first diminished and then increased, the operator had the differential weight already on the pan of the balance before the pressure was brought to bear on the finger, and made the reverse changes at the command of the subject. The subject then stated his judgment and also his degree of confidence, whereupon the total pressure was at once removed by the cam, and the card that had been used to direct the change was placed face down or face up according as the answer was right or wrong, and with corner indicating the degree of confidence in a determinate position. By means of these trifling devices the important object of rapidity was secured, and any possible psychological guessing of what change the operator was likely to select was avoided. A slight disadvantage in this mode of proceeding arises from the long runs of one particular kind of change, which would occasionally be produced by chance and would tend to confuse the mind of the subject. But it seems clear that this disadvantage was less than that which would have been occasioned by his knowing that there would be no such long runs if any means had been taken to prevent them. At the end of each set the results were of course entered into a book.*

The following tables show the results of the observations for each day:

Date.	Ratios of pressures. [Subject: Mr. Peirce.]						
	1.100	1.080	1.060	1.050	1.040	1.030	1.015
December 10	2 errors.			13 errors.			
December 13		4 errors.	8 errors.		15 errors.		
December 17			11			20 errors.	
December 20			7			16	21 errors.
January 3			14			20	28
January 15			15			29	28
January 22			12			16	20
January 24			6			15	22
Means	2	4	10.4±1.0	13	15	19.3±1.4	21.6±1.1
Calculated from probable error=0.051	4.6±1.0	7.2±1.6	10.7±0.8	12.7±2.1	14.9±2.2	17.2±0.9	21.0±1.1
Average confidence.							
Observed	1.9	0.9	0.7	0.8	0.3	0.3	0.2
Calculated	1.3	1.0	0.7	0.6	0.5	0.3	0.2

The numbers in the columns show the number of errors in fifty experiments. With the average number of errors in a set of fifty we compare the theoretical value of this average as calculated by the method of least squares. The number .051 thus obtained in this case best satisfies the mean number of errors. The numbers affixed with a sign denote, in the upper row the observed (*a posteriori*) probable error of the mean value as given, in the lower row the calculated (*a priori*) probable error. The last two lines give the average confidence observed and calculated with each variation of the ratios of pressure. It will be seen that the correspondence between the real and theoretical numbers is close, and closest when the number of sets is large. The probable errors also closely correspond, the observed being, as is natural, slightly larger than the calculated probable errors.

---

* In the experiments of December, 1883, and January, 1884, the method as above described was not fully perfected the most important fault being that the total weight instead of being removed and replaced by a mechanical device, was taken off by the operator pressing with his finger upon the beam of the balance.

SMALL DIFFERENCES OF SENSATION.

The following is a similar table for Mr. Jastrow as subject:

Date.	Ratios of pressures.								
	1.100	1.080	1.060	1.050	1.040	1.030	1.020	1.015	1.010	1.005
December 10	5			19						
December 13		9	15		15					
December 17			14			23				
December 20			10			17		25		
January 3			8			14		24		
January 10			7			13		17		
January 15			12			6		22		
January 22			11			10		16		
January 24			4			11		18		
February 11			1			7		18		
February 17			2			10		17		
February 18			2			11		17		
February 24			2			8		15		
March 4							13	16		
March 5							13	17		
March 18							14	19	25	
March 19							11	21	18	
March 23							14	17	18	
March 25							12	16	18	
March 30							11	16	21	
March 31							10	15	21	
April 2							11	17	21	
April 3							9	18	20	
April 6							12	15	21	
April 7			0			5	7	14	15	17
Means	5	9	6.6	19	15.0	11.6	11.4	18.9	16.8	20.5

It would obviously be unfair to compare these numbers with any set of theoretical numbers, since the probable error is on the decrease throughout, owing to effects of practice, etc. For various reasons we can conveniently group these experiments into four groups. The first will include the experiments from December 10 to January 22, inclusive; the second from January 24 to February 24, inclusive; the third from March 4 to March 25, inclusive; the fourth from March 30 to the end of the work.

The mean results for the different groups are exhibited in the following tables:

First group.

[Probable error=0.05.]

Ratios of pressures.	Number of sets of 50.	Average number of errors.		Average confidence.	
		Observed.	Calculated from probable error.	Observed.	Calculated.
1.100	1	5	4.4±1.4	0.9	1.5
1.080	1	9	7.0±1.7	0.9	1.2
1.060	7	11.0±0.7	10.4±0.7	0.85	0.9
1.050	1	19	12.5±2.1	0.35	0.7
1.040	1	15	14.7±2.2	0.3	0.6
1.030	6	13.8±1.5	17.0±0.9	0.5	0.4
1.015	5	20.8±1.1	21.0±1.1	0.3	0.2

Second group.

[Probable error=0.0235.]

1.060	5	2.2±0.3	2.1±0.4	1.0	1.2
1.030	5	9.4±0.6	9.6±0.8	0.55	0.6
1.015	5	17.0±0.3	16.6±1.0	0.3	0.3

Third group.
[Probable error=0.02.]

Ratios of pressures.	Number of sets of 50.	Average number of errors.		Average confidence.	
		Observed.	Calculated from probable error.	Observed.	Calculated.
1.020	6	12.8±0.3	12.5±0.8	0.12	0.12
1.010	6	17.7±0.6	18.3±0.9	0.07	0.06
1.005	4	20.7±1.7	21.6±1.2	0.00	0.03

Fourth group.
[Probable error=0.0155.]

1.060	1	0	0.8±0.6	1.6	....
1.030	1	5	4.8±1.4	0.5	0.4
1.020	6	10.0±0.5	9.6±0.8	0.1	0.2
1.015	1	14	12.8±2.1	0.1	0.13
1.010	6	16	16.5±0.9	0.05	0.12
1.005	6	20.8±0.4	20.6±1.0	0.00	0.06

The tables show that the numbers of errors follow, as far as we can conveniently trace them, the numbers assigned by the probability curve,* and therefore destroy all presumption in favor of an *Unterschiedsschwelle*. The introduction and retention of this false notion can only confuse thought, while the conception of the mathematician must exercise a favorable influence on psychological experimentation.†

The quantity which we have called the degree of confidence was probably the secondary sensation of a difference between the primary sensations compared. The evidence of our experiments

---

* In the tables of the third and fourth groups, there is a marked divergence between the *a priori* and *a posteriori* probable error, for the average number of errors in 50, making the observed probable error too small. This can only be partly accounted for by the fact that the subject formed the unconscious habit of retaining the number of each kind of experiment in a set and answering according to that knowledge. In point of fact the plus errors and minus errors separately do not exhibit the singular uniformity of their sums, for which we are quite unable to account. Thus in the fourth group we have:

Number of + and − errors.

Date.	1.020	1.010	1.005
March 30	−4, + 7	−6, +10	−13, + 8
March 31	−7, + 3	−5, +10	− 6, +15
April 2	−1, +10	−8, + 9	− 8, +13
April 3	−4, + 5	−4, +14	−10, +10
April 6	−6, + 6	−8, + 7	−10, +11
April 7	−5, + 9	−8, + 7	− 8, + 9

† The conclusions of this paper are strengthened by the results of a series of experiments on the color sense, made with the use of a photometer by Mr. Jastrow. The object was to determine the number of errors of a given magnitude, and compare the numbers thus ascertained with the theoretical numbers given by the probability curve. A thousand experiments were made. Dividing the magnitude of the errors from 0 to the largest error, made into 5 parts, the number of errors, as observed and calculated, that occur in each part are as follows:

Observed	199	181	217	213	190
Calculated	213	197	209	181	200

These numbers would be in closer accordance if the probable error were the same throughout, as it is not owing to the effects of practice, &c. Moreover, the experiments were made on different colors—300 on white and 100 each on yellow, blue, dove, pink, green, orange, and brown. These experiments were not continuous.

seems clearly to be that this sensation has no *Schwelle*, and vanishes only when the difference to which it refers vanishes. At the same time we found the subject often overlooked this element of his field of sensation, although his attention was directed with a certain strength toward it, so that he marked his confidence as *zero*. This happened in cases where the judgments were so much affected by the difference of pressures as to be correct three times out of five. The general fact has highly important practical bearings, since it gives new reason for believing that we gather what is passing in one another's minds in large measure from sensations so faint that we are not fairly aware of having them, and can give no account of how we reach our conclusions about such matters. The insight of females as well as certain "telepathic" phenomena may be explained in this way. Such faint sensations ought to be fully studied by the psychologist and assiduously cultivated by every man.

SIMON NEWCOMB, PERCY DANIELL, AND THE HISTORY OF ROBUST ESTIMATION

Stephen M. Stigler

Simon Newcomb, Percy Daniell, and the History of Robust Estimation 1885–1920

STEPHEN M. STIGLER*

This article reviews some of the history of robust estimation, emphasizing the period 1885–1920 and the work of Newcomb, Edgeworth, Sheppard, and Daniell. Particular attention is paid to lines of development which have excited recent interest, including linear functions of order statistics and mixtures of normal densities as models for heavy-tailed populations.

1. INTRODUCTION

In the eighteenth century, the word "robust" was used to refer to someone who was strong, yet boisterous, crude, and vulgar. By 1953 when Box first gave the word its statistical meaning, the evolution of language had eliminated the negative connotation: robust meant simply strong, healthy, sufficiently tough to withstand life's adversities. The subject of robust inference, just like the word "robust," has a long and varied history. It is the aim of this present study to examine a part of this history and its relationship to current work.

The scope of this article will be rather narrow—we shall only be concerned with the *mathematical* background and development of robust estimation up to 1920. Thus we shall be less concerned with the first appearances of estimators such as the median and trimmed mean than with the first mathematical analyses of their behavior and properties. The main emphasis will be on the period 1885–1920, and particular attention will be given to work which is not widely known, yet is relevant to modern lines of thought. Section 3 discusses the contributions of Simon Newcomb to robust estimation, and to the use of normal mixtures as models for heavy-tailed distributions; Section 4 is concerned with the history of the mathematical analysis of order statistics in relation to robust estimation, with due attention to the works of Laplace, Sheppard, and Percy Daniell; and Section 5 contains some brief remarks on "M-estimators."

The reader may be as surprised as the author was to find to what extent priorities in these areas have been misassigned. While many other points will be touched upon in the article, our major findings are as follows: Laplace [46, 1818] and Sheppard [70, 1899a] seem to have been the first to present a large sample theory for one or two order statistics. Simon Newcomb [54, 1886] provided the first sound, modern approach to robust estimation, including the first use of mixtures of normal densities as representing heavy-tailed distributions. Percy Daniell [8, 1920] should be credited with the first mathematical analysis of the class of estimators which are linear functions of order statistics, including the derivation of the optimal weighting functions for estimating scale and location parameters (the so-called "ideal" linear estimators) and the first mathematical treatment of the trimmed mean. Some of Newcomb's work has been commented on recently by Huber [39, 1972], but much of the remainder of the work discussed in this article, including that due to Edgeworth, Galton, Laplace, Sheppard, and Daniell, has been largely ignored in recent years.

We shall begin with a brief overview of the situation prior to 1885.

2. THE SITUATION BEFORE 1885

Scientists have been concerned with what we would call "robustness"—insensitivity of procedures to departures from assumptions, particularly the assumption of normality—for as long as they have been employing well-defined procedures, perhaps longer. For example, in the first published work on least squares, Legendre [47, 1805] explicitly provided for the rejection of outliers:

> If among these errors are some which appear too large to be admissible, then those equations which produced these errors will be rejected, as coming from too faulty experiments, and the unknowns will be determined by means of the other equations, which will then give much smaller errors.

Yet most of the early work in *mathematical* statistics was obsessed with "proving" the method of least squares, either starting with the assumption that the sample mean is the best estimate of the mean and deriving the normal distribution, as Gauss did in his first proof in 1809, or starting with the Central Limit Theorem, as did Laplace in 1811. The first mathematical work on robust estimation seems to have been that of Laplace [46, 1818] on the distribution of the median. We shall defer a discussion of Laplace's work until Section 4 where it will

---

* Stephen M. Stigler is associate professor, Department of Statistics, University of Wisconsin, Madison, Wis. 53706, and is presently on leave (1972-3) at the Department of Statistics, University of Chicago. This research was partially supported by Research Grant No. NSF GP 32037x from the Division of Mathematical, Physical and Engineering Sciences of the National Science Foundation at the Department of Statistics, University of Chicago, and by the U.S. Navy through the Office of Naval Research under Contract No. ONR-N00014-67A-0128-0017 at the Department of Statistics, University of Wisconsin. The author wishes to thank William Kruskal, Churchill Eisenhart, and Oscar B. Sheynin for a number of references and helpful comments.

be considered with later work on linear functions of order statistics.

The next statistical problem connected with robust estimation to receive mathematical treatment was the rejection of outliers. In 1852, the first proposal of a criterion for the determination of outliers was published by Benjamin Peirce, the Harvard mathematician-astronomer and father of logician-philosopher C. S. Peirce. Peirce's paper and most others on this subject[1] were not really about robust estimation, as their authors did not concern themselves with the properties of the resulting estimators; rather, they implicitly assumed that after the outlier test was performed the estimation could be done with no thought given to what had gone before, nor what information might be lost. This narrowness of view did not go unnoticed at the time. The first paper proposing an outlier criterion (Peirce [66, 1852]) was soon followed by the first paper criticizing the use of outlier criteria (Airy [1, 1856]). Airy, the Astronomer Royal, wrote:

> And I have, not without surprise to myself, been led to think that the whole theory is defective in its foundation, and illusory in its results; that no rule for the exclusion of observations can be obtained by any process founded purely upon a consideration of the discordance of these observations.

A lively debate ensued, with the participants not always expressing themselves with Airy's restraint. For example, Glaisher [32, 1872] wrote "Professor Pierce's [sic] criterion for the rejection of doubtful observations seems to me to be destitute of scientific precision." At one point an exchange in print between the mathematician Glaisher and the astronomer Stone became so heated that one of Glaisher's papers was itself rejected by the *Monthly Notices of the Royal Astronomical Society* due to the personal nature of his comments (see Glaisher [34, 1874]).

One of the more interesting papers of this time (and one of the most unusual statistical papers of all time) appeared in the *Report of the Superintendent of the U.S. Coast Survey for 1870*. It is by C.S. Peirce, written while he was an Assistant to the Coast Survey (at the time his father was Superintendent of the Survey!). In the paper Peirce presented the then standard material of the theory of errors, but in the language and notation which he had developed for the logic of relations, for which he later became famous. Thus we find, regarding averages,

> Since [m] denotes all men, we may naturally write [m]/m to denote what all men become when that factor is removed which makes [m] refer to *men* rather than to anything else; that is to say, to denote the *number* of men. We may write this for short [m] with heavy brackets. Then t being a relative term ("a tooth of,") by [t] will be denoted the total number of teeth in the universe. But [t] will be used as equivalent to [tl]/[l], or the average number of teeth that anything has.

Peirce included a sensible—one is tempted to say "logical"—defense of his father's outlier criterion in the paper (p. 210). By 1885 a number of rejection criteria were in use, often only by the proposer and his employees.[2]

But techniques other than simply "reject outliers, then use the sample mean" were also employed. A variety of weighted means had been used prior to 1885. For example, in 1763 James Short (an English astronomer and noted manufacturer of telescopes) had estimated the sun's parallax based on observations of the transit of Venus of 1761 by averaging three means: the sample mean, the mean of all observations with residuals less than one second, and the mean of those with residuals less than half a second. The median and the midrange had appeared even earlier (Eisenhart [23, 1971]).

By the last half of the nineteenth century, weighted least squares had become a standard topic in the literature of the theory of errors, and it was a frequent practice (at least in astronomical investigations) to weight observations differently, depending upon the scientist's (often subjective) estimate of the "probable error"[3] of the observation. The estimate of the probable error was supposed to be based solely on external evidence: scientists were warned of the possible biases if the magnitude of the observation were allowed to influence its weight (see, for example, Jevons [44, 1874, p. 450]), but it is doubtful that this advice was faithfully adhered to. We shall discuss the use of these weighted means further in the next section, in connection with the contributions of Simon Newcomb.

Other estimators were proposed in this period. In particular, De Morgan [11, 1847, p. 456] had outlined a scheme for discounting the more extreme observations. This method, more fully developed by Glaisher [33, 1873], amounted to starting with the sample mean, then assigning different probable errors to the different observations based on the value of the likelihood function at those observations, and iterating this process. Glaisher's estimate was criticized by both Stone [75, 1873] and Edgeworth [14, 1883], who both (independently) proposed an alternative based on looking for a maximum of the likelihood function (without assuming equal probable errors).[4] Edgeworth later became disenchanted with this alternative [17, 1887a].

At about this time, Francis Galton was making much use of the median [27, 1875], although his motivation was less suspicion of the normal distribution, which he considered a good representation of many real phenomena, than an appreciation of the simplicity, ease of calculation, and ease of interpretation of the median. Many of these same features were also cited in the apparently independent work of Fechner [26, 1878]; see Walker, [76, 1929, p. 84–6]. Also, various formulas for

---

[1] See Anscombe [3, 1960] and Rider [69, 1933] for historical surveys of outlier techniques.

[2] Outlier rejection rules for multivariate data seem to have not been considered prior to 1899, when Sheppard [71, 1899b] proposed a multivariate extension of Chauvenet's [6, 1863] rule.

[3] The probable error of a symmetric distribution is half the interquartile range; for normal distributions p.e. = (.6745)σ.

[4] Actually, the equations both Stone and Edgeworth derived by differentiation would lead to local minima of the likelihood function. This problem was recently discussed by Mantel [49, 1972].

index numbers were developed during this period; these included weighted averages and geometric means, each designed for a specific purpose.

However, it can still be said that by 1885, the conventional wisdom (but by no means the unanimous view) was that for purposes of estimation, the cautious use of the sample mean was recommended—sometimes weighted, sometimes after discarding outliers, but still the sample mean.

3. SIMON NEWCOMB AND MIXTURES OF NORMAL DENSITIES

1885 can be conveniently taken as the start of one of the most active and innovative periods in the history of mathematical statistics. The story of the development of mathematical statistics into a subject in its own right through the work of such men as Edgeworth, Karl Pearson, Gosset, and Fisher has been told by E.S. Pearson [62, 1967]. Our present, rather narrow purpose is to describe how the modern theory of robust estimation developed over this period. To this end, we shall place particular emphasis on the introduction of mixtures as models for the heavy-tailed distributions which scientists had encountered in practice, and on the use of linear functions of order statistics as robust estimators of location parameters.

Simon Newcomb appears to have been the first to introduce a mixture of normal densities as a model for a heavy-tailed distribution, and to exploit this model to get an estimator of location which was more robust than the sample mean. (Francis Galton and Karl Pearson had modeled measurements of natural populations by normal mixtures about the same time, but with a completely different object in mind, namely to demonstrate how a single population could be broken down into components.) While Newcomb's name may be unfamiliar to present day statisticians, it should not be so to astronomers, applied mathematicians, and economists.

Simon Newcomb (1835–1909) was born in Nova Scotia, attended Harvard, and spent most of his adult life (1861–1897) as a professor of mathematics in the U.S. Navy, working for the U.S. Nautical Almanac Office. He is generally regarded as the greatest American astronomer of the nineteenth century, and was responsible for many of the determinations of astronomical constants which are still accepted today. In addition, he was a powerful applied mathematician, and co-founded and for many years edited the *American Journal of Mathematics*. As an avocation he wrote *Principles of Political Economy* [53, 1885], a book which has established him as a major American economic theorist, and which contains one of the earliest modern mathematical statements of the quantity theory of money.

As was the practice in astronomy at the time, Newcomb made frequent use of weighted means in his estimation of astronomical constants. The relative weights were usually thought of in terms of "probable errors," and were assigned somewhat subjectively on the basis of Newcomb's judgment of the relative accuracy of the process which produced the observation. For example, after assessing some data on eclipses collected by Ptolemy in the second century A.D., he remarked [51, 1878, p. 41]:

> the [assigned] probable errors are the result of judgment from the terms of [Ptolemy's] description rather than of calculation; they were estimated without any knowledge of the way the comparison with theory would come out, and are printed without subsequent alteration.

With more contemporary data, Newcomb would base his choice of weights upon "the quality of the image and the generally satisfactory way in which the image was kept on the crosswires" [55, 1891a, p. 170] in the case of an experiment he was personally involved with, and upon the number of observers, general opinion of the reporting observatory, and "number and force of the doubtful circumstances "[56, 1891b, p. 383], in cases involving combination of others' measurements. He was apparently aware of criticism of the subjective nature of these assignments, but he maintained that

> Opinions may doubtless differ as to whether a judicious system of weights has always been applied, but it is not likely that any unbiased reassignment would materially affect the result [58, 1898, p. 211].

Newcomb also rejected outliers when necessary, but usually only based on external evidence or really huge deviations.

With this experience in dealing with observations made with differing degrees of precision, it is not surprising that, when faced with a collection of non-normal observations for which there was no satisfactory way to weight them individually, he should consider a mixture of normal densities with different variances as a model. For, having observed that a collection of 684 residuals based on observations of the transits of Mercury had much heavier tails than the corresponding normal distribution (even with excessive deviations ignored), he wrote [52, 1882, p. 382]:

> It is evident that if we have a collection of observations of different degrees of probable error, in which, however, there is no way of distinguishing those of great probable error from those of small probable error, the law of the errors will not be that usually adopted, but there will be a comparative excess of large residuals. It is also evident that in such a case the arithmetical mean does not necessarily give the most probable result. For, in the case of an observation of large residual, there is evidently a preponderance of probability that it belongs to a class with large probable error, and therefore should be assigned least weight. . . . That any general collection of observations of transits of Mercury must be a mixture of observations with different probable errors was made evident to the writer by his observations of the transit of May 6, 1878, which may be here described as an illustration of the subject.

Four years after writing this, Newcomb published a remarkable paper in his own journal, the *American Journal of Mathematics*, in which he used this model to arrive at a more robust estimator of location than the sample mean. In this paper [54, 1886],[5] after criticizing the overuse of outlier criteria and presenting his mixture model, he proceeded to develop an estimator upon the principles of Bayesian decision theory that gave "less weight to the more discordant observations." Adopting squared error as a loss function (Newcomb's word for loss was "evil"), he demonstrated that in general the posterior mean minimizes the expected mean square error, and he suggested the following procedure. 1) Calculate the residuals based on the sample mean, and, using trial and error, fit a mixture of a finite number of normal densities with zero means to these residuals. 2) Take this fitted mixture and, considering the location family it generates, estimate the desired mean by the posterior mean with respect to a uniform prior, given the original observations. Newcomb realized that this procedure presented practical difficulties and gave a number of simplifying approximations to arrive at a usable estimator. He illustrated its use with the data on the transits of Mercury.[6]

As an interesting sidelight, we note that in this article and in a later work, Newcomb made an early use of a simple version of Tukey's sensitivity function (see Andrews et al. [2, 1972, p. 96]). In a paper [60, 1912, p. 212] discussing the unsatisfactory nature of outlier criteria, he wrote that if all observations with large residuals are rejected (and the mean estimated from the remaining observations), then the final result

> becomes a discontinuous function of the residual of the rejected observation, the continuity being broken at the point regarded as the limit of normal error. A simple example will make the case clear. If we have three observed results a, b, c of which the mean is to be taken, and if c be the result which may be abnormal, then so long as c is retained we shall have
>
> mean $= \frac{1}{3}(a + b + c)$;
>
> the mean will then continuously increase with c. When c passes the normal limit, the mean changes per saltum to
>
> $\frac{1}{2}(a + b)$.

In the same posthumous paper [60, 1912, p. 214], Newcomb also proposed a very simple estimator in the spirit of his 1886 paper: weight the observation X_i by $w_i = c/\max(|X_i - \bar{X}|, c)$, where c is a constant to be specified, stating that when $|X_i - \bar{X}| > c$, "This will lead to practically the same result as if we substituted for the actually observed quantity another quantity corresponding to the residual c." Peter J. Huber has pointed out to me that the fixed point of the iterative version of this estimator, where one replaces $\bar{X}$ by the weighted mean with weights w_i and iterates, is identical to one of Huber's favorite M-estimators (see Section 5), with $\phi(x) = \max[-c, \min(x, c)]$!

4. LAPLACE, SHEPPARD, DANIELL, AND LINEAR FUNCTIONS OF ORDER STATISTICS

With few exceptions, statisticians were quite late in coming to consider any but the simplest linear functions of order statistics as estimators of means. By a linear function of order statistics we shall mean any weighted linear combination of observations where the weights depend *only* on their order, not on their magnitudes or the size of their residuals. The median and the midrange, two members of this class, evidently have a long history (Eisenhart [22, 1964; 23, 1971]), but perhaps the first extensive mathematical analysis to be published involving order statistics was by Laplace. In the second supplement [46, 1818] to his monumental *Théorie Analytique des Probabilités*, Laplace considered the problem we would now call linear regression through the origin: $a_i = p_i y + x_i$, a_i, p_i known, y to be estimated, where the errors x_i were assumed to have an arbitrary continuous, symmetric distribution. By looking for that estimator which minimized the sum of the absolute values of the residuals, he was led to consider an estimator of y which reduces to the median of the a_i's in the case $p_i = 1$. Laplace derived the density of this estimator, showed that this density approaches the normal density as the sample size increases, and gave the necessary and sufficient condition on the error distribution that the median have a smaller asymptotic variance than the sample mean.[7] Laplace's proof is easily adapted to any sample percentile and asymmetrical populations, as was in fact later noted by Edgeworth [15, 1885; 16, 1886]. In addition, Laplace derived the joint asymptotic density of the sample mean and median, and used it to find which linear combination of these estimators has the smallest asymptotic variance. (As the weights depend upon the unknown error distribution, he termed this result "impracticable," but noted that if the error distribution were normal, the best linear combination was the sample mean alone.)[8] Two years before Laplace's investigation, Gauss [31, 1816], considering the problem of estimating the probable error of a normal distribution, had suggested the use of the median of the absolute values of the residuals, and stated (without proof) the asymptotic probable error of the median for this special case. Gauss apparently never published or circulated a proof, for 16 years later Encke [25, 1832], who had corresponded with Gauss, felt it necessary to provide one, attributing it to Dirichlet. It seems likely that Dirichlet's proof for this special case was simply an adaptation of Laplace's, as Dirichlet was quite familiar with Laplace's work, the second supplement in particular (see Dirichlet [12, 1836]).

Later in the nineteenth century, Galton [27, 1875] and particularly Edgeworth [15, 1885; 18, 1887b; 19,

---

[5] Some of his arguments also appear in Newcomb [57, 1895, p. 81–86].
[6] Ogorodnikoff [61, 1928] provided a different simplification of Newcomb's estimator based on a Charlier expansion of the posterior distribution. The relationship between Newcomb's simplified estimator and the maximum likelihood estimator was discussed by Hulme and Symms [40, 1939].

[7] Laplace actually carried through his entire investigation in the more general regression situation, comparing the general estimator with the least squares estimator for this situation. For other views of Laplace's work and its historical context, see Eisenhart [21, 1961] and Stigler [74, 1973].
[8] The possibility of a linear function of two estimators outperforming both has been more fully exploited in the recent Princeton robustness study (see Andrews et al. [2, 1972, p. 132]).

1888], touted the use of the median in situations where heavier tails than the normal could be expected. Specifically, Edgeworth [19, 1888] used Laplace's results to conclude that the median may well be better than the mean when the population distribution is one of Newcomb's mixtures of normal distributions. Also, Edgeworth [16, 1886] seems to have been the first to realize that the median may possess an advantage over the sample mean for serially correlated data.

More complicated linear estimators began to appear in 1889, when Galton (in a footnote on p. 61–62 of *Natural Inheritance*, [28, 1889]) suggested estimating the mean and standard deviation of a normal distribution by what amounts to taking

$$\hat{\mu} = \frac{\xi_p X^{(nq)} - \xi_q X^{(np)}}{\xi_p - \xi_q},$$

$$\hat{\sigma} = \frac{X^{(np)} - X^{(nq)}}{\xi_p - \xi_q},$$

where ξ_p and ξ_q are the p and q percentiles of the standard normal distribution, $X^{(np)}$ and $X^{(nq)}$ are the sample $100p$ and $100q$ percentiles, and p and q are arbitrary but fixed ($0 < p < q < 1$). In 1899 in a long paper on the multivariate normal distribution and its applications, Sheppard proved the joint asymptotic normality of Galton's estimators when the population is normal. He also showed the joint asymptotic normality of $X^{(np)}$ and $X^{(nq)}$, and gave analogues to $\hat{\mu}$ and $\hat{\sigma}$ based on any finite number of sample percentiles [70, 1899, p. 131–2]. Sheppard's (sketchy) proof, which is based on an implicit use of the probability integral transformation, can be easily adapted to any regular distribution.[9]

Sheppard's paper also represented the first attempt since Laplace to optimize performance within a class of linear functions of order statistics. He both showed how the best choice (for normal populations) of p and q can be made [70, 1899, p. 135] and found which linear combination of the three quartiles has the smallest asymptotic variance (again for normal populations) [70, 1899, footnote, p. 134]. Such functions of the three quartiles had been considered earlier by Edgeworth [1893], who neglected the quartiles' correlation and erroneously claimed the estimator with weights in proportions 5:7:5 to be superior to the sample mean for normal populations. Recent work, however, seems to bear out Edgeworth's claim that such an estimator is to be recommended on grounds of robustness (for example, see Gastwirth [30, 1966] and Andrews et al. [2, 1972]).

The next mathematical work to appear on order statistics was Karl Pearson's [63, 1902] examination of the Galton difference problem. In this paper, which was inspired by an inquiry of Galton's [29, 1902] as to the most suitable proportion between the values of first and second prizes, Pearson gave the joint density of any two consecutive order statistics and found their expected difference. He remarked in a footnote that

> I propose on another occasion to consider the application of Galton's problem to a new theory for the rejection of outlying individuals.

This proposal was later carried out by J.O. Irwin [41, 1925].

In 1920, a remarkable paper appeared in the *American Journal of Mathematics* (the journal Simon Newcomb co-founded) by the English mathematician P.J. Daniell. This paper, "Observations Weighted According to Order," has been all but totally overlooked since it's publication. It could in fact be claimed that Daniell was at least thirty years ahead of his time, for it took that long for his major results to be rediscovered. While his paper itself is a model of clarity and rigor, its relevance to modern work is such that it merits a short summary, in his own notation.

The work was apparently inspired by a reading of Poincaré's *Calcul des Probabilités* [68, 1912]. After remarking how Poincaré had suggested discarding extreme observations (when normality is suspect) before taking the mean, Daniell wrote:

> Besides such a discard-average [i.e., the trimmed mean] we might invent others in which weights might be assigned to the measures according to their order. In fact the ordinary average or mean, the median, the discard-average, the numerical deviation (from the median, which makes it a minimum), and the quartile deviation can all be regarded as calculated by a process in which the measures are multiplied by factors which are functions of order. It is the general purpose of this paper to obtain a formula for the mean square deviation of any such expression. This formula may then be used to measure the relative accuracies of all such expressions.

Daniell's analysis proceeded as follows: First he explicitly introduced the probability integral transformation (apparently the first time this was done)[10] and explained how it could be used to find the moments of any function of order statistics. Then, he assumed the population density $p(t)$ was regular (and indefinitely differentiable), and he expanded the inverse of the distribution function in a Taylor series to derive asymptotic expressions for the mean of an order statistic t_r and the mean product of any two. He thus duplicated some of Sheppard's [70, 1899a] results, but in a much more rigorous manner.

Daniell then considered statistics of the form $\bar{t} = \sum_{r=1}^{n} f_r t_r$, where he assumed that the weight f_r associated with the rth order statistic t_r was given by

$$f_r = \frac{1}{n} f\left(\frac{r}{n+1}\right),$$

and put things together to obtain the (now standard)[11] expression for the asymptotic variance of $\bar{t}$,

$$S^2 = \int_{-\infty}^{\infty} \phi^2(t) p(t) dt,$$

---

[9] Twenty years later, Karl Pearson [64, 1920] presented part of Sheppard's proof in more detail, made the obvious step to more general distributions than normal, and much more fully examined the consequences of the result.

[10] The next being Karl Pearson [65, 1931].
[11] See Chernoff, Gastwirth, and Johns [7, 1967].

where $\phi(t)$ is the indefinite integral of $f[x(t)]$,

$$x(t) = \int_{-\infty}^{t} p(u)du.$$

If he was less than specific as to why the remainder terms are uniformly negligible, his standard of rigor was nonetheless far above that of the statistical literature of the time.

In the third section of the paper, Daniell gave the conditions on f under which the asymptotic mean of $\hat{t}$ is the population mean or standard deviation, and defined the "accuracy" of $\hat{t}$ as the ratio of the asymptotic variance of the sample mean (or sample standard deviation, as the case may be) to that of $\hat{t}$. (He also derived the asymptotic variance of the sample standard deviation here.) In the fourth section, Daniell gave the optimal weight function f—that which minimizes S^2—for both the location and scale cases, using standard results from the calculus of variations, and noted that the optimal estimate of σ for the normal case is as accurate as the sample standard deviation in this case. These results were not to appear in print again until Jung [45, 1955], after the papers of Godwin [35, 1949] and Lloyd [48, 1952] had renewed interest in this subject, although they are in Bennett's [4, 1952] unpublished thesis.

The final two sections were concerned with applications. Daniell gave special attention to the "discard-average" (the trimmed mean), presenting the (now standard) expression for its asymptotic variance and evaluating its performance for various Pearson densities, including Student's t. He also gave conditions under which the quartile-discard average is superior to the sample mean. The paper ended with a number of applications to other estimators of location and scale,[12] with numerical results. Daniell did not derive the asymptotic normality of $\hat{t}$, nor did he try to state minimal regularity conditions (indeed, some of his regularity conditions were implied rather than stated). However, taken altogether it is a thoroughly modern paper which almost appears to have been gleaned from the literature of the 1950's and 1960's.

How could such a paper have gone unnoticed for all these years?[13] To see why, we need to learn something of Daniell's life. Percy John Daniell (1889–1946) received a B.A. degree at Cambridge in 1910 (and an M.A. in 1914), where his honors included Senior Wrangler in Mathematics (1909), First Class Physics Tripos (1910), and the Raleigh Prize (1912). His stay at Cambridge would have overlapped R.A. Fisher's, but they were at different colleges and may not have met. After graduation (and brief stays at Göttingen and Liverpool), Daniell went to Rice Institute in Houston, Texas in 1912 as a travelling fellow. He remained at Rice until 1923, becoming a full professor in 1920. It was at Rice that he did his most important work, principally on the theory of integration (including the development of what is now known as the Daniell integral.) In 1924 he returned to England to the University of Sheffield, where he remained until his death at the age of 57. In the latter part of his life he published occasional papers on applied mathematics, on such topics as flame motion, potentials, and quadrature formulas.

The paper [8, 1920], written at Rice, seems to have been Daniell's only related work in statistics. This fact, together with his isolation from active statistical research (both at Rice and Sheffield), was largely responsible for the obscurity of the paper. Daniell's death before his results were rediscovered and widely discussed, and Wilks' overlooking his work in the survey paper of 1948 also served to delay recognition of his priority. As a further irony, these circumstances have helped relegate to obscurity another important paper of Daniell's, "Integral Products and Probability" [9, 1921], in which he presents one of the earliest mathematical treatments of continuous time Markov processes, including the Chapman-Kolmogorov equation (ten years before Kolmogorov) and a short treatment of the Wiener process (two years before Wiener).

5. M-ESTIMATORS

Recently, much attention has been given to a class of robust estimators which Huber calls "M-estimators," M for maximum-likelihood type (see Huber [39, 1972]). T is said to be an M-estimator corresponding to a function ϕ if T is a solution to $\sum \phi(X_i - T) = 0$. Each choice of ϕ determines an estimator; if $\phi = p'/p$, T is the maximum likelihood estimator for the location parameter of the population with density $p(t - \theta)$. As the first appearance of these estimators in the context of robustness seems to be in the work of Jeffreys after 1920 (see Jeffreys [42, 1932] and [43, 1939], in particular), and as this work is outside the scope of this study, we shall not dwell on this subject. However, we cannot resist calling attention to an early reference in which the class of M-estimators is introduced and their consistency claimed.

In a paper examining the various "proofs" of the method of least squares, Ellis [24, 1844] began with Gauss's first proof. Letting x_i's denote observed values, "a" the quantity to be estimated, and $e_i = x_i - a$, Ellis questions Gauss's a priori designation of the arithmetic mean (the solution to $\sum(x_i - a) = 0$) as the most probable value.

> It [the arithmetic mean] is not the only rule to which these considerations might lead us. For not only is $\sum e = 0$ ultimately, but $\sum fe = 0$, where fe is any function such that $fe = -f(-e)$; and therefore we should have
>
> $$\sum f(x - a) = 0,$$
>
> as an equation which ultimately would give the true value of x when the number of observations increases *sine limite*, and which therefore for a finite number of observations may be looked on

---

[12] Including the "discard-deviation," where the *inner* quartiles are discarded.
[13] A fairly complete review of the literature reveals only two published citations, Dodd [13, 1922] and Greenberg [37, 1968], and the descriptions there are superficial and misleading. Daniell's paper came to my attention as the result of a systematic inspection of the *American Journal of Mathematics*.

in precisely the same way as the equation which expresses the rule of the arithmetic mean. There is no discrepancy between these two results. At the limit they coincide: short of the limit both are approximations to the truth. Indeed we might form some idea how far the action of fortuitous causes had disappeared from a given series of observations by assigning different forms of f, and comparing the different values thus found for a.

No satisfactory reason can be assigned why, setting aside mere convenience, the rule of the arithmetic mean should be singled out from other rules which are included in the general equation $\sum f(x - a) = 0$.

Thus Ellis has claimed (without proof or regularity conditions) the consistency of M-estimators,[14] and even suggested the class may be useful for judging to what degree an estimated value depends on the choice of estimator, a stability test. Of course Ellis was not really concerned with robustness, only with illuminating the arbitrary nature of Gauss's proof, but his comments are of interest nonetheless.[15]

[Received January 1973. Revised June 1973.]

REFERENCES

[1] Airy, G.B., (1856), Letter from Professor Airy, Astronomer Royal, to the editor, *Astronomical Journal*, 4, 137–38.
[2] Andrews, D.F., Bickel, P.J., Hampel, F.R., Huber, P.J., Rogers, W.H. and Tukey, J.W., (1972), *Robust Estimates of Location: Survey and Advances*, Princeton: Princeton University Press.
[3] Anscombe, F.J., (1960), "Rejection of Outliers," *Technometrics*, 2, No. 1, 123–47.
[4] Bennett, C.A., (1952), "Asymptotic Properties of Ideal Linear Estimators," Unpublished dissertation, University of Michigan.
[5] Box, G.E.P., (1953), "Non-Normality and Tests on Variances," *Biometrika*, 40, 318–35.
[6] Chauvenet, W., (1863), *A Manual of Spherical and Practical Astronomy, Vol. 2*, Philadelphia: J. B. Lippincott Co., 558–66.
[7] Chernoff, H., Gastwirth, J. and Johns, M.V., (1967), "Asymptotic Distribution of Linear Combinations of Functions of Order Statistics with Applications to Estimation," *Annals of Mathematical Statistics*, 38, No. 1, 52–72.
[8] Daniell, P.J., (1920), "Observations Weighted According to Order," *American Journal of Mathematics*, 42, 222–36.
[9] ———, (1921), "Integral Products and Probability," *American Journal of Mathematics*, 43, 143–62.
[10] David, H.A., (1970), *Order Statistics*, New York: John Wiley and Sons, Inc.
[11] De Morgan, A., (1847), "Theory of Probabilities," *Encyclopedia of Pure Mathematics* (Part of *Encyclopedia Metropolitana*), 393–490.
[12] Dirichlet, G.L., (1836), "Über die Methode der kleinsten Quadrate," In *G. Lejeune Dirichlet's Werke, Vol. I*, 281–2. Berlin; Reimer, 1889.
[13] Dodd, E.L., (1922), "Functions of Measurements under General Laws of Error," *Skandinavisk Aktuarietidskrift*, 5, 133–58.
[14] Edgeworth, F.Y., (1883), "The Method of Least Squares," *Philosophical Magazine*, 16, Ser. 5, 360–75.

[15] ———, (1885), "Observations and Statistics. An Essay on the Theory of Errors of Observation and the First Principles of Statistics," *Transactions of the Cambridge Philosophical Society*, 14, 138–69.
[16] ———, (1886), "Problems in Probabilities," *Philosophical Magazine*, 22, Ser. 5, 371–84.
[17] ———, (1887a), "On Discordant Observations," *Philosophical Magazine*, 23, Ser. 5, 364–75.
[18] ———, (1887b), "The Choice of Means," *Philosophical Magazine*, 24, Ser. 5, 268–71.
[19] ———, (1888), "On a New Method of Reducing Observations Relating to Several Quantities," *Philosophical Magazine*, 25, Ser. 5, 184–91.
[20] ———, (1893), "Exercises in the Calculation of Errors," *Philosophical Magazine*, 36, Ser. 5, 98–111.
[21] Eisenhart, C., (1961), "Boscovich and the Combination of Observations," Chapter 7, in L.L. Whyte, ed., *R.J. Boscovich Studies of His Life and Work*, London: Allen and Unwin. (Reissued 1963 by Fordham University Press, New York).
[22] ———, (1964), "The Meaning of 'Least' in Least Squares," *Journal of the Washington Academy of Sciences*, 54, 24–33. Reproduced in Precision Measurement and Calibration. Selected NBS Papers on Statistical Concepts and Procedures (ed. H.H. Ku). National Bureau of Standards (US) Special Publication 300, Volume 1. Washington, D.C.: U.S. Government Printing Office, 1969.
[23] ———, (1971), "The Development of the Concept of the Best Mean of a Set of Measurements from Antiquity to the Present Day," 1971 A.S.A. Presidential Address.
[24] Ellis, R.L., (1844), "On the Method of Least Squares," *Transactions of the Cambridge Philosophical Society*, 8, 204–19.
[25] Encke, J.F., (1832), "Über die Methode der Kleinsten Quadrate," *Berliner Astronomisches Jahrbuch für 1834*, 249–312. Translated from the German in R. Taylor, ed., *Scientific Memoirs, Selected from the Transactions of Foreign Academies of Science and Learned Societies and from Foreign Journals, Vol. 2*, 1841, (reissued by Johnson Reprint Corporation, New York and London, 1966).
[26] Fechner, G.T., (1878), "Ueber den Ausgangswerth der Kleinsten Abweichungssumme dessen Bestimmung, Verwendung, und Verallgemeinerung," *Abhandlungen der Mathematisch-Physischen Classe der Königlich Sächsischen Gesellschaft der Wissenschaften*, Leipzig, 11, 1–76.
[27] Galton, F., (1875), "Statistics by Intercomparison, with Remarks on the Law of Frequency of Error," *Philosophical Magazine*, 49, Ser. 4, 33–46.
[28] ———, (1889), *Natural Inheritance*. London: Macmillan and Co., Ltd.
[29] ———, (1902), "The Most Suitable Proportion Between the Values of First and Second Prizes," *Biometrika*, 1, 385–90.
[30] Gastwirth, J., (1966), "On Robust Procedures," *Journal of the American Statistical Association*, 61 (December 1966), 929–48.
[31] Gauss, C.F., (1816), "Bestimmung der Genauigkeit der Beobachtungen," in *Carl Friedrich Gauss Werke*, Band 4, 109–17, Göttingen: Königliche Gesellschaft der Wissenschaften (1880). Summarized in Section 103 of E.T. Whittaker and G.R. Robinson, *The Calculus of Observations* (London and Glasgow: Blackie and Son, Ltd., 1924; 4th edition (1944) reissued 1967 by Dover Publications, Inc., New York).
[32] Glaisher, J.W.L., (1872), "On the Law of Facility of Errors of Observations, and on the Method of Least Squares," *Memoirs of the Royal Astronomical Society*, 39, Part II, 75–124.
[33] ———, (1873), "On the Rejection of Discordant Observations," *Monthly Notices of the Royal Astronomical Society*, 33, 391–402.
[34] ———, (1874), "Note on a Paper by Mr. Stone, 'On the Rejection of Discordant Observations'," *Monthly Notices of the Royal Astronomical Society*, 34, 251.

---

[14] Huber proved this in [38, 1964].
[15] In addition to those works cited below, many other works were consulted for references and general information. The information on the life of Simon Newcomb came principally from the *Encyclopedia Britannica*, the *International Encyclopedia of the Social Sciences*, and Newcomb's autobiography [59, 1903]. The information on the life of Percy Daniell came from various editions of *Who's Who* and *American Men of Science*, and Stewart [73, 1947]. Merriman's [50, 1877] bibliography on least squares was quite useful for the period prior to 1877. Good bibliographies of work since 1920 can be found in H.A. David's *Order Statistics* [10, 1970], and in Govindarajulu and Leslie [36, 1972].

[35] Godwin, H.J., (1949), "On the Estimation of Dispersion by Linear Systematic Statistics," *Biometrika*, 36, 92–100.
[36] Govindarajulu, Z. and Leslie, R.T., (1972), *Annotated Bibliography on Robustness Studies of Statistical Procedures*, U.S. Department of H.E.W. DHEW Publication No. (HSN), U.S. Government Printing Office, 72-1051.
[37] Greenberg, B.G., (1968), "Nonparametric Statistics: Order Statistics," *International Encyclopedia of the Social Sciences*, New York: The Macmillan Company and the Free Press.
[38] Huber, P.J., (1964), "Robust Estimation of a Location Parameter," *Annals of Mathematical Statistics*, 35, No. 1, 73–101.
[39] ———, (1972), "Robust Statistics: A Review," *Annals of Mathematical Statistics*, 43, No. 4, 1041–67.
[40] Hulme, H.R. and Symms, L.S.T., (1939), "The Law of Error and the Combination of Observations," *Monthly Notices of the Royal Astronomical Society*, 99, 642–9.
[41] Irwin, J.O., (1925), "On a Criterion for the Rejection of Outlying Observations," *Biometrika*, 17, 238–50.
[42] Jeffreys, H., (1932), "An Alternative to the Rejection of Observations," *Proceedings of the Royal Society*, 137, Ser. A, 78–87.
[43] ———, (1939), "The Law of Error and the Combination of Observations," *Philosophical Transactions of the Royal Society of London*, 237, Ser. A, 231–71.
[44] Jevons, W.S., (1874), *The Principles of Science I*, London: Macmillan, 2nd ed., (1877) reissued 1958 by Dover Publications, Inc., New York.
[45] Jung, J., (1955), "On Linear Estimates Defined by a Continuous Weight Function," *Arkiv For Matematik*, Band 3, nr. 15, 199–209.
[46] Laplace, P.S. de, (1818), *Deuxieme Supplement a la Théorie Analytique des Probabilités*. Paris: Courcier. (pp. 569–623 in *Oeuvres de Laplace*, 7, Paris: Imprimerie Royale (1847); pp. 531–80 in *Oeuvres Complétes de Laplace*, 7, Paris: Gauthier-Villars (1886).)
[47] Legendre, A.M., (1805), "On the Method of Least Squares." Translated from the French in D.E. Smith, ed., *A Source Book in Mathematics*, 576–79, New York: Dover Publications, Inc., 1959.
[48] Lloyd, E.H., (1952), "Least Squares Estimation of Location and Scale Parameters Using Order Statistics," *Biometrika*, 39, 88–95.
[49] Mantel, N., (1972), Letter to the editor, "Another Maximum Likelihood Oddity," *The American Statistician*, 26, No. 5 (December 1972), 45.
[50] Merriman, M., (1877), "A List of Writings Relating to the Method of Least Squares, with Historical and Critical Notes," *Transactions of the Connecticut Academy of Arts and Sciences*, 4, 151–232.
[51] Newcomb, S., (1878), "Researches on the Motion of the Moon, I," *Washington Observations for 1875*—Appendix II (published by the U.S. Naval Observatory, Washington.)
[52] ———, (1882), "Discussion and Results of Observations on Transits of Mercury from 1677 to 1881," *Astronomical Papers 1*, 363–487. Published by the U.S. Nautical Almanac Office.
[53] ———, (1885), *Principles of Political Economy*, New York: Harper and Brothers, Reissued in 1966 by Augustus M. Kelley, New York.
[54] ———, (1886), "A Generalized Theory of the Combination of Observations so as to Obtain the Best Result," *American Journal of Mathematics*, 8, 343–66.
[55] ———, (1891a), "Measures of the Velocity of Light Made under the Direction of the Secretary of the Navy during the Years 1880 to 1882," *Astronomical Papers*, 2, 107–230. Published by the U.S. Nautical Almanac Office.
[56] ———, (1891b), "Discussion of Observations of the Transits of Venus in 1761 and 1769," *Astronomical Papers*, 2, 259–405. Published by the U.S. Nautical Almanac Office.
[57] ———, (1895), *The Elements of the Four Inner Planets and the Fundamental Constants of Astronomy* (Supplement to the American Ephemeris and Nautical Almanac for 1897). Washington: U.S. Government Printing Office.
[58] ———, (1898), "Catalogue of Fundamental Stars for the Epochs 1875 and 1900 Reduced to an Absolute System," *Astronomical Papers*, 8, 77–403. Published by the U.S. Nautical Almanac Office.
[59] ———, (1903), *The Reminiscences of an Astronomer*, Boston: Houghton Mifflin. Reissued in 1972 by Somerset Publishers, New York.
[60] ———, (1912), "Researches on the Motion of the Moon, II," *Astronomical Papers*, 9, 1–249. Published by the U.S. Nautical Almanac Office.
[61] Og[o]rodnikoff, K., (1928), "On the Occurrence of Discordant Observations and a New Method of Treating Them," *Monthly Notices of the Royal Astronomical Society*, 88, 523–32.
[62] Pearson, E.S., (1967), "Studies in the History of Probability and Statistics XVII: Some Reflexions on Continuity in the Development of Mathematical Statistics, 1885–1920," *Biometrika*, 54, No. 2, 341–55.
[63] Pearson, K., (1902), "Note on Francis Galton's Problem," *Biometrika*, 1, 390–99.
[64] ———, (1920), "On the Probable Errors of Frequency Constants, III," *Biometrika*, 13, 113–32.
[65] ——— and Pearson, M.V., (1931), "On the Mean Character and Variance of a Ranked Individual, and on the Mean and Variance of the Intervals, between Ranked Individuals, I: Symmetrical Distributions (Normal and Rectangular)," *Biometrika*, 23, 364–97.
[66] Peirce, B., (1852), "Criterion for the Rejection of Doubtful Observations," *Astronomical Journal*, 2, 161–3.
[67] Peirce, C.S., (1873), "On the Theory of Errors of Observations," *Report of the Superintendent of the United States Coast Survey* (1870), 200–24.
[68] Poincaré, H., (1912), *Calcul des Probabilités*, Paris: Gauthier-Villars.
[69] Rider, P.R., (1933), "Criterion for Rejection of Observations," *Washington University Studies—New Series, Science and Technology*, No. 8.
[70] Sheppard, W.F., (1899a), "On the Application of the Theory of Error to Cases of Normal Distribution and Normal Correlation," *Philosophical Transactions of the Royal Society of London*, 192, Ser. A, 101–67.
[71] ———, (1899b), "On the Statistical Rejection of Extreme Variations, Single or Correlated (Normal Variation and Normal Correlation)," *Proceedings of the London Mathematical Society*, 31, 70–99, London, 1900.
[72] Short, J., (1763), "Second Paper Concerning the Parallax of the Sun Determined from the Observations of the Late Transit of Venus; in Which This Subject Is Treated of More at Length, and the Quantity of the Parallax More Fully Ascertained," *Philosophical Transactions of the Royal Society of London*, 53, 300–45.
[73] Stewart, C.A., (1947), "P.J. Daniell," *Journal of the London Mathematical Society*, 22, 75–80.
[74] Stigler, S.M., (1973), "Laplace, Fisher, and the Discovery of the Concept of Sufficiency," *Biometrika*, 60, (December).
[75] Stone, E.J., (1873), "On the Rejection of Discordant Observations," *Monthly Notices of the Royal Astronomical Society*, 34, 9–15.
[76] Walker, H.M., (1929), *Studies in the History of Statistical Method*, Baltimore: Williams and Wilkins.
[77] Wilks, S.S., (1943), "Order Statistics," *Bulletin of the American Mathematical Society*, 5, 6–50.

A TABLE SHOWING THE PROBABILITY
OF THE DURATION, THE DECREMENT,
AND THE DURATION OF LIFE,
IN THE STATES OF MASSACHUSETTS
AND NEW HAMPSHIRE

Edward Wigglesworth

XIII. *A Table shewing the Probability of the Duration, the Decrement, and the Expectation of Life, in the States of Massachusetts and New-Hampshire, formed from sixty two Bills of Mortality on the files of the American Academy of Arts and Sciences, in the year* 1789. *By* EDWARD WIGGLESWORTH, D. D. F. A. A.

ON examination of the bills of Mortality, on the files of the Academy, it appears that the society are under obligation to a considerable number of gentlemen, in different parts of the commonwealth, for the attention which they have paid to this subject. Since their formula has been dispersed through the state, many gentlemen have communicated bills of their respective towns, or parishes, with a topographical description of the same, which will lead to an investigation of the natural causes, which produce them, whenever it appears that particular diseases are endemial to any places.

Returns have been made from towns, scattered along the sea coast, from Nantucket on the south, to Portland in Casco-bay, on the north east, and through the counties of Middlesex, Worcester, and Hampshire, in a western direction. From Hingham,

Hingham, Ipswich, East Kingston, Dover, Portland, Edgarton, Waltham, Ashburnham, Brookfield, and Brimfield, they have been made for a long course of years. And though those, which have been made from other places, are for a shorter term; yet, as they are from places very distant from one another, it is presumed that the result, from a combination of the bills, will give a very just representation of the increasing population of this state and New-Hampshire, and the longevity of their inhabitants. At least, it will exhibit more authentick evidence on the last of these subjects, than has, as yet, been laid before the publick. Should there be any errour in the deductions, there will be at least an approximation towards the truth; and the table will give the probability, decrement, and expectation of life, among the inhabitants of this state and New-Hampshire, with more accuracy, than has yet been done. Any errours in the deductions may be corrected, by taking the present result for the basis of a new combination with the bills, that may hereafter be communicated.

The whole number of deaths on the bills is 4893, which happened at the following periods of life.

Under	5	years	1942
Between	5 and	10	236
	10	15	136
	15	25	425
	25	35	382
	35	45	349
	45	55	270
	55	65	270
	65	75	372
	75	80	185
	80	85	171
	85	90	103
	90	95	36
	95	100	16

Deductions from Bills of Mortality.

From these elements the table is formed, by taking the number of deaths as the radix of calculation. This would have given the proportional numbers of persons living, and dying at every age, from the birth to the latest extremity of life, had the annual number of deaths been equal to the births. But by the bills it appears, that the births are annually double to the deaths. Therefore, the number of persons of each age, as given by the table, is less, than is actually in life together, from an annual excess of 4893 births. Consequently the expectation of life is less than just, especially in the early periods of life.

The TABLE.

Ages.	Persons living.	Decrem. of life.	Expectation of life. yrs. dec.	Ages.	Persons living.	Decrem. of life.	Expectation of life. yrs. dec.
At birth	4893	1264	28.15	20	2365	43	34.21
1	3629	274		21	2322	42	
2	3355	188		22	2280	42	
3	3167	132		23	2238	42	
4	3035	84		24	2196	42	
5	2951	58	40.87	25	2154	40	32.32
6	2893	55		26	2114	38	
7	2838	47		27	2076	38	
8	2791	40		28	2038	38	
9	2751	36		29	2000	38	
10	2715	28	39.23	30	1962	38	30.24
11	2687	27		31	1924	38	
12	2660	27		32	1886	38	
13	2633	27		33	1848	38	
14	2606	27		34	1810	38	
15	2579	42	36.16	35	1772	35	28.22
16	2537	43		36	1737	35	
17	2494	43		37	1702	35	
18	2451	43		38	1667	35	
19	2408	43		39	1632	35	

Deductions from Bills of Mortality.

Ages.	Persons living.	Decrem. of life.	Expectn. of life. yrs. dec.	Ages.	Persons living.	Decrem. of life.	Expectat. of life. yrs. dec.
40	1597	35	26.04	70	698	37	10.06
41	1562	35		71	661	37	
42	1527	35		72	624	37	
43	1492	35		73	587	38	
44	1457	34		74	549	38	
45	1423	27	23.92	75	511	37	7.83
46	1396	27		76	474	37	
47	1369	27		77	437	37	
48	1342	27		78	400	37	
49	1315	27		79	363	37	
50	1288	27	21.16	80	326	35	5.85
51	1261	27		81	291	34	
52	1234	27		82	257	34	
53	1207	27		83	223	34	
54	1180	27		84	189	34	
55	1153	27	18.35	85	155	21	4.57
56	1126	27		86	134	21	
57	1099	27		87	113	21	
58	1072	27		88	92	20	
59	1045	27		89	72	20	
60	1018	27	15.43	90	52	8	3.73
61	991	27		91	44	7	
62	964	27		92	37	7	
63	937	27		93	30	7	
64	910	27		94	23	7	
65	883	37	12.43	95	16	6	1.62
66	846	37		96	10	5	
67	809	37		97	5	3	
68	772	37		98	2	1	
69	735	37		99	1	1	

The whole number of inhabitants, according to this table, is 140182; of which 48183 are persons under 16 years of age; and 91999, above sixteen years of age. By the enumeration

meration of the inhabitants of Maffachufetts, the whole number of free males under fixteen was 95453; and 87189, above fixteen. Therefore, 35851 perfons under fixteen, muft be added to thofe in the table under fixteen, to make the table accord with the enumeration; which will give 176033 inhabitants, produced by an excefs of 4893 annual births. This addition will raife the expectation of a child juft born, from 28 years and $\frac{15}{100}$ to 35 years and $\frac{47}{100}$; of a child of 5 years of age, from 40 years $\frac{87}{100}$ to 48 years and $\frac{46}{100}$; of a perfon of 10 years, from 39 years and $\frac{23}{100}$ to 43 years and $\frac{23}{100}$; of a perfon of 15 years, from 36 years and $\frac{16}{100}$ to 36 years and $\frac{10}{100}$.

The annual excefs of 4893 births above the deaths, on a ftock of 176033 inhabitants, determines the period of duplication to be twenty five years and three tenths of a year. At this rate, the inhabitants of the five New England ftates are probably increafing at this time by natural population, without any confideration being had either to foreign or American acceffions.

Similar bills, kept in the other ftates with the fame accuracy that they have been kept here, would determine their natural population, with a degree of accuracy, which would be of utility to the publick; and would afford entertainment to perfons of a philofophical difpofition, both in Europe and America.

NOTE ON C.S. PEIRCE'S EXPERIMENTAL DISCUSSION OF THE LAW OF ERRORS

Edwin B. Wilson

NOTE ON C. S. PEIRCE'S EXPERIMENTAL DISCUSSION OF THE LAW OF ERRORS

BY EDWIN B. WILSON AND MARGARET M. HILFERTY

DEPARTMENT OF VITAL STATISTICS, HARVARD SCHOOL OF PUBLIC HEALTH

Read before the Academy November 19, 1928

Incident to the preparation of the official biography of C. S. Peirce for this Academy the senior author came across Peirce's experimental discussion of the law of errors[1] wherein the conclusion was that the normal law was, on the whole, verified. This is in accord with the dictum of Poincaré[2] that everybody believes in the law of errors: the mathematicians because they think it empirically demonstrated by experimenters and the experimenters because they think the mathematicians have proved it *a priori*. The series of observations given by Peirce is long, consisting of about 500 records each day for 24 different days of the time elapsed between the making of a sharp sound and the record of reception of the sound by an observer. According to our previous experience such long series of observations generally reveal marked departures from the normal law, and it seems interesting to examine this material of Peirce's from the point of view of the modern theory of frequency functions.[3] Accordingly we have entered in the tables, (1) the median with its standard deviation, (2) the mean with its standard deviation, (3) the semi-interquartile range, (4) two-thirds of the standard deviation, (5) the ratio of probable error as defined by the semi-interquartile range to the probable error as defined by two-thirds or more accurately 0.6745 of the standard deviation, (6) the arithmetic mean error about the median as origin, (7) the number of negative errors greater than 3 σ and the number of positive errors of similar magnitude and their sum, (8) the number of observations within 0.25 σ of the mean and the number expected according to the normal law, (9) the percentage excess of the observed over the expected number, (10) the second moment μ_2 and its standard deviation, (11) the third moment μ_3 and its standard deviation, (12) the Pearsonian constant $\sqrt{\beta_1} = \mu_3/\sigma^3$

(Charlier's measure of skewness), (13) the Pearsonian skewness sk, (14) Yule's measure of skewness as the sum of the quartiles less twice the median divided by the semi-interquartile range, (15) the Pearsonian kurtosis $\beta_2 - 3$, (16) the Pearsonian frequency type, with the appended sign $+$ meaning that the sampling error of the fourth moment is infinite and with $++$ meaning that the third moment also has an infinite sampling error for samples drawn from the universe defined by the curve.

The following observations may be made on the table:

1. There is on each of the 24 days a greater concentration of small errors in the vicinity of the mean or median than the normal law, fitted with the computed value of the standard deviation, predicts. This may be seen in a number of ways: (a) the quotient of the probable error defined in column (3) divided by the probable error in column (4), given in column (5), is uniformly less than 1, varying from a maximum of 0.98 which presumably does not depart significantly from 1 to 0.51 which does greatly depart therefrom, with a mean or median value of about 0.765 which cannot in any way be reconciled with the theoretical value 1. (b) The arithmetic mean error in column (6) is never so high as 0.8 σ as required by the normal curve but varies from 0.76 σ to 0.50 σ about a median value of 0.7 σ. (c) The number of observations within 0.25 σ of the mean on either side as observed in column (8) is uniformly in excess of the theoretical number by anywhere from 4 to 94%.

2. There is a considerable excess of large errors as compared with the normal law. For example, the normal law would indicate on the basis of 500 observations that one error should exceed 3.1 σ. In column (7) the observed numbers are entered. There is only one day on which the number of errors above 3.1 σ is less than 3 and on five days there are 8. The mean number of negative errors greater than 3.1 σ is 1.7 instead of 0.5 and the mean number of positive errors greater than 3.1 σ is 3.9 instead of 0.5, making the mean value of the total 5.6 instead of 1. When factors of safety of 3 to 8 are observed, can one say that the theoretical distribution is substantiated? When one calculates the ratio of the largest error to the value of σ for each day one finds numbers running up to over 15; and such errors are inconceivably rare on the basis of the normal law.

As a matter of fact, is it not generally true of errors of observation that they exhibit an excess of small and an excess of large errors as compared with the normal law?

3. The curves exhibit positive skewness. Just how to formulate the precise amount or even the sign of this phenomenon is difficult to say. If the ratio $\beta = \mu_3/\sigma^3$ is taken, column (12), we know that according to the books the value of β for samples drawn from any normal universe is $\sqrt{6/n} = 0.11$ when $n = 500$. There is only one value of β lying between -0.11 and $+0.11$, and the values run up to 10.9 with a median of 1. Two

values are negative and 22 are positive. On the basis of Pearson's skewness, column (13), of which the standard deviation is 0.055, there are 4 values less than 0.055 and 20 values greater; there are 10 values less than

	1	2	3	4	5	6	7			8	
			PROBABLE ERROR							NO. WITHIN 0.25σ	
DAY	MEDIAN	MEAN	$\frac{Q_3-Q_1}{2}$	0.6745σ	$\frac{Q_3-Q_1}{2(0.6745\sigma)}$	A.D.	ERRORS $>3.1\sigma$			OF THE MEAN	
							NEG.	POS.	TOTAL	OBS'D.	EXPECTED
1	468.0 ±3.0	475.6 ±4.2	57.9	62.1	0.932	70.1	1	3	4	110	98
2	236.8 ±2.4	241.5 ±2.1	26.6	31.6	0.842	35.7	1	0	1	113	97
3	199.9 ±1.8	203.1 ±2.0	27.5	30.4	0.905	33.7	0	7	7	113	97
4	201.3 ±1.2	205.6 ±1.8	19.4	26.6	0.730	26.9	1	7	8	134	99
5	147.0 ±2.1	148.5 ±1.6	21.4	23.5	0.912	26.7	0	4	4	110	97
6	172.4 ±1.8	175.6 ±1.8	20.0	26.9	0.744	28.0	0	6	6	119	97
7	184.0 ±1.6	186.9 ±2.2	24.6	32.7	0.753	32.3	0	6	6	132	98
8	193.9 ±1.2	194.1 ±1.4	17.2	20.5	0.840	22.0	2	6	8	120	97
9	195.4 ±1.4	195.8 ±1.6	17.9	23.7	0.756	24.0	2	4	6	132	98
10	215.1 ±1.5	215.5 ±1.3	16.4	19.3	0.850	21.0	2	1	3	120	99
11	213.1 ±1.8	216.6 ±1.7	19.7	25.2	0.782	25.9	1	5	6	135	99
12	232.7 ±1.9	235.6 ±1.7	17.0	22.4	0.759	23.4	3	5	8	103	78
13	244.4 ±1.3	244.5 ±1.2	16.3	17.7	0.922	19.9	6	1	7	101	97
14	236.3 ±1.2	236.7 ±1.8	14.7	27.8	0.529	21.6	2	3	5	192	99
15	253.3 ±1.0	236.0 ±1.4	14.3	21.6	0.662	20.8	4	4	8	162	98
16	233.3 ±1.4	233.2 ±1.7	15.6	25.5	0.612	21.7	4	2	6	162	98
17	264.0 ±1.7	265.5 ±1.7	20.6	26.0	0.792	27.6	3	5	8	123	100
18	254.0 ±1.3	253.0 ±1.1	15.8	16.5	0.959	18.6	0	4	4	114	98
19	255.0 ±0.9	258.7 ±1.8	13.9	27.7	0.502	20.4	1	3	4	187	99
20	252.6 ±1.3	255.4 ±2.0	15.2	29.2	0.521	21.6	0	3	3	179	98
21	245.2 ±1.6	245.0 ±1.2	14.6	18.5	0.790	19.2	3	4	7	120	99
22	255.0 ±1.5	255.6 ±1.4	14.8	21.5	0.688	19.0	2	4	6	142	99
23	252.0 ±1.1	251.4 ±1.6	15.0	24.6	0.610	19.8	0	3	3	158	98
24	244.2 ±0.9	243.4 ±1.1	11.6	15.9	0.730	16.8	3	3	6	113	98
							1.7	3.9	5.6		

	9	10	11	12	13	14	15	16
	EXCESS OBS. OVER				SKEWNESS			PEAR-
DAY	EXPECTED	$\mu_2 \pm \sigma_{\mu_2}$	$\mu_3 \pm \sigma_{\mu_3}$	CHARLIER'S μ_3/σ^2	PEARSON'S sk	YULE'S	β_2-3	SONIAN TYPE
	%							
1	12	8475 ±856	918,117 ±312,000	1.18	0.417	+0.107	3.1	IV
2	16	2197 ±168	41,338 ± 14,000	0.43	0.158	+0.192	0.9	IV
3	16	2046 ±219	100,824 ± 37,000	1.09	0.314	−0.029	3.6	IV
4	35	1552 ±237	111,274 ± 64,000	1.82	0.407	+0.108	9.7	IV+
5	13	1210 ± 99	14,511 ± 11,800	0.39	0.126	+0.252	1.3	IV
6	22	1596 ±209	91,972 ± 38,600	1.48	0.375	0	6.4	IV+
7	35	2353 ±547	336,594 ±248,000	2.96	0.592	+0.114	24.9	IV++
8	23	926 ±101	13,856 ± 12,000	0.48	0.100	+0.070	4.1	IV++
9	35	1233 ±220	73,081 ± 59,000	1.71	0.268	−0.056	13.8	IV++
10	21	818 ±120	18,569 ± 23,600	0.84	0.129	−0.012	8.8	IV++
11	36	1394 ±214	83,461 ± 48,900	1.69	0.352	+0.152	9.8	IV++
12	32	1108 ±143	22,133 ± 17,000	0.63	0.118	+0.312	4.7	IV++
13	4	692 ± 67	4,894 ± 5,800	−0.22	−0.046	+0.037	2.6	IV+
14	94	1702 ±616	401,764 ±207,000	5.74	1.595	−0.034	63.6	VI+
15	65	1030 ±252	55,666 ± 25,000	1.68	0.222	−0.042	27.9	IV++
16	65	1426 ±614	345,877 ±328,000	6.39	1.445	+0.006	90.6	VI+
17	23	1486 ±165	14,176 ± 23,700	0.21	0.039	+0.126	4.3	IV++
18	16	596 ± 52	3,879 ± 3,700	0.27	0.080	−0.139	1.8	IV+
19	89	1689 ±910	758,187 ±512,000	10.94	112.0	+0.288	143.9	I
20	81	1884 ±818	632,595 ±427,000	7.71	3.496	+0.006	91.4	VI
21	21	752 ±107	4,724 ± 16,000	0.23	0.030	−0.055	8.2	IV++
22	43	1024 ±383	171,371 ±156,000	5.27	1.08	−0.148	68.1	VI++
23	61	1333 ±343	144,188 ±135,000	2.73	0.420	−0.220	31.1	IV++
24	15	559 ± 68	−342 ± 6,600	−0.02	−0.006	−0.241	5.4	IV++

$3\sigma = 0.165$ but 14 values greater. Still we have two negative and 22 positive values. On the basis of Yule's value, column (14), we find 10 negative, 1 zero and 13 positive values for the skewness; the comparison of the signs and numerical magnitudes in column (14) and in columns (12) and (13) is interesting.

4. The values of the kurtosis are impossible to reconcile with the notion of random sampling from a normal universe. For the value of $\beta_2 - 3$, of column (15), is, as ordinarily given, $0 \pm \sqrt{24/n} = 0 \pm 0.22$, and there is

no value of $\beta_2 - 3$ which is less than 4 times this amount; indeed, the observed values seem to run up almost indefinitely.

The failure of β to lie within the limits ± 0.11 and of $\beta_2 - 3$ to lie within the limits ± 0.22 justifies the statement that on no single day of the 24 was the distribution of errors such that its values of β and $\beta_2 - 3$ could have arisen by random sampling from a normal universe.

5. On account of the large kurtosis the usual formula for the standard deviation of the standard deviation, $\sigma_\sigma = \sigma \sqrt{2n}$, based on the normal curve cannot be applied; we must use $\sigma_\sigma = \sigma \sqrt{(\beta_2 - 1)/4n}$.

6. The third moments as calculated from the data are not reliable

because the sampling errors as calculated from the data, using sixth moments, are so large—see column (11).

7. The type of Pearsonian frequency curve which fits the data is IV on 19 of the 24 days, is VI on 4 days and I on one day. But it should be observed that on 18 of the 24 days the fitted Pearsonian curve has infinite eighth moments μ_8 and that therefore if the universe of errors of observation were really as thus defined on any one of these days the sampling errors of the values of μ_4 would be infinite. The erratic conduct of $\beta_2 - 3$, observed just above, is possibly fair evidence that in very truth the universe of errors of observation in the case of this whole series is heterotypic and that the values of μ_4 or β_2 are in no way to be depended upon. (In 12 of the 24 days the fitted Pearsonian curve has infinite sixth moments so that the reliability of the third moments must be nearly nil—and judging from the sampling errors as given in column (11) it is not alone in these 12 days that the moments are unreliable.)

8. The distributions cannot be represented by the Charlier A-type expansions using the third and fourth moments (but no higher moments) as computed. The departures from the normal curve are in fact so large that the A-type expansion becomes negative in some ranges of the variable less than 3σ removed from the mean in 20 of the 24 days.

The upshot of this all is that Peirce had observations which could show as completely as one might desire that the departures of the errors from the normal law was for his series uniformly great.

Let us turn from the frequency distribution to the statistical constants. The ordinary statement based on the normal law is that the determination of the median is 25% worse than that of the mean. A comparison of the standard deviations of the median and mean in columns (1) and (2) shows that for these observations the median is better determined than the mean on 13 days, worse determined on 9 days, and equally well determined on 2 days. Roughly speaking this means that mean and median are on the whole about equally well determined.

The changes in the mean from day to day are significant in the statistical sense. The value of the mean gives the correction for the "personal equation" of the observer for this type of observation. For scientific purposes in applying the corrections and in estimating the inexactness of the correction we must include in that estimate of inexactness all variations of the personal equation which are not subject to known correction. As the standard deviation of the mean is $\sigma/\sqrt{n}$ or about 1.7 we may not expect better exactness than this. However, the real (as yet uncontrolled and unexplained) variation of the mean is very much greater. Just how to estimate it from the data is impossible to say because there are obviously progressive changes in the mean. On the first day when the observer is "finding himself" the mean is very large; on the second day it drops to about the terminal value, but continues to drop for three days more; thereafter it rises to a maximum on the 17th day and then falls off.

The actual changes in the mean for the last half of the series are

$$+9, -8, -1, -3, +32, -12, +5, -3, -10, +11, -5, -8.$$

There is little to suggest anything but statistical fluctuations in these differences. It is evident that whatever may be the proper estimate of the standard deviation of the mean even during the last half of the series when the observer may be considered to have settled down, the value must very greatly exceed the 1.7 obtained by the formula $\sigma/\sqrt{n}$, i.e., a reasonably large Lexian ratio is involved. This illustrates the principle that we must have a plurality of samples if we wish to estimate the variability of some statistical quantity, and that reliance on such formula as $\sigma/\sqrt{n}$ is not scientifically satisfactory in practice, even for estimating unreliability of means.

With respect to the variation of the standard deviation from day to day it is clear that the first day is quite imcomparable with the subsequent days, that beginning with the second day there is a gradual reduction of the value of σ of total extent of perhaps 4 or 5 units in 23 days, superposed upon which there are fluctuations of very considerable size. The standard deviation of σ is indeed about 6.2. The formula $\sigma/\sqrt{2n}$ gives about 0.8 for σ_σ on the basis of a mean value $\sigma = 36$. But, as seen above, the kurtosis of the frequency curves is such as to make necessary the use of $\sigma\sqrt{(\beta_2-1)/4n}$ in place of $\sigma/\sqrt{2n}$. The root mean square of $(\beta_2 - 1)/2$ is 3.7 and if it be appropriate to use this as a factor we should give σ_σ the value 3.0 which still is less than half the observed value 6.2.

It appears, therefore, that the sampling theory is not applicable to the results of the identical experiment repeated on 24 days (when the first day is rejected) even when the analysis is confined to the last half of the series. The mean and standard deviation vary much more than is predicted.

[1] On the "Theory of Errors of Observations" by Assistant C. S. Peirce, *Rep. Super. U. S. Coast Survey* (for the year ending Nov. 1, 1870), Washington, Gov't. Printing Office, 1873, Appendix No. 21, pp. 200–224 and Plate No. 27.

[2] H. Poincaré, *Calcul des Probabilités*, chap. X, p. 171, 1912.

[3] The only reference to the series that we have found in the literature is one by M. Fréchet, "Sur la loi des erreurs d'observation," *Rec. Soc. Math., Moscow*, 32, 1924, in which the author states that at his instigation one of his pupils (Samama) had adjusted by the normal law and by the first Laplacian law $e^{-a|x|}$ several of Peirce's curves and had found that, though the two laws present satisfactory adjustments, the first gives the better result.

ON PROFESSOR AIRY'S OBJECTIONS TO PEIRCE'S CRITERION

Joseph Winlock

THE ASTRONOMICAL JOURNAL.

No. 91.

ON PROFESSOR AIRY'S OBJECTIONS TO PEIRCE'S CRITERION.*

By JOSEPH WINLOCK.

The conclusion to which Professor Airy has been led, in his endeavors to make himself acquainted with Professor Peirce's *Criterion for the Rejection of Doubtful Observations*,—that is, "that the whole theory is defective in its foundation, and illusory in its results,"—may well surprise, not only the Astronomer Royal himself, but mathematicians generally, who are acquainted with the character of Professor Peirce's numerous and valuable mathematical investigations, or who have given any attention to this one.

The *Criterion* has already come into use, to some extent, in this country; it has been employed in several astronomical researches, and in its extended form, prepared by Dr. Gould (*Astronomical Journal*, No. 83), is now used altogether in the discussion of observations in the United States Coast Survey. It might, then, be safely assumed, that American mathematicians are sufficiently aware of its correctness and value. But as it is well known that improvements in methods of calculation are slowly and cautiously adopted by practical computers, even when their utility and accuracy are unquestioned, it is but reasonable to suppose, that the adverse opinion of so eminent a mathematician as the Astronomer Royal, emphatically and authoritatively expressed, would add not a little to this natural opposition to what is new. And it might tend, in no unimportant degree, to retard the general adoption of the *Criterion*, especially among that class of astronomers who would not be sufficiently attracted to it as a theoretical discussion, to take the trouble requisite to satisfy themselves of the soundness of the theoretical basis on which it rests. For this reason, and not that I think his objections of sufficient importance, otherwise, to require an answer, I have thought I might render some service to science by calling attention to what I consider the unsoundness of his arguments, and the falsity of his conclusions.

His objections go back to the first principles of the Calculus of Probabilities, and his difficulties are confined to obvious elementary considerations, such as would present themselves at once to any one in a preliminary examination of the subject; they could not have escaped a mathematician of Professor Peirce's sagacity, and were no doubt readily disposed of by him before approaching the real difficulties of the problem.

An objection to which he seems to attach much importance is contained in the following paragraph:—

"It appears to me that, in the whole process to which I am alluding, there is a fault of reasoning of a vital character; namely, the assumption that we can apply the reasoning of antecedent probability to the discussion of observations which are recorded as having been actually made. In ordinary life, instances perpetually occur (some of a permanent, some of a transient character) which, considered beforehand, are in the highest degree improbable, but which, as matters of history, are absolutely certain."

If this could be urged as an objection of a vital character against the *Criterion*, it would prove equally fatal to any theory of combination of discordant observations. The Professor certainly accepts the *Method of Least Squares*; but, in using that, is he not applying a "theory of antecedent probability" to a set of observations for the purpose of obtaining a result which he receives as the most probable, although it may be at variance with the evidence of each and every one of the observations which are recorded as having been actually made?

Why should there be, *a priori*, any greater objection to approaching a series of observations with a theory of antecedent probability for the purpose of detecting abnormality in some of them, than for the purpose of ascertaining any other peculiarity or property that they may possess, whose existence or magnitude has to be received or rejected by us on an estimate of the probabilities for or against it? But we cannot speak with propriety of the residual errors as being actually made, until we know with certainty the quantity observed.

While it is true that, in ordinary life, and in matters of history, things are perpetually occurring which are beforehand in the highest degree improbable, it is equally true that, on an

---

* "Letter from Professor Airy, Astronomer Royal, to the Editor," *Astronomical Journal*, Vol. IV. pp. 137, 138.

estimate of the probabilities against events said to have occurred, we refuse our belief in them altogether, or consider them so improbable that it would be unwise for us to take them into account, or to allow them any influence in forming our opinions or directing our actions. More especially would this be the case when the question to be decided by us in regard to them was, whether they were produced by the operation of ordinary or of extraordinary causes, which is the question in the case of doubtful observations.

He urges against it another argument, based on its assumed failure in a special case, — the observed zenith-distances of *Polaris*. But since he seems to have applied it here with no other data than the error of the doubtful observation, we are justified in doubting whether he has any right to pronounce this a case of failure. Nor can we attach any consequence to his assertion that it would fail in the instance which he has cited in his note, of observations for determining the figure of the earth, until it has been actually applied in accordance with its own rules and conditions.

It is not easy to see how the enumeration of the separate sources of error, in this connection, strengthens his position. If the effect of any one of them is known, the observations must first be corrected for it; if the manner of its action is known to be such that we cannot assume that the probability of an error produced by it may be expressed by the function $\frac{h}{\sqrt{\pi}} e^{-h^2 \Delta^2}$, any objection on this account against the *Criterion* would lie also against the application of the *Method of Least Squares* for any purpose; for this method supposes that the residual errors are not affected by anything inconsistent with the law of error upon which it is based. There is nothing in the *Criterion* which would prevent our using any knowledge we may have of the errors arising from different causes. When we know nothing to the contrary, we are authorized to assume, whatever may be those separate causes, whether "bad definition of the star, failure of the nervous system of the observer's eye, failure of the nervous or muscular system of the fingers, or the irregular blasts of wind on the spokes of the circle," that each one and all together produce errors in accordance with the law of error which the *Method of Least Squares* supposes, and that, therefore, the *Criterion* is applicable to them, if it has been logically deduced from that method.

But granting, in the case he has cited, all that he claims, it amounts to nothing more than this, — that the *Criterion* might reject, in an extreme case, a good observation; that it might fail to distinguish between the most improbable and an abnormal observation. Here, and throughout his paper, he seems to hold the opinion that it is just as injurious to reject a good observation as it is to retain a bad one, provided the latter is within the limit of the greatest possible chance error. In connection with the case cited above, of the zenith-distances of *Polaris*, he says : "But is it not injurious to retain such a result? I cannot say; but it is quite as probable that its retention will be beneficial." How do we know that retaining an observation affected by an error, which requires the concurrence of the maxima of all the constituent errors with the same sign, will affect the mean of a series beneficially quite as often as injuriously? Before we can be prepared to say that, in a given group, there is an even chance of this being the case, we must know the number of observations, and the number of distinct sources of error.

We have a right to expect such large errors in an extended series, where their effect is unimportant, and counterbalanced by similar errors with opposite signs; but when we find them irregularly in small groups, we have reason to fear that their effect may be prejudicial to the mean. But we are not required to use the same care to avoid the rejection of a good observation (that is, one affected only by ordinary causes of error) as the retention of a bad one. When we reject a good observation, we do not necessarily introduce error, we simply decline to use all the evidence we have, and our conclusions are of proportionally less value; but retaining one which is abnormal vitiates the result. If the observation is simply doubtful, the doubt goes to depreciate the probability of the correctness of our conclusions, and in this way has the same effect as rejecting one known to be good. Then we are not bound, as Professor AIRY contends, to be absolutely certain that extraordinary causes, such as the slip of a clamp, have operated, before we are authorized to reject; but we must reject whenever an observation is so doubtful, that retaining it makes our conclusions less reliable than they would be if its evidence had not been used. The *Criterion* not only shows us when this is the case, but, which is one of its greatest merits, it enables us to estimate how much the doubtful character of the rejected observations, even after their rejection, detracts from the weight of results obtained from the series in which they were found; and thus furnishes us the only method we have of ascertaining the probable error of conclusions based on some observations of a series of which the rest have been rejected as defective or doubtful.

Finally, he concludes, "that we can hope for no rule to guide us in the rejection of observations, and we must abandon the mathematical theory altogether, and rely upon the application of our mental reasoning to the independent examination of the special cases which may come before us." This rests on the following argument : —

The mathematical theory of probabilities fails in all questions applying to errors of extreme magnitude. In applying the theory to doubtful observations, the large errors are those under consideration; then, in such cases, we are using the failing part of the theory; and, therefore, our conclusions must necessarily be worthless.

The difficulty that arises from the fact that theory admits the possibility of errors of any magnitude in an infinitely extended series of observations, and therefore in any group, however small, regarded as a part of such series, is one of the first that presents itself in the consideration of the problem. And it seems to me that the failure referred to is, in this case, as in all others for which the *Method of Least Squares* is used, of no practical importance. It is purely theoretical; represented geometrically, — A curve which should meet a line can, theoretically, only approach it within a distance less than any assignable quantity. Nor is the fact that large errors generally occur even more frequently than the theory requires, of any consequence to our present purpose. If the quantity observed were already known with certainty, and we should propose to decide which of a small group of observations were affected by extraordinary causes, when their number was too small for us to expect the distribution of their errors to exhibit conformity to theory, and they were supposed to be taken at random from a series of indefinite extent, the failure of the theory would render impossible any satisfactory solution of the problem. And this seems to be the case that Professor AIRY had in his mind when he gave up the problem. But even in this case, if the observations were to be used to determine an unknown quantity, the discordances might authorize us to doubt whether it would be safe to admit the evidence of some of them. But when, as is always the case in practice, our knowledge of the quantity observed rests on the evidence of the observations, and is affected by their errors, and we must aim to find that value from the evidence before us to which the least doubt can be attached, it is of no consequence to us that an observation is *theoretically possible*, if its discordance renders it so doubtful that it would make our result less reliable than if we rejected its evidence altogether.

The residual errors furnish us a legitimate means, and, in most cases, the only means we have, of estimating the degree of confidence due to any single observation, or to a result obtained by a combination of many; and, regarding the probability of an error as a function of its magnitude, we are enabled to find the probability of any system of residual errors, and by the comparison of the systems of errors before and after rejection in accordance with the rule of the *Criterion*, we can decide, within safe limits, whether the probability of our final result is lessened by retaining the doubtful observations, without troubling ourselves to know whether they are theoretically or even practically possible or not.

In conclusion, I will examine his *résumé*, which contains the points on which he wishes especially to insist.

1. " *The mathematical theory of probabilities fails in all questions applying to errors of extreme magnitude.*"

This failure is of no practical importance, and cannot be urged as an objection to the *Method of Least Squares*, nor to anything logically deduced from that method.

2. " *No consideration of the magnitudes of residual errors per se will justify us in rejecting a result.*"

Any theory of probabilities applied to observations must rest upon the principle, that the probability of any error is a function of its magnitude. No. 2, then, is equivalent to the assertion, that no consideration of the probabilities against an observation *per se* can justify us in rejecting it.

3. " *We are justified in rejecting a result only when, from the best estimate that we can form of the extent of action of the various causes which can produce error, we find that the combination of those causes of error cannot possibly produce the discordance in question;* —

4. " *And when we perceive that other causes may have intervened, whose nature is such that they cannot be recognized as occurring in the ordinary series of observations.*"

In practice it is impossible to know all the possible causes which may produce error, and we could form no estimate of the maximum or probable effect of such causes as irregular blasts of wind on the spokes of a circle, failure of the nervous or muscular system of the fingers, or of the eye, even if their effects could be separated from the others, without a consideration of the residual errors *per se*.

No. 3, then, if it could ever be a practical guide to any one, could only be used by the observer himself, before *all* the circumstances of the observations had escaped his memory; and it might then lead him right in extraordinary cases, such as that recorded of General COLBY.

No. 4 adds another practical difficulty to the application of the criterion of " mental reasoning " to special cases. Taken in connection with No. 3, it serves to present in a strong light Professor AIRY's positions: — that observations are not to be judged by their discordances, but by a knowledge of instrumental defects, the observer's liability to error, and extraneous accidental influences, independent of that furnished by the observations themselves; and that we must have certainty of the existence of abnormal error, not only by being convinced that no ordinary causes could have possibly produced the discordance in question, but we must, in addition to this, perceive that extraordinary influences may have intervened, before we are authorized to reject; or that, in forming our judgment in doubtful cases, we must lean to the side of error.

ON THE DEVELOPMENT OF FORMULAE
FOR GRADUATION BY LINEAR COMPOUNDING,
WITH SPECIAL REFERENCE
TO THE WORK OF ERASTUS L. DE FOREST

H.H. Wolfenden

On the Development of Formulae for Graduation by Linear Compounding, With Special Reference to the Work of Erastus L. De Forest.

BY

HUGH H. WOLFENDEN

[This paper was read in abstract before the International Mathematical Congress, Toronto, in August, 1924, and is here published in full by permission of the Editorial Committee of the Congress.]

1. The problem of "graduation" by "linear compounding" with which actuaries and vital statisticians have to deal concerns the replacement of an observed value* u_r (of a true value U_r) by a linear compound, v_r, of u_r and terms $u_{r+1}, u_{r+2}, \ldots u_{r-1}, u_{r-2}, \ldots$ on either side of it, on the assumption that differences of U beyond a certain order j may be neglected. Any such graduation formula, therefore, will be of the form

$$v_r = l_r u_r + (l_{r+1} u_{r+1} + l_{r-1} u_{r-1}) + (l_{r+2} u_{r+2} + l_{r-2} u_{r-2}) + \ldots$$
$$\ldots + (l_{r+n} u_{r+n} + l_{r-n} u_{r-n}) \ldots (1)$$

for a range of $2n+1$ terms; and where the l's are symmetrical, so that $l_{r+t} = l_{r-t} = b_t$, say, this becomes

$$v_r = b_0 u_r + b_1 u_{r \pm 1} + b_2 u_{r \pm 2} + \ldots + b_n u_{r \pm n} \ldots \ldots (2)$$

where $u_{r \pm t}$ is written for $u_{r+t} + u_{r-t}$. The fundamental conditions under which such replacements of u_r by linear compounds of the u's are legitimate are well set out in the papers of Dr. W. F. Sheppard; and it will be sufficient here to say that if e_r

---

* The notation and terminology used herein follow closely those of Dr. W. F. Sheppard, with such modifications as are necessary to include the formulae of other writers. Dr. Sheppard's papers, to which a number of references are given herein, are: "Reduction of Errors by means of Negligible Differences" (5th Int. Cong. Mathematicians, 1912, ii. 348); "Fitting of Polynomial by Method of Least Squares" (Proc. London Math. Soc., 2nd Ser., XIII. 97); "Graduation by Reduction of Mean Square of Error" (J. I. A. XLVIII, 171 and 390, and XLIX, 148); and "Reduction of Error by Linear Compounding" (Phil. Trans., Ser. A, vol. 221, p. 199). The J. I. A. papers will generally be referred to here, as they embody the essential features pertinent to this discussion.

be the error (or departure) of the observed from the true value, so that $u_r = U_r + e_r$, it may be assumed that the e's are independent, and that the mean square of each is (say) e^2.

2. The determination of the l's or b's may be effected by (I) Interpolation, (II) Fitting, or (III) Reduction of Error. In (I) the unadjusted data are made the subject of direct interpolation, by using (a) single values, (b) groups of values, (c) combinations of interpolations from single values, or (d) combinations of interpolations from groups of values; and such formulae, which are not made to satisfy any criterion of "fit" or "smoothness" other than the assumption that certain of the original values (or combinations of them) shall be reproduced exactly and that differences beyond order j may be neglected, may also be produced directly in "summation" form without any preliminary selection of interpolations. In (II) a polynomial of degree j is "fitted" to the u's—the "best" formula, according to the usual criterion of the method of least squares, being taken to be that which makes the sum of the squares of the departures of the graduated from the ungraduated values a minimum. In (III) the mean square of error of v_r, or the mean square of error of $\Delta^n v$, may be made a minimum; and since differences beyond order j are assumed to be negligible several cases arise in practice according to the value assumed for j, and depending also upon whether the mean square of error of $\Delta^{j+1} v$, or of $\Delta^j v$, or of $\Delta^{j-1} v, \ldots$ is made a minimum.

3. The formulae of (III) which secure the maximum reduction of mean square of error in v itself clearly give, on the assumptions of par. 1, the same results as those of (II) which give the "best fitting" polynomial on the above criterion of "best" (see par. 23). Since therefore such formulae reduce the error in v itself as much as possible, they may be said to adjust the original series of u's with the greatest possible "accuracy"—and the term "accuracy" will therefore be used in that sense. Similarly, the other formulae of (III), which effect the maximum reduction of mean square of error in $\Delta^n v$, may be said to secure the maximum "smoothness" in the resulting series. These formulae give the greatest reductions of error possible, on the assumptions made; and consequently the same criteria of "accuracy" and "smoothness" may be applied to test also the

lesser accuracy and smoothing power of the interpolation formulae of (I), as will be seen hereafter (see pars. 13-14).

4. By (1) and (2), any graduation formula deduced by the above methods may be stated as a linear compound of the u's. Alternatively, since the assumption that (say) $j=3$ means that $U_x = A + Bx + Cx^2 + Dx^3$, the u's may be stated in terms of A, B, C, D,...and powers of x (cf. par. 12). Furthermore, since u_{r+t} may be expressed in terms of u_r and its differences, the graduation formulae may also be expressed in this way; or since a linear compound may be put in terms of successive sums, they may be given in that form, as in Dr. Sheppard's papers. The process of "summation" may also be conveniently applied in the manner usually adopted by actuaries, by which the formulae are put in terms of sums $[p]$ $[q]$ $[r]$...where $[p]u_s$ denotes the sum of p u's of which u_s is the middle term. For the particular purposes of this paper, however, the differences between the various formulae will be seen most clearly by stating them all as linear compounds of the u's; and that method of presentation will therefore generally be used throughout.

5. The earliest application of the interpolation methods was that of Griffith Davies in connection with the mortality experience of the Equitable Society (1834); the least square formulae of (II) were indicated briefly by C. L. Landré in 1901, and have been fully worked out by Dr. Sheppard in his valuable series of papers (1912-15); and the formulae under (III) were indicated by G. F. Hardy in 1909, fully examined by Dr. Sheppard's papers (1912-15), and some of them afterwards were rediscovered independently by R. Henderson (1916) and J. R. Larus (1918). It is not generally known, however, that much of the theory of (I), the least square formulae of (II), and those of (III) which make the mean square error in $\Delta^4 v$ a minimum had previously been discussed very fully by Erastus L. De Forest, M.A., Ph.B. (Yale) in a most invaluable series of papers which appeared in the Smithsonian Reports of 1871 and 1873, a pamphlet "Interpolation and Adjustment of Series" (1876), and The Analyst (Des Moines) 1877-80. It is the purpose of this paper, therefore, to trace the development of the above methods, with particular reference to De Forest's work and the recognition which should be accorded to it; and this at the same

time will permit of a classification being made of the various formulae in a manner which will be clearer, to the actuary and statistician, than a theoretical discussion of the general problem.

(I). INTERPOLATION.

6. *Method* (*A*). As stated in par. 2, the interpolations may be performed from (*a*) *single* values of the unadjusted data; and the assumption that differences above order j may be neglected clearly amounts to assuming that $U_x = A + Bx + Cx^2 + \ldots + Jx^j$. Consequently if we take, as in Griffith Davies' early application to the Equitable Experience, decennial values u_0, u_{10}, u_{20}, u_{30}, with $j=3$, we have

$$\left.\begin{array}{l} u_0 = A \\ u_{10} = A + 10B + 100C + 1000D \\ u_{20} = A + 20B + 400C + 8000D \\ u_{30} = A + 30B + 900C + 27000D \end{array}\right\} \ldots\ldots\ldots\ldots (3)$$

from which A, B, C, and D follow at once without any further conditions beyond the assumption that differences above the third may be neglected. This obvious process, which of course may be varied in a number of ways by taking different ranges of the data, and different values of j, is illustrated in the Institute of Actuaries' Text-Book, Part II, p. 435, and all works on interpolation; and it is included here because it is the first step in the development, and has been advocated by James Meikle ("On a Mode of Graduating a Table of Mortality," Trans. Actuarial Soc. of Edinburgh, IV, No. 18—1901) many years after Davies' application of it.

7. This method, however, is open to the objection that the selected points among which the interpolations are made themselves remain unadjusted in the final results. Consequently in 1865 G. W. Berridge (J. I. A. XII, 220 "On a Method of Graduation applied to the Peerage Mortality . . .," and J. I. A. XIV, 244) employed ordinary interpolation from (*b*) *groups* of values, which were taken in six consecutive decennial sums S_1, $S_2, \ldots S_6$ with $j=5$, and thus covered a range of 60 terms. The formulae are given in his papers, and in the Institute of Actuaries' Text-Book, Pt. II, p. 465. The effect of the use of sums in this manner is that the consecutive sums of the adjusted and

unadjusted data are the same; and by using $\Sigma \log p_x$ in decennial sums, i.e. $\log {}_{10}p_x$, the original decennial probabilities of living are undisturbed, while the yearly values p_x are redistributed in a smooth curve (see diagram, J. I. A. XII, 224-5).

8. The above method as used by Berridge, with six consecutive decennial sums and $j=5$, may of course be shortened or extended by taking $j=1$, or 2, or 3...6, or 7...; and the corresponding 2, or 3, or 4,.... 7, or 8,...sums need not necessarily be consecutive, nor need they all contain the same number of terms. While such formulae have been superseded by others (see later) as a means of performing complete graduations, nevertheless they are required in certain preliminary adjustments of population data; and a very clear and complete treatment of them is given by De Forest* (I, 277-289, 293-300, 304-315, 327-329; II, 319-327; III, 43-50). He uses a general method of areas, for which the formulae for any particular case are very easily derived, as follows:

The assumption that differences beyond order j are negligible means that

$$U_x = A + Bx + Cx^2 + Dx^3 + Ex^4 + Fx^5 + Gx^6 + Hx^7 + Ix^8 \ldots + Jx^j \ldots (4).$$

Taking x as the abscissa of the middle point, the sum S of any n terms in a group is therefore $\int_{x-\frac{1}{2}n}^{x+\frac{1}{2}n} U_x \, dx$, which from (4) is

$$\begin{aligned}
S = n\Big[& A + \frac{C}{12}n^2 + \frac{E}{80}n^4 + \frac{G}{448}n^6 + \frac{I}{2304}n^8 \ldots + \Big(B + \frac{D}{4}n^2 + \frac{F}{16}n^4 + \frac{H}{64}n^6 \ldots\Big)x \\
& + \Big(C + \frac{E}{2}n^2 + \frac{3}{16}Gn^4 + \frac{I}{16}n^6 + \ldots\Big)x^2 + \Big(D + \frac{5}{6}Fn^2 + \frac{7}{16}Hn^4 + \ldots\Big)x^3 \\
& + \Big(E + \frac{5}{4}Gn^2 + \frac{7}{8}In^4 + \ldots\Big)x^4 + \Big(F + \frac{7}{4}Hn^2 + \ldots\Big)x^5 + \Big(G + \frac{7}{3}In^2 + \ldots\Big)x^6 \\
& + (H + \ldots)x^7 + (I + \ldots)x^8 + \ldots \Big]
\end{aligned} \quad (5)$$

If, therefore, we take any $j+1$ groups and their corresponding values by (5), there will be $j+1$ equations from which the

---

* In referring to his various papers, of which a list is given in the Appendix, the Roman numerals denote the corresponding papers as there listed.

$j+1$ constants A, B, C,...may be completely determined. The general formulae for each value of j from 1 to 8 for consecutive groups of n_1 terms each are given by De Forest, loc. cit., together with the still more general expressions for non-consecutive and unequal groups for $j=2$ and 3.

9. Omitting the unimportant case of $j=1$, the general formulae for $j=2$ and $U_x=A+Bx+Cx^2$ are thus stated by De Forest (I, 277-279) :† "Assume any three areas S_1, S_2, S_3, so situated that the middle ordinates of S_1 and S_3 shall fall respectively to the left and right of the middle ordinate of S_2, which is taken as the axis of Y. Let n_1, n_2, n_3, be....the bases of these areas, and let a_1 and a_3 be the portions of the axis of X intercepted between the axis of Y and the middle ordinates of S_1 and S_3 respectively.

Then

$$\left. \begin{array}{l} S_1 = \int_{-a_1-\frac{1}{2}n_1}^{-a_1+\frac{1}{2}n_1} U_x dx = n_1 \left[A - Ba_1 + C\left(a_1^2 + \frac{1}{12} n_1^2\right) \right] \\[2mm] S_2 = \int_{-\frac{1}{2}n_2}^{+\frac{1}{2}n_2} U_x dx = n_2 \left(A + \frac{C}{12} n_2^2 \right) \\[2mm] S_3 = \int_{a_3-\frac{1}{2}n_3}^{a_3+\frac{1}{2}n_3} U_x dx = n_3 \left[A + Ba_3 + C\left(a_3^2 + \frac{1}{12} n_3^2\right) \right] \end{array} \right\} \quad \ldots\ldots(6)$$

Let S be a fourth area whose base is n, and let x' be the abscissa corresponding to its middle ordinate; then

$$S = \int_{x'-\frac{1}{2}n}^{x'+\frac{1}{2}n} U_x dx = n \left[A + Bx' + C\left(x'^2 + \frac{n^2}{12}\right) \right] \text{"}$$

Eliminating A, B, and C from these four equations, we may thus "find the magnitude S of an area whose position only is given, when the three other areas S_1, S_2, S_3 are given both in

---

† It should be noted that in the symmetrical cases considered in pars. 9-11 here, the formulae are the same for $j=0$ or 1, 2 or 3, 4 or 5, etc., as pointed out in par. 23. Formula (8), for example, can be obtained not only as in par. 9, but also on the assumption that $j=3$ as in par. 10; and likewise (13) emerges by taking five consecutive symmetrical groups and $j=5$ as well as from those groups and $j=4$ as in par. 11. The formulae (8) and (13), which are here found for $j=2$ and 4 respectively, are therefore actually correct to third and fifth differences of U.

magnitude and position". For example, in the special case of three consecutive groups of n_1 terms each (so that $n_3=n_2=n_1$, and $a_3=a_1=n_1$) we get—dropping the accent from x'—

$$A = \frac{1}{24n_1}[26S_2-(S_1+S_3)]$$

$$B = \frac{1}{2n_1^2}[S_3-S_1] \qquad \qquad \qquad \qquad (7)$$

$$C = \frac{1}{2n_1^3}[(S_1+S_3)-2S_2]$$

$$\text{and } S = n\left[A + \frac{C}{12}n^2 + Bx + Cx^2\right]$$

Taking now $n_1=5$, there will be three quinquennial groups. The central one will be given by putting $n=1$ and $x=0$ in S, giving

$$\int_{-\frac{1}{2}}^{+\frac{1}{2}} U_x\, dx = A + \frac{C}{12} = .216S_2 - .008(S_1+S_3)\ldots\ldots\ldots(8)$$

Thus, from the populations or deaths in the groups 30-35 (S_1), 35-40 (S_2), and 40-45 (S_3) we can compute the number for the central year of age 37-38 (the origin being at age 37½). This is the formula used by Mr. George King in finding graduated quinquennial "pivotal values" in his well-known method of constructing life-tables. His proof, however, is based on u's and finite differences, instead of areas; and the formula when so deduced is, for the range $u_0\ldots u_{14}$,

$=.2w_5-.008\triangle^2 w_0$ where $w_n = \sum_{x=n}^{x=n+4} u_x$ and $\triangle$ refers to interval 5 ....(8a)

$=.2w_5-.008[w_{10}-2w_5+w_0]\ldots\ldots\ldots\ldots\ldots\ldots\ldots\ldots\ldots\ldots\ldots$(8b)

$=.216w_5-.008(w_0+w_{10})\ldots\ldots\ldots\ldots\ldots\ldots\ldots\ldots\ldots\ldots\ldots\ldots\ldots$(8c)

Putting this now as a linear compound of the u's, and taking the range $u_{-7}\ldots u_{+7}$ instead of $u_0\ldots u_{14}$ as above, we find (using v_0 as in par. 1) that v_0

$.216u_0+.216u_{\pm 1}+.216u_{\pm 2}-.008u_{\pm 3}-.008u_{\pm 4}-.008u_{\pm 5}-.008u_{\pm 6}-.008u_{\pm 7}$ .. (8d)

From (7) the formula for any other term emerges at once. Thus to get the middle term of the group S_1 we take $n=1$ and $x=-5$ in S, giving

$$\int_{-5\frac{1}{2}}^{-4\frac{1}{2}} U_x\,dx = A - 5B + \frac{301}{12}C = .192S_1 + .016S_2 - .008S_3 \quad\ldots(9)$$

or, as in (8a), $u_2 = .2w_0 - .008\Delta^2 w_0 \ldots\ldots\ldots\ldots\ldots\ldots\ldots(9a)$
which is the non-central formula (for the range $u_0\ldots u_{14}$) deduced in the U. S. Abridged Life Tables, 1919-20, p. 34.

10. When $j=3$ we use four areas $S_1\ldots S_4$; and the general formulae are given by De Forest, I, 282. With consecutive groups of n_1 terms each they are

$$\left.\begin{array}{l} A = \dfrac{1}{12n_1}[7(S_2+S_3)-(S_1+S_4)] \\[1ex] B = \dfrac{1}{12n_1^2}[15(S_3-S_2)-(S_4-S_1)] \\[1ex] C = \dfrac{1}{4n_1^3}[(S_1+S_4)-(S_2+S_3)] \\[1ex] D = \dfrac{1}{6n_1^4}[(S_4-S_1)-3(S_3-S_2)] \\[1ex] \text{and } S = n\left[\left(A+\dfrac{C}{12}n^2\right) + \left(B+\dfrac{D}{4}n^2\right)x + Cx^2 + Dx^3\right] \end{array}\right\}\ldots(10)$$

When $n_1=5$, and $n=1$ and $x=0$ in S, we get

$$\int_{-\frac{1}{2}}^{+\frac{1}{2}} U_x\,dx = A + \frac{C}{12} = .1165(S_2+S_3) - .0165(S_1+S_4)\ldots\ldots\ldots(11)$$

which enables the populations or deaths in (say) the year of age $39\frac{1}{2}$ to $40\frac{1}{2}$ to be computed from those in the four groups 30-35, 35-40, 40-45, and 45-50. This is the formula obtained by Mr. Henderson in "Mortality Laws and Statistics", par. 131, pp. 74-5, and applied (p. 95) in the construction of a mortality table for the Northeastern States of the U. S. The alternative forms as in (8a)-(8d) above, and the non-central formulae such as (9) and (9a), may be deduced at once in the same way.

11. With $j=4$ and five consecutive groups $S_1\ldots S_5$ of n_1 terms each the formulae are (De Forest, I, 284):

$$\left.\begin{aligned}A &= \frac{1}{1920n_1}[2134S_3+9(S_1+S_5)-116(S_2+S_4)]\\ B &= \frac{1}{48n_1^2}[34(S_4-S_2)-5(S_5-S_1)]\\ C &= \frac{1}{16n_1^3}[12(S_2+S_4)-22S_3-(S_1+S_5)]\\ D &= \frac{1}{12n_1^4}[(S_5-S_1)-2(S_4-S_2)]\\ E &= \frac{1}{24n_1^5}[6S_3+(S_1+S_5)-4(S_2+S_4)]\\ \text{and } S &= n\left[\left(A+\frac{C}{12}n^2+\frac{E}{80}n^4\right)+\left(B+\frac{D}{4}n^2\right)x+\left(C+\frac{E}{2}n^2\right)x^2+Dx^3+Ex\right]\end{aligned}\right\}\ldots(12)$$

When $n_1=5$, and $n=1$ and $x=0$ in S, we get

$$\int_{-\frac{1}{2}}^{+\frac{1}{2}} U_x\,dx = A+\frac{C}{12}+\frac{E}{80} = .221376S_3 - .011584(S_2+S_4)$$
$$+.000896(S_1+S_5)\ldots\ldots(13)$$

which would give the populations or deaths in (say) the year of age 42-43 from the groups 30-35, 35-40, 40-45, 45-50, and 50-55. This, like (8), is the central term of an odd number of groups (and terms); and it may also be found easily by finite differences (with Δ referring to interval 5) in the equivalent form (with range $u_0\ldots u_{24}$)

$$u_{12}=.2w_{10}-.008\Delta^2 w_5+.000896\Delta^4 w_0\ldots\ldots\ldots(13a)$$

as shown by King, J. I. A. XLIII, 114. The non-central formulae for any other term is immediately obtainable from (12).

12. All the above methods are based strictly upon the principle of interpolation, i.e. the calculation of intermediate values from certain given values which are assumed to be correct, and which are so chosen that the unknowns $A, B, C,\ldots$ may be obtained at once by direct solution. These given values, being single u's as in par. 6 or groups of u as in pars. 7-11, are exactly

reproduced; and the differences of the new v's above order j are zero *so long as each v is found from the same range of u's.* For example, when $j=2$, formulae (8)-(8d) give the central v_0 from the 15 terms $u_{-7}\ldots u_{+7}$ taken in three quinquennial groups; but all the other v's from $v_{-7}\ldots v_{+7}$ follow similarly— like (9) and (9a)—by putting $n_1=5$, $n=1$, and $x=-7, -6,\ldots +6, +7$ in (7), which may be written, in terms of S's and powers of x,

$$S = v_x = \frac{1}{250}[S_1(-2-5x+x^2)+S_2(54-2x^2)+S_3(-2+5x+x^2)] \ldots (14)$$

whence the complete series (of which (8) and (9) for v_0 and v_{-5} are particular cases) is

$$\left.\begin{array}{ll} & v_0 = -.008S_1 + .216S_2 - .008S_3 \\ v_{-1}=.016S_1+.208S_2-.024S_3 & v_1 = -.024S_1+.208S_2+.016S_3 \\ v_{-2}=.048S_1+.184S_2-.032S_3 & v_2 = -.032S_1+.184S_2+.048S_3 \\ v_{-3}=.088S_1+.144S_2-.032S_3 & v_3 = -.032S_1+.144S_2+.088S_3 \\ v_{-4}=.136S_1+.088S_2-.024S_3 & v_4 = -.024S_1+.088S_2+.136S_3 \\ v_{-5}=.192S_1+.016S_2-.008S_3 & v_5 = -.008S_1+.016S_2+.192S_3 \\ v_{-6}=.256S_1-.072S_2+.016S_3 & v_6 = +.016S_1-.072S_2+.256S_3 \\ v_{-7}=.328S_1-.176S_2+.048S_3 & v_7 = +.048S_1-.176S_2+.328S_3 \end{array}\right\} \ldots (15)$$

That is to say, from the three quinquennial groups $S_1\left(=\sum_{-7}^{-3} u_r\right)$, $S_2\left(=\sum_{-2}^{2} u_r\right)$, and $S_3\left(=\sum_{3}^{7} u_r\right)$ each new v from v_{-7} to v_{+7} may be found by the above formulae. The quinquennial sums of these v's will exactly reproduce the original quinquennial sums of the u's, as may be seen from the fact that in the above series

$$\sum_{-7}^{-3} v_r = 1.000 S_1 + S_2(.248-.248) + S_3(.064-.064) = S_1 = \sum_{-7}^{-3} u_r,$$ and similarly $\sum_{-2}^{2} v_r = \sum_{-2}^{2} u_r$ and $\sum_{3}^{7} v_r = \sum_{3}^{7} u_r$; but, of course, there is no equality for other quinquennial sums—for example, $\sum_{1}^{5} v_r \neq \sum_{1}^{5} u_r$.

If we now difference the above v's we get (taking, for example, the leading differences of v_0),

$$\left. \begin{array}{l} \Delta v_0 = v_1 - v_0 = -.016 S_1 - .008 S_2 + .024 S_3 \\ \Delta v_1 = v_2 - v_1 = -.008 S_1 - .024 S_2 + .032 S_3 \\ \Delta v_2 = v_3 - v_2 = -.040 S_2 + .040 S_3 \end{array} \right\} \dots \dots \dots (16)$$

from which

$$\left. \begin{array}{l} \Delta^2 v_0 = \Delta v_1 - \Delta v_0 = .008 S_1 - .016 S_2 + .008 S_3 \\ \Delta^2 v_1 = \Delta v_2 - \Delta v_1 = .008 S_1 - .016 S_2 + .008 S_3 \end{array} \right\} \dots \dots \dots (17)$$

and therefore $\Delta^3 v_0 = \Delta^2 v_1 - \Delta^2 v_0 = 0$, which is correct since $j=2$. Thus the new values, v, which replace the original u's, exactly reproduce the original selected sums of the u's; and at the same time the third differences of the new v's are exactly zero, although the third differences of the original u's may not have been zero. These new zero third differences (and their constant second differences) are therefore perfectly smooth; that is, since the assumption that $j=2$ means that the original third differences $\Delta^3 u$ could be neglected and therefore presumably arose from "errors", it may be said that the process has resulted not merely in a reduction but in a complete removal of those errors, on the assumptions made.

13. The principle of the above, as already stated, is that of interpolation from consecutive sums; and since those consecutive prearranged sums are reproduced, the process may be viewed as one of "fitting" in which the criterion of "fit" is the identity of the graduated and ungraduated consecutive sums. It now becomes of importance to examine also the effect of the method on the errors of the original series, on the usual assumption (pars. 1 and 2) that each observed u_r is affected by an error e_r, the mean square of which is e^2. The linear compound form (1) may be used to illustrate this. For on this assumption it follows from (1), by the theory of combination of errors, that the mean square error of v_r is $(\Sigma l_r^2) e^2$; and therefore, since the process of linear compounding replaces the original u_r with its mean square error e^2 by the new v_r with its mean square error $(\Sigma l_r^2) e^2$, we see that the mean square error will have been reduced to

$\frac{(\Sigma l_r^2) \ e^2}{e^2}$, or Σl_r^2, of its original amount, and this reduction of error in the original series will therefore measure the "accuracy" of the graduation. In the case, for example, of formula (8d), which is the central one of the complete series (15), we find that $\Sigma l_r^2 = 5(.216)^2 + 10(-.008)^2 = .23392$; and the reduction of error effected by each of the other v's of (15) may be similarly calculated.

This theory of reduction of error is, of course, well known. In actuarial literature it seems to have been used first in 1878 by W. S. B. Woolhouse (J. I. A. XXI, 61) and again in 1887 (J. I. A. XXVI, 421), and then by G. F. Hardy in 1888 (J. I. A. XXVII, 323) and 1896 (J. I. A. XXXII, 376)—the latter being frequently referred to as the originator of its application to the graduation formulae of actuaries (see Lidstone, J. I. A. XLII, 114, 124, 129, and Spencer J. I. A. XLI, 386); while a number of more recent papers, such as those of Dr. W. F. Sheppard (J. I. A. XLVIII and XLIX) are based upon it. It does not seem to be known, however, that precisely the same theory was used extensively by De Forest in 1871 (I, 325) and 1873 (II, 327); and in his subsequent works he gives many examples —such as a very clear explanation in III, 43, of the significance and calculation of Σl_r^2 for each v resulting from (7) when $n_1 = 4$ instead of 5 as above (see also par. 26 here), as well as important applications to other cases which will be considered later in sections II and III of this paper.

14. This same theory may, of course, be applied similarly to investigate the effect of the adjustment on the errors of the differences of the series. Thus, on the same assumption that each original u_r is affected by an error e_r with a mean square of e^2, the mean square error of the original $\Delta u_r \equiv u_{r+1} - u_r$ is, according to the theory of errors, $[1^2 + (-1)^2]e^2 = 2e^2$, while that of the new Δv_r is obtainable similarly by taking the sum of the squares of the coefficients of u in Δv_r multiplied by e^2. For example, from (16), the mean square error of Δv_0 is thus $[5(-.016)^2 + 5(-.008)^2 + 5(.024)^2]e^2 = .00448e^2$; and therefore the proportionate reduction of error is $\frac{.00448e^2}{2e^2} = .00224$. Similarly, from (17), the proportionate reduction of error in

$\Delta^2 v_0$ is $[5(.008)^2+5(-.016)^2+5(.008)^2]e^2 \div [1^2+(-2)^2+1^2]e^2$ $=.00032$; while the original mean square error in $\Delta^3 u_0$ is reduced in $\Delta^3 v_0$ to $0 \div [1^2+(-3)^2+(3)^2+(-1)^2]$, that is, as stated in par. 12, it is made to disappear.

In order to keep clear the distinction between these cases, we shall write R_n^2 for the proportion which the mean square error in the new $\Delta^n v$ bears to that in the original $\Delta^n u$ (thus using Dr. Sheppard's R^2 with the addition of the subscript n)—so that in the particular case above, $R_0^2 = .23392$, $R_1^2 = .00224$, $R_2^2 = .00032$, and $R_3^2 = 0$. Consequently, since in the theory of errors the square root of the mean square error is proportionate to the "average error" (irrespective of sign), R_n will represent similarly the proportion to which that "average error" is reduced; and this is the method of comparison actually used by De Forest, Woolhouse, and Hardy, and shown also by Sheppard. Finally, since R_n thus represents the proportion to which the original mean square error is reduced, and since the original errors are supposed to have been accidental errors of observation, it may be said that the mean square error of the new v has been reduced to the value which would result from the u of an unadjusted experience $\dfrac{1}{R_n^2}$ times as great; and likewise the "average error" will have been reduced to the value which would be expected in the u of an unadjusted experience $\dfrac{1}{R_n}$ times as great (cf. Woolhouse, J. I. A. XXI, 61; Hardy, XXXII, 377, and also his "Lectures," p. 28).

15. *Method (B)*. In the preceding paragraphs it is supposed that all the new v's are to be obtained *from the same original u's*—for example, in par. 12 we are given u_{-7} to u_{+7}, and by (15) the new series v_{-7} to v_{+7} is constructed; and it has been pointed out how the method of that paragraph is in effect a method of curve fitting in which the reproduction of consecutive sums of u is the criterion of fit, and how the method secures that the differences of v beyond order j shall actually be zero as assumed. This last assumption, moreover, is that U_x may be represented by a parabolic curve *with the same constants throughout*. This assumption is sufficiently near the truth for

comparatively short ranges; but for long ranges, as in Berridge's process which used no less than 60 terms (in six groups of ten terms each with $j=5$, or—as applied also by De Forest, I, 286—in five groups of twelve terms with $j=4$) the assumption is less defensible—as may be appreciated from the fact that it can hardly be held that u_{60}, u_{59}, and other terms at the end should have much influence in determining v_0, v_1, etc. Consequently, instead of employing the same long range of u's throughout to give each v—which will thus occupy a variable position in relation to the original u's—we may employ *successive short ranges of u's* from each of which one v—occupying as nearly as possible the same position in the range each time—may be determined. For example, in the case of par. 12, v_0 could be found centrally from $u_{-7}\ldots u_{+7}$, as was done there, by formula (8); and then, by successive applications of *this same formula* over the successive ranges $u_{-6}\ldots u_{+8}$, $u_{-5}\ldots u_{+9}$, $u_{-4}\ldots u_{+10}$, etc., we could get v_1, v_2, v_3, and so on. This is the principle now generally adopted in such practical "graduations" as those of mortality tables, where the assumption that U_x may be represented by a parabolic curve with (say) $j=3$ is legitimate only over comparatively short ranges of u.

The effect of this process, however, is clearly different from that of Method (A) of the preceding pars. 12-14. For now, although differences of the true U's above order j are still assumed to be negligible, the differences of the new v's will not be exactly zero as before, because the progressive application of the same formula over successive ranges now means that each v is determined from different values of S and consequently involves different constants A, B, C,...; and similarly there will now be no reproduction of the original consecutive sums. For example, in the case of formula (8) we should now have, instead of the series (14)-(15) for v_{-7} to v_{+7}, the same formula $v_x=$

$-.008 S_1 + .216 S_2 - .008 S_3$ in which $S_1 = \sum_{x-7}^{x-3} u_r$, $S_2 = \sum_{x-2}^{x+2} u_r$, and $S_3 = \sum_{x+3}^{x+7} u_r$,

and x takes successively the values $-7, \ldots +7$; that is to say, formula $(8d)$ is applied progressively.

The effect on the errors, therefore, is easily seen from formula $(8d)$. For R_0^2 will now be $5(.216)^2 + 10(-.008)^2 = .23392$ for

DEVELOPMENT OF FORMULAE FOR GRADUATION   95

each v from v_{-7} to v_{+7}, instead of being obviously different for each v in series (15). In the case of $\triangle v_0 \equiv v_1 - v_0 =$

[.216 $\{u_1+(u_2+u_0)+(u_3+u_{-1})\} - .008 \{(u_4+u_{-2})+(u_5+u_{-3})$
$+ (u_6+u_{-4})+(u_7+u_{-5})+(u_8+u_{-6})\}] - [.216(u_0+u_{\pm 1}+u_{\pm 2})$
$- .008(u_{\pm 3}+u_{\pm 4}+u_{\pm 5}+u_{\pm 6}+u_{\pm 7})]$

$= + .008u_{-7} - .224u_{-2} + .224u_3 - .008u_8$, R_1^2 is (see par. 14)
$$\frac{2(.008)^2+2(.224)^2}{2} = .05024.$$

Similarly for $\triangle^2 v_0$ we find at once by differencing again (from right to left) the detached coefficients in $\triangle v_0$ (cf. J. I. A. XXXII, 377, and XLII, 115) that the significant coefficients are .008, —.008, —.224, .224, .224, —.224, —.008, .008, whence

$$R_2^2 = \frac{4(.008)^2 + 4(.224)^2}{6} = .033493;$$ and likewise $R_3^2 = .030144$,

$R_4^2 = .028709$, and so on. These values show how in Method (B) the new third differences are merely reduced in error instead of being made to vanish entirely as in (A); but beyond that fact these particular values are merely illustrative, and are not to be taken as measuring the relative advantages of Methods (A) and (B), because formula (8) is merely the first step in improved formulae which may be evolved for Method (B).

16. These other formulae of Method (B) may be approached, as in par. 2, either by taking other direct interpolations from (a) single values or (b) groups of values, or by using (c) combinations of interpolations from single values or (d) combinations of interpolations from groups.

The formulae under (a) or (b) are easily obtainable from the general formulae (6), (7), (10), (12), etc., and a number of them are given by De Forest. For example with 5 terms, taking "the three middle ones as the middle group and the first one and last one as the two lateral groups" so that $n_2 = 3$, $n_3 = n_1 = 1$, $a_3 = a_1 = 2$ in (6) we get (De Forest, I, 289)

$$v_0 = .4u_0 + .4u_{\pm 1} - .1u_{\pm 2} \dots\dots\dots\dots\dots\dots\dots (18)$$

and with seven terms and $n_2 = 5$, $n_3 = n_1 = 2$, $a_3 = a_1 = \frac{5}{2}$ —"five in the middle group and two in each lateral one, so that the second and sixth terms belong to two groups each"—

$$v_0 = \frac{1}{45}[13u_0 + 13u_{\pm 1} + 8u_{\pm 2} - 5u_{\pm 3}] \quad\quad\quad\quad\quad (19)$$

These formulae are not satisfactory in themselves; but they may be used under (d) to give improved formulae—see par. 19.

17. Under (c)—combinations of interpolations from single values—an obvious principle will be simply to take a number of ordinary interpolations from single values as in Davies' method of par. 6, and average them. The first formula of this type was that used by W. S. B. Woolhouse in the graduation of the Institute of Actuaries Life Tables in 1869 (J. I. A. XV, 389), as follows: Using 15 terms, interpolate separately for u_0 from the five sets of three original values (i) u_{-7}, u_{-2}, u_3 (ii) u_{-6}, u_{-1}, u_4 (iii) u_{-5}, u_0, u_5 (iv) u_{-4}, u_1, u_6, and (v) u_{-3}, u_2, u_7, thus obtaining

$$\left.\begin{array}{ll}(i) & u_0 = -.12u_{-7} + .84u_{-2} + .28u_3 \\ (ii) & u_0 = -.08u_{-6} + .96u_{-1} + .12u_4 \\ (iii) & u_0 = \quad\quad\quad\quad u_0 \\ (iv) & u_0 = \ .12u_{-4} + .96u_1 - .08u_6 \\ (v) & u_0 = \ .28u_{-3} + .84u_2 - .12u_7 \end{array}\right\} \quad (20)$$

from which, taking the average,

$$\left.\begin{array}{l}v_0 = \frac{1}{125}[25u_0 + 24u_{\pm 1} + 21u_{\pm 2} + 7u_{\pm 3} + 3u_{\pm 4} - 2u_{\pm 6} - 3u_{\pm 7}] \\ \ = .2u_0 + .192u_{\pm 1} + .168u_{\pm 2} + .056u_{\pm 3} + .024u_{\pm 4} \\ \quad\quad\quad\quad\quad\quad\quad\quad\quad\quad\quad - .016u_{\pm 6} - .024u_{\pm 7}\end{array}\right\} \ (21)$$

For this formula, by the method of par. 15, $R_0^2 = .1793$, as shown by Woolhouse, J. I. A. XXVI, 422, and $R_0 = .423$; $R_2 = .18$ (given also in Hardy's Lectures, p. 28); and $R_3 = .0655$.

18. This principle of simply averaging a number of interpolations may, of course, be applied over various ranges of u's besides u_{-7} to u_{+7} as in Woolhouse's formula; and in each interpolation the selected u's need not be three in number, nor need the interval between them be 5. General formulae are considered by C. D. Higham, J. I. A. XXVI, 50, and R. Todhunter, XXXII, 387. For practical purposes, however, a clearer treatment is given by C. H. Wickens in a little-known paper "An

Extension of the Principle underlying Woolhouse's Method of Graduation" (Trans. Actuarial Society of New South Wales, 1912). Keeping firstly Woolhouse's principle of sets of three u's with interval 5, his scheme as in (20) above is there extended from 5 to 11 sets, commencing obviously with u_{-10}, u_{-5}, u_0 and ending with u_0, u_5, u_{10}; from which—giving the first and last sets the weighting ½—the 19-term formula is

$$v_0 = .200u_0 + .168u_{\pm 1} + .132u_{\pm 2} + .092u_{\pm 3} + .048u_{\pm 4} \\ - .008u_{\pm 6} - .012u_{\pm 7} - .012u_{\pm 8} - .008u_{\pm 9} \Bigg\} \cdot \cdot (22)$$

for which R_3 is .0265.

Varying now the interval 5, and using each time—as for (22)—the complete scheme in which the first set ends with u_0 and the last set begins with u_0, Wickens gives a complete table of the resulting formulae for intervals 1 to 10, embracing 1 to 39 terms. The 15-term formula so constructed is, for example,

$$v_0 = .250u_0 + .199u_{\pm 1} + .141u_{\pm 2} + .074u_{\pm 3} - .012u_{\pm 5} \\ - .016u_{\pm 6} - .011u_{\pm 7} \Bigg\} \cdot \cdot (23)$$

for which R_3 is .0415.

Wickens also gives a similar table of the formulae resulting from interpolations from sets of 4 u's instead of 3; and their "summation" forms, for sets of both three and four u's, show their relationships to Woolhouse's very clearly. The values of R_3 are smaller with four u's than with three; and those of R_3 for such formulae as (22) and (23) above are smaller than Woolhouse's (.0655)—in fact, all Wickens' formulae "for intervals of 4 and upwards may be expected to give much smoother results than Woolhouse's original formula."

19. Under (d)—combinations of interpolations from groups —several systems have been used. Thus De Forest, I, 291, took (19) as above and (18) in the form $10kv_0 = k[4u_0 + 4u_{\pm 1} - u_{\pm 2}]$, from which, by addition,

$$v_0 = \left(\frac{1}{45+10k}\right)[(13+4k)(u_0+u_{\pm 1})+(8-k)u_{\pm 2}-5u_{\pm 3}];$$

and then, making the coefficients in arithmetical progression so that $(13+4k)-(8-k)=(8-k)+5$, he found $k=\dfrac{4}{3}$ and hence

$$v_0 = \frac{1}{35}[11u_0 + 11u_{\pm 1} + 4u_{\pm 2} - 3u_{\pm 3}] \dots \dots \dots \dots (24)$$

or, increasing the coefficient of u_0 by 7,

$$v_0 = \frac{1}{42}[18u_0 + 11u_{\pm 1} + 4u_{\pm 2} - 3u_{\pm 3}] \quad\quad\quad\quad (25)$$

in which all the coefficients are in A. P. He also gave (I, 292 and II, 341) the following similar formulae with the coefficients in A. P. for 9, 11, and 15 terms:

$$v_0 = \frac{1}{30}[10u_0 + 7u_{\pm 1} + 4u_{\pm 2} + u_{\pm 3} - 2u_{\pm 4}]. \quad\quad\quad (26)$$

$$v_0 = \frac{1}{165}[45u_0 + 34u_{\pm 1} + 23u_{\pm 2} + 12u_{\pm 3} + u_{\pm 4} - 10u_{\pm 5}] \quad (27)$$

$$\left.\begin{array}{l} v_0 = \frac{1}{140}[23u_0 + 23u_{\pm 1} + 18u_{\pm 2} + 13u_{\pm 3} + 8u_{\pm 4} + 3u_{\pm 5} \\ \quad\quad\quad\quad - 2u_{\pm 6} - 7u_{\pm 7}] \\ = .2u_0 + .164u_{\pm 1} + .129u_{\pm 2} + .093u_{\pm 3} + .057u_{\pm 4} \\ \quad\quad + .021u_{\pm 5} - .014u_{\pm 6} - .050u_{\pm 7} \end{array}\right\} (28)$$

Of these formulae, Nos. (25)-(27) were "presented to the Smithsonian Institution" in 1868 (see II, 341) and published in 1871, while (28) was first given in 1873 in II, 341. Fourteen years later No. (26) was rediscovered by J. A. Higham (J. I. A. XXIII, 336) with apparently no knowledge of De Forest's work. Both authors, however, recognized the limited smoothing power of such formulae—De Forest basing his conclusion (II, 331) on the value of R_4 (=.0818) for No. (27) in comparison with other formulae which he deduced, and both De Forest (I, 322) and Higham properly objecting to the fact that the coefficients, being in A. P., are in straight lines (see par. 32 here).

20. Instead of combining them as in De Forest's method, however, another process under (d) is, of course, to take the simple average of a number of interpolations from groups, as in pars. 17-18 for single values. Thus just as Woolhouse's formula may be viewed as the average of a number of interpolations based on Davies' method (par. 6), we may construct a formula as the average of a number of interpolations on Berridge's principle of groups—the resulting formula for 3rd differences and quinquennial groups being, as shown by G. King, J. I. A. XLI, 554 (see also Wickens, loc. cit., p. 11) :—

$$\left.\begin{array}{l} v_0 = .2u_0 + .1728u_{\pm 1} + .1344u_{\pm 2} + .0896u_{\pm 3} + .0432u_{\pm 4} \\ \quad - .0096u_{\pm 6} - .0128u_{\pm 7} - .0112u_{\pm 8} - .0064u_{\pm 9} \end{array}\right\} (29)$$

King also gives the similar formulae for groups of 3 instead of 5, and for fifth as well as third differences.

Osculatory, instead of ordinary, interpolation may also be employed. For groups of 5 the resulting formulae for 3rd and 5th differences were first given by Dr. Johannes Karup (Trans. 2nd Int. Cong. Actuaries, p. 78), and subsequently were rediscovered by George King in his paper mentioned above. The formulae for groups of 3 are also worked out by Mr. King, who also states all of them in "summation" form, and indicates very clearly their probable utility.

21. All the above methods, however, are based upon the preliminary selection of certain interpolations. While, therefore, a great number of such formulae may be produced by varying the scheme, it is not easy to pick out interpolations with any certainty that the resulting formulae will show marked improvement. Furthermore, actuaries generally prefer the formulae in "summation" form—although, with modern calculating machines, they may now be applied easily and rapidly in the "linear compound" form (cf. Larus, T. A. S. A. XIX, 18-20 and 298). The direct production of formulae in summation form has therefore received a great deal of attention; and the methods, which are well known and need not be detailed here, are given in the J. I. A. in a series of papers commencing in 1882 by J. A. Higham (vols. XXIII to XXVII, and XXXI), in vol. XXXII by A. Levine, G. F. Hardy, and R. Todhunter, and in vols. XLI-XLII by G. King, J. Spencer, and G. J. Lidstone, while convenient resumés are available in J. B. Maclean's paper T. A. S. A. XIV, 256, R. Henderson's Mortality Laws and Statistics, and the Actuarial Society's Study No. 4 on Graduation (also by Mr. Henderson). In these papers R_3—the reduction of error in the third difference—is generally referred to as the "smoothing coefficient," since it may be taken as a measure of the smoothness of the graduation; and, following the nomenclature of the theory of errors, $\frac{1}{R_0^2}$ —being the reciprocal of the mean square error of the graduated value—is the "weight" of v_0, and may be regarded (like R_0^2) as a measure of the

accuracy of the graduation, in order to compare the formulae with the "best" formulae of (II) below which, as shown in par. 23, effect the greatest reduction of error in v, i.e., are those of maximum weight.

(II). Fitting.

22. The interpolation formulae of (I) above, when tested by these criteria of accuracy and smoothness, give in many cases very satisfactory results; but they are not the "best" formulae as judged by those criteria. These "best" formulae in the case of the "accuracy" are clearly obtainable, as stated in par. 3, by making R_0^2 —the mean square error of v_0—a minimum; and, as shown in the next paragraph, they are the same as those obtained by "fitting" a polynomial by the usual method of least squares. The first detailed treatment of such formulae seems to have been given in 1871 by De Forest, I, 301, 316, 326-7; II, 331; III, 43-50; and IV, 79-82; although, as stated by him (I, 335) they had been considered independently in Schiaparelli's work "Sul modo di ricavare la vera espressione delle leggi della natura dalle curve empiriche" (1867), while it may be noted that they were also indicated by Woolhouse in 1865 (J. I. A. XII, 143 et seq.).

As in (I), formulae may be given for (a) single values or (b) groups of values; and in each of these subdivisions they may be applied by Method (A)—in which all the v's are obtained from the same original u's—or by Method (B), in which the same formula is applied progressively.

23. Taking first the case of (a) single values, the problem is to fit equation (4) by the method of least squares. For example, when $j=3$ and the range is u_{-3} to u_{+3} we have from (4) the 7 observation equations:

$$\left.\begin{array}{l} A-3B+9C-27D = u_{-3} \\ A-2B+4C-8D = u_{-2} \\ A-B+C-D = u_{-1} \\ A = u_0 \\ A+B+C+D = u_1 \\ A+2B+4C+8D = u_2 \\ A+3B+9C+27D = u_3 \end{array}\right\} \quad \ldots\ldots\ldots\ldots\ldots (30)$$

from which the normal equations are, by the usual rule of least squares,

$$\left.\begin{array}{l} 7A+28C-(u_{-3}+u_{-2}+u_{-1}+u_0+u_1+u_2+u_3)=0 \\ 28B+196D+(3u_{-3}+2u_{-2}+u_{-1}-u_1-2u_2-3u_3)=0 \\ 28A+196C-(9u_{-3}+4u_{-2}+u_{-1}+u_1+4u_2+9u_3)=0 \\ 196B+1588D+(27u_{-3}+8u_{-2}+u_{-1}-u_1-8u_2-27u_3)=0 \end{array}\right\}..(31)$$

whence $A = v_0 = \dfrac{1}{21}[7u_0 + 6u_{\pm 1} + 3u_{\pm 2} - 2u_{\pm 3}]$ ....(32)

and the other values for v_1, v_{-1}, etc., emerge similarly from the values of B, C, and D resulting from (31).

This straightforward method is illustrated a number of times by De Forest, who gives the formulae for 5, 7, 9, 11, and 12 terms.*

Another process, however, which will be useful subsequently, is to express the condition "$j=3$" directly in terms of the unknowns instead of implicitly by equation (4), and thence to find the minimum for R_0^2 subject to that condition, either by

---

* Woolhouse, however, attacks the problem with a slightly different objective; for he sets out to determine the fourth difference corrections which are involved when the u's are adjusted by a curve of the third degree by the method of least squares. In doing so he uses, incidentally, the method of moments, which in the case of a linear equation is equivalent to the method of least squares. Thus for 7 terms from u_0 to u_6 we have, by the usual

$$u_n = u_0 + n\Delta u_0 + {}^nC_2 \Delta^2 u_0 + {}^nC_3 \Delta^3 u_0 + \ldots {}^nC_n \Delta^n u_0$$

$$= (\text{say}) \ [A+nB+({}^nC_2)C+({}^nC_3)D] + \sum_{x=0}^{x=n-4} \frac{\overline{(n-x+1)}\,\overline{(n-x+2)}\,\overline{(n-x+3)}}{6} \Delta^4 u_x$$

the following strictly correct observation equations:

$u_0 = A$
$u_1 = A + B$
$u_2 = A + 2B + C$
$u_3 = A + 3B + 3C + D$
$u_4 = A + 4B + 6C + 4D + (\Delta^4 u_0)$
$u_5 = A + 5B + 10C + 10D + (4\Delta^4 u_0 + \Delta^4 u_1)$
$u_6 = A + 6B + 15C + 20D + (10\Delta^4 u_0 + 4\Delta^4 u_1 + \Delta^4 u_2)$

and now determining A, B, C, and D by the method of least squares (or moments) we obtain for, say, the central u_3 the formula

$$\frac{1}{21}[(-2u_0+3u_1+6u_2+7u_3+6u_4+3u_5-2u_6)+(2\Delta^4 u_0+5\Delta^4 u_1+2\Delta^4 u_2)]$$

which is formula (32) above with the fourth difference correction as given by Woolhouse J. I. A. XII, 157.

direct elimination or by the method of undetermined multipliers. Thus by (4) the general formula (1) for v_0 may be written

$$\left.\begin{aligned}v_0 =& A[l_0 + (l_1+l_{-1}) + (l_2+l_{-2}) + \ldots + (l_n+l_{-n})]\\&+ B[(l_1-l_{-1}) + 2(l_2-l_{-2}) + \ldots + n(l_n-l_{-n})]\\&+ C[1^2(l_1+l_{-1}) + 2^2(l_2+l_{-2}) + \ldots + n^2(l_n+l_{-n})]\\&+ D[1^3(l_1-l_{-1}) + 2^3(l_2-l_{-2}) + \ldots + n^3(l_n-l_{-n})]\\&+ \ldots\ldots\ldots\ldots\ldots\\=& \text{(say)}\ c_0A + c_1B + c_2C + c_3D + \ldots\ldots\ldots\ldots\end{aligned}\right\}\ldots(33)$$

Consequently when $j=0$, so that $U_x=A$, we must have $c_0=1$; when $j=1$ and $U_x=A+Bx$, it is necessary and sufficient that $c_0=1$ and $c_1=0$; and generally when differences above order j are negligible the conditions are that $c_0=1$, $c_1=0$, $c_2=0,\ldots$, $c_j=0$. These conditions cover both unsymmetrical and symmetrical cases; and for the latter, since $l_n=l_{-n}$, they reduce to $c_0=1$, $c_2=0$, $c_4=0$, etc., and are the same when $j=0$ or 1, 2 or 3, 4 or 5,.... (cf. De Forest, VII, 131, and VIII, 145).

In the above illustrative symmetrical case $j=3$, and $v_0 = b_0u_0 + b_1u_{\pm 1} + b_2u_{\pm 2} + b_3u_{\pm 3}$; and therefore, in order to make R_0^2 or Σl_r^2 (par. 13) a minimum subject to the conditions $c_0=1$, $c_1=0$, $c_2=0$, $c_3=0$ we have to make $b_0^2 + 2(b_1^2 + b_2^2 + b_3^2)$ a minimum subject to the conditions $b_0+2b_1+2b_2+2b_3=1$ and $b_1+4b_2+9b_3=0$ (cf. Sheppard, J. I. A. XLVIII, 173). Eliminating b_0 and b_1 these give $b_1=-4b_2-9b_3$ and $b_0=1+6b_2+16b_3$, so that $b_0^2 + 2(b_1^2 + b_2^2 + b_3^2)$ becomes $1 + 12b_2 + 32b_3 + 70b_2^2 + 336b_2b_3 + 420b_3^2$; and differentiating with regard to b_2 and b_3 and equating to zero, this expression will be a minimum when $12+140b_2+336b_3=0$ and $32+336b_2+840b_3=0$, whence $b_3=-\frac{2}{21}$ and $b_2=\frac{3}{21}$ which, with $b_1=\frac{6}{21}$ and $b_0=\frac{7}{21}$ from $c_1=0$ and $c_0=1$, is the same formula (32) as that found by fitting by the method of least squares. Or, as stated above, the same result may be obtained by the method of undetermined multipliers, as used by C. L. Landré in his paper on these formulae in 1901 in Ehrenzweig's Assekuranz-Jahrbuch, vol. XXII, and pp. 91-94 of his book "Mathematisch-Technische Kapitel zur Lebensversicherung" (see also C. Dizler, 5th Int. Cong. Actuaries, 1906, II, 293; R. Risser, 6th Congress, 1909, II, 576; Review, J. I. A. XXXVII, 108; and Davis, T. A. S. A. XIX, 301-2).

The above analysis shows, as pointed out by Dr. Sheppard, that the process of finding the "best" formula by least squares gives the same result as that of determining the l's so as to give the smallest possible mean square error—i.e. the greatest "weight," being the minimum value of R_0^2—in v.

24. These formulae are frequently referred to as the "least square formulae"; and the one for 19 terms, for which $R_0 = .345$, is sometimes called "Landré's formula" (see Risser and Dizler, loc. cit.). A complete table of them up to 25 terms for $j=2$ or 3, with R_0 and R_4, was given by De Forest in 1877 (IV, 81)—the b's being given in decimal form. More recently tables of them up to 31 terms for $j=2$ or 3, and 4 or 5, have been given in Dr. Sheppard's papers (J. I. A. XLVIII, 181-2) in terms of differences or sums—from which the formulae in terms of u are easily derived; and the latter are also set out still more fully, by including also the formulae up to 21 terms for $j=6$ or 7, and 8 or 9, in Catherine W. M. Sherriff's paper "On a Class of Graduation Formulae" (Proc. Royal Soc. of Edinburgh, 1919-20, XL, 112). These tables give only the symmetrical formulae; the unsymmetrical ones, which would be required under Method (A), and would be useful under Method (B) for completing the end terms which the symmetrical formulae will not reach under that method, may be deduced similarly, and are indicated by De Forest (I, 326), Landré and Dizler (loc. cit.), and Sheppard (J. I. A. XLIX, 155, and his other papers there mentioned).

These "least square formulae" give the "best" v's, according to the above criterion of "best"; and this is seen at once by comparing their values of R_0^2 or R_0 with those of any other formulae. But—as was shown and emphasized by De Forest (I, 317, 325, 327, 335; II, 331; III, 81 et seq.)—when we compare the values of R_1^2, R_2^2, etc.,* it is found that the effect on

---

* Under Method (A) the formulae for Δv, $\Delta^2 v$, etc., emerge very easily from the scheme of observation equations as in (30). Thus $\Delta v_0 \equiv v_1 - v_0 = B + C + D$; the central $\Delta^2 v_{-1} \equiv v_1 - 2v_0 + v_{-1} = 2C =$
$$\frac{1}{42}[5u_{\pm 3} - 3u_{\pm 1} - 4u_0]$$
which is the formula resulting from Sheppard's Table 4, J. I. A. XLVIII, 406 and par. 15, p. 394; and so on—whence R_1^2, R_2^2, etc., emerge at once. For Method (B) the process of par. 15 here—differencing with detached coefficients—gives the required expressions.

the errors of the differences of the series is not satisfactory; and consequently the formulae do not possess satisfactory smoothing power and are desirable only when it is required to adjust the u's themselves with the greatest possible accuracy. This is due, as De Forest makes very clear, to "the great want of continuity" between the b's of the terms included by the formulae and those beyond (see par. 32 here); and this feature has been pointed out subsequently also by R. Henderson, T. A. S. A. XV, 151, M. Davis, T. A. S. A. XIX, 302, and Sherriff, p. 118.

25. Coming now to (b), groups of values, the "least square formulae" have been suggested for use instead of formulae (8)-(13) in the preliminary determination of the pivotal values in Mr. King's method of constructing life tables, because that is a case where the best v's themselves are required (J. I. A. LI, 368). Since the use of groups necessarily means that the coefficients of every u within a group are the same, the linear compound becomes for, say, quinquennial groups,

$$v_0 = \ldots r'(u_{-12}+\ldots+u_{-8})+q'(u_{-7}+\ldots+u_{-3}) \\ +p(u_{-2}+\ldots+u_2)+q(u_3+\ldots+u_7)+r(u_8+\ldots+u_{12})+\ldots \quad \ldots(34)$$

and (33) and its resulting conditions are modified similarly, so that any required formula is easily derivable by the methods indicated above. For three symmetrical quinquennial groups we have $q=q'$ and $r=r'=0$; and the conditions $c_0=1$ and $c_2=0$ (when $j=3$) become $5p+10q=1$ and $5p+135q=0$, whence $p=.216$ and $q=-.008$ as already found in formula (8). With five symmetrical groups ($r=r'$ and $q=q'$) the conditions $c_0=1$ and $c_2=0$ can give a complete solution only when the three unknowns p, q, and r are reduced to two; and, as pointed out in J. I. A. LI. 368, if this is done by taking $p=q$, so that the conditions become $10r+15p=1$ and $140p+510r=0$, we obtain—without, it is to be noted, any introduction of the method of least squares—Mr. Henderson's formula (Actuarial Study No. 4, p. 22)

$$v_0 = \frac{1}{625}[51w_0 + 51w_{\pm 5} - 14w_{\pm 10}] \quad\quad\quad\quad\quad (35)$$

If, however, $p \neq q$, there are three unknowns and the two conditional equations

$$\left. \begin{array}{l} 5p + 10q + 10r = 1 \\ 5p + 135q + 510r = 0 \end{array} \right\} \quad\quad\quad\quad (36)$$

so that the "best" formula as above will be that which makes Σl^2, or $5p^2 + 10q^2 + 10r^2$ a minimum subject to (36)—the formula, as in J. I. A. LI, 368, being

$$v_0 = \frac{1}{7}[.696w_0 + .488w_{\pm 5} - .136w_{\pm 10}] \quad\quad\quad (37)$$

By an analogous process the unsymmetrical formulae, like (9a), can also be obtained.

26. While it is usual in practice to use equal groups as a matter of convenience, there is no reason why the most advantageous grouping should not be adopted wherever values for each age are available; and this question, which is referred to in J. I. A. LI, 369, was investigated by De Forest, III, 43-50. Taking a range of 12 terms from $u_{-\frac{11}{2}}$ to $u_{+\frac{11}{2}}$ he investigates R_0 for each v of the range (thus using Method (A)) by the following methods:—(i) Three equal groups of four terms each, (ii) Tchebicheff's principle (De Forest, I, 310, and IV, 113) of six terms in the middle group and three in each end group, (iii) The "best" number, n, of terms in each of two end groups, and the remaining ($12-2n$) in the middle—n being found to be .224 times the whole number of terms N ("a little less than the number $\frac{N}{4}$ adopted...under Tchebicheff's principle") by determining n so as to make the mean of R_0^2 over the whole range a minimum, and (iv) Individual terms by the 12-term "least square formulae". The results may be summarized thus:—

Distance from centre of range = x	Values of R_0 for v_x by			
	Three equal groups	Tchebicheff's grouping	De Forest's "best" grouping	Individual values
$\pm \frac{1}{2}$	.534	.475	.469	.432
$\pm \frac{3}{2}$	.488	.445	.437	.411
$\pm \frac{5}{2}$	.423	.410	.411	.389
$\pm \frac{7}{2}$	.423	.429	.434	.411
$\pm \frac{9}{2}$	.586	.567	.569	.526
$\pm \frac{11}{2}$	.901	.823	.816	.739
Square root of mean of squares of above	.582	.544	.541	.500

This table shows clearly the manner in which R_0 varies in each case with the distance from the central term, and that the results obtained from using each individual value are preferable to grouping—as would be expected from the fact that any grouping as in (34) necessarily restricts the values of l by imposing the condition that they shall not vary within each group.

(III). REDUCTION OF ERROR.

27. The process of "fitting" used above for the "least square formulae" of (II) gives the same results as that of determining the l's so as to give the greatest possible reduction of error—i.e., the minimum value of R_0—in v, as shown in par. 23. The "smoothness" of the results may, however, as already stated, be more important than the "accuracy" of v itself. This essential distinction between accuracy and smoothness was repeatedly emphasized by De Forest; for example in I, 325, he remarks that R_0 "is not conclusive as regards smoothness of

adjustment—the comparative regularity of the graduation of two series obtained by using different adjustment formulae will be best ascertained by comparing their corresponding orders of differences", and—as we shall see hereafter—he gave three series of formulae satisfying the criterion of smoothness in a greater or less degree which seem to antedate, in their fundamental conception, all the formulae subsequently considered in section (III) of this paper.

In accordance with the theory of pars. 13-15 it therefore becomes of importance, first of all, to investigate the formulae which shall give the greatest possible reduction of error in $\Delta^n v$, that is, the smallest possible values of R_n, rather than the greatest reduction of error in v, that is, the minimum values of R_0, as in (I). Omitting, for brevity, the unimportant case of first differences, we may therefore examine the formulae—like those for R_0 in (I)—which will give minimum values of $(i) R_2$, $(ii) R_3$ and $(iii) R_4$ under Method (B);* and it will be sufficient, as is customary, to take $j=3$.

28. (i). *Minimum Values of R_2*. The formulae for this particular case were directly suggested first by G. F. Hardy in 1909 (Lectures, p. 28), although he dismissed them as "somewhat theoretical" because they would not generally be reducible to "summation" form—an objection which need no longer be considered fatal (see par. 21). The application of the principle may be illustrated (to compare with the case of R_3 given in full in T. A. S. A. XIX, 26-28) by applying the second method of par. 23—with $j=3$, and hence $c_0=1$ and $c_2=0$ in the symmetrical case—to find the 9-term formula for v_0 which shall have the minimum value for R_2 (or R_2^2) under Method (B). Taking therefore

$$v_0 = b_0 u_0 + b_1 u_{\pm 1} + b_2 u_{\pm 2} + b_3 u_{\pm 3} + b_4 u_{\pm 4}$$

---

* For practical purposes it is not necessary to proceed farther, to R_5, etc. It may be noted, however, that De Forest gives (II, 332) the formula

$$v_0 = \frac{1}{3059}[1959 u_0 + 825 u_{\pm 1} - 330 u_{\pm 2} + 55 u_{\pm 3}] = .640 u_0 + .270 u_{\pm 1}$$
$$- .108 u_{\pm 2} + .018 u_{\pm 3}$$

for the case of a minimum value of R_4 when $j=5$, which may be used "when the given series varies so rapidly that any five consecutive terms cannot be regarded as approximating closely to the form of a curve of the third degree".

and differencing the coefficients twice as in par. 15 we get
$$R_2^2 = 2[(b_4)^2 + (b_3 - 2b_4)^2 + (b_2 - 2b_3 + b_4)^2 + (b_1 - 2b_2 + b_3)^2 \\ + (b_0 - 2b_1 + b_2)^2 + (b_0 - b_1)^2]e^2 \div 6e^2$$
so that we have to make the expression in square brackets, that is,
$$6b_4^2 + 6b_3^2 - 8b_3b_4 + 6b_2^2 - 8b_2b_3 + 2b_2b_4 + 7b_1^2 - 8b_1b_2 + 2b_1b_3 \\ + 3b_0^2 - 8b_0b_1 + 2b_0b_2,$$
a minimum, subject to the conditions $c_0=1$, that is (from (33)), $b_0+2(b_1+b_2+b_3+b_4)=1$, and $c_2=0$, being $b_1+4b_2+9b_3+16b_4=0$. Eliminating b_0 and b_1 as in par. 23 the expression to be made a minimum is therefore
$$8338b_4^2 + 2475b_3^2 + 9064b_3b_4 + 462b_2^2 + 2112b_2b_3 + 3894b_2b_4 \\ + 70b_2 + 168b_3 + 308b_4 + 3$$
from which, by differentiation with regard to b_2, b_3, and b_4 we get
$$462b_2 + 1056b_3 + 1947b_4 + 35 = 0$$
$$1056b_2 + 2475b_3 + 4532b_4 + 84 = 0$$
$$1947b_2 + 4532b_3 + 8338b_4 + 154 = 0$$
and from these, together with $c_0=1$ and $c_2=0$, the b's are found and hence the formula
$$v = \frac{1}{429}[135u_0 + 112u_{\pm 1} + 56u_{\pm 2} - 21u_{\pm 4}] \quad \ldots \ldots \ldots (38)$$
This formula, those for 11, 17, 21, and 29 terms, and also the general form, were first given in Dr. Sheppard's papers—see J. I. A. XLVIII, 396-403 (pars. 18-23).

29. (ii). *Minimum Values of R_3*. This is the case which has received particular attention from actuaries, because, as stated in par. 21, the smallness of R_3 has generally been used as the standard of smoothness (in preference to R_2 as above or R_4 below), so that R_3 in actuarial literature has become known as the "smoothing coefficient." As in the case of R_2 the direct determination of the b's corresponding to the minimum value for R_3 seems to have been suggested first by G. F. Hardy (Lectures, p. 27-29); and again the first complete statement of the formulae was given in 1915 by Dr. Sheppard, J. I. A. XLIX, 148-152 (pars. 24-26)—although, like Hardy, he considered that

they were "not convenient for practical purposes" on account of their complicated nature when expressed in terms of successive sums. Later, in 1916, the general formulae were rediscovered independently by R. Henderson (T. A. S. A. XVII, 43) and again in 1918 by J. R. Larus (T. A. S. A. XIX, 14)— the latter paper showing the practicability of using the formulae in linear compound form with modern calculating machines, so that the once fatal objection of inconvenience may now be considered to have been removed (see also T. A. S. A. XIX, 298, and J. I. A. LII, 401). The formulae are obtainable in exactly the same way as those for R_2 above; and the demonstration therefore need not be given here, especially as it is set out in full in Larus' paper and Mervyn Davis' discussion thereof (T. A. S. A. XIX, 300-1). The general formulae, and the values of $\dfrac{1}{R_3}$ up to 31 terms, are given by Sheppard; and Larus gives a complete table of the coefficients from 7 to 21 terms, with the "smoothing coefficient" R_3.

30. (iii). *Minimum Values of R_4.* As already stated, the criterion of "smoothness" is necessarily arbitrary. Either the "R_2 formulae" of par. 28 or the "R_3 formulae" of par. 29— which produce a new series of v's in which the mean square error of $\Delta^2 v$, or of $\Delta^3 v$, is a minimum—may be held to give the theoretically "smoothest" result according to whether the smallness of the mean square error of $\Delta^2 v$ or $\Delta^3 v$ is adopted as the criterion; and, since j has been taken as 3, it is to be noted in the "R_2 formulae" that, as stated by Dr. Sheppard (J. I. A. XLVIII, 396), we must "assume that, even if $j=3$, we are not concerned about special accuracy of 3rd differences". The fundamental assumption that $j=3$, however, means that the fourth differences of the true U's may be assumed to be zero. In Method (A), as shown in par. 12, the new values of $\Delta^4 v$ are made to conform exactly with this assumption, and are themselves precisely zero—even though the original u's may not have had their fourth differences zero. But in Method (B), which is here under consideration, the new values of $\Delta^4 v$ are not zero unless the original u's had their fourth differences zero; and since, when $j=3$, U_r is a polynomial of the 3rd degree and,

as usual, $u_r = U_r + e_r$, it follows that the values of $\triangle^4 v$ in Method (B) will arise entirely from the errors e, whereas $\triangle^3 v$, $\triangle^2 v$, and $\triangle v$ arise partly from the true differences of U and partly from the errors. Consequently, as the object is to find the "best" formula for reducing the errors by linear compounding of the u's, it is logical to make the mean square error in $\triangle^4 v$ as small as possible—for the series will then have been smoothed by operating directly upon that order of differences which, on account solely of the errors e, is known to need adjustment.*

These formulae were first given by De Forest (I, 332-334; II, 327-331 and 351-2; III, 50-2; IV, 82-84); and having thus been published so long ago as 1873 (the publication date of I) they seem, as already stated, to have antedated in their fundamental conception all the important formulae considered in section (III) of this paper. De Forest obtains them by a process very similar to that of pars. 28 and 29 above, and, in another demonstration (I, 334, and II, 352) from the fact that "the whole series of weights [the b's] taken together with the eight nearest zero weights, constitutes a series of the tenth order". Adopting here exactly the same method as that in pars. 28-29, we should have for, say, the 5-term formula (for brevity of illustration), $R_4^2 = 2[(b_2)^2 + (b_1 - 4b_2)^2 + (b_0 - 4b_1 + 6b_2)^2 + (7b_1 - 4b_2 - 4b_0)^2 + (3b_0 - 4b_1 + b_2)^2]e^2 \div 70e^2$ together with the conditions (since $j=3$), $b_0 + 2b_1 + 2b_2 = 1$, and $b_1 + 4b_2 = 0$; from which the expression in square brackets to be made a minimum becomes $6435 (b_2)^2 + 924 b_2 + 35$, and thence by differentiation we find b_2 and hence the formula

$$v_0 = \frac{1}{195}[111 u_0 + 56 u_{\pm 1} - 14 u_{\pm 2}] \dots \dots \dots \dots \dots (39)$$

De Forest gives a complete table of these "R_4 formulae", for $j=3$, up to 25 terms, together with their values of R_0 and R_4, as follows:—

---

* It may be noted that this, of course, is a different procedure from that used by Dr. Sheppard (J. I. A. XLIX, 153) for obtaining "improved values for the fourth differences" in order to test whether 4th differences are really negligible when $j=3$. For his formulae, like those for $\triangle^2 v_{-1}$ in the footnote to par. 24, are those for $\triangle^4 v_{-2}$, that is, $24E$, when $U_x = A + Bx + Cx^2 + Dx^3 + Ex^4$ is fitted by least squares—giving, for example, $24E = \triangle^4 v_{-2} = \frac{1}{11}[6u_0 + u_{\pm 1} - 7u_{\pm 2} + 3u_{\pm 3}]$ for 7 terms.

DEVELOPMENT OF FORMULAE FOR GRADUATION

COEFFICIENTS OF THE u's IN THE "R_4 FORMULAE"

Number of Terms	5	7	9	11	13	15	17	19	21	23	25
b_0	.56	.42	.34	.290	.252	.222	.200	.182	.166	.1534	.1424
b_1	.29	.29	.27	.245	.222	.202	.184	.170	.157	.1463	.1368
b_2	−.07	.05	.11	.135	.145	.146	.143	.137	.132	.1262	.1205
b_3		−.05	.02	.026	.056	.076	.087	.093	.096	.0969	.0962
b_4			−.03	−.028	.007	.015	.034	.047	.057	.0635	.0678
b_5				−.023	−.027	−.018	−.004	.010	.022	.0317	.0395
b_6					−.015	−.022	−.020	−.012	−.003	.0065	.0151
b_7						−.010	−.017	−.018	−.015	−.0092	−.0024
b_8							−.007	−.013	−.016	−.0151	−.0119
b_9								−.005	−.010	−.0133	−.0140
b_{10}									−.003	−.0077	−.0110
b_{11}										−.0025	−.0060
b_{12}											−.0018
R_0	.70	.60	.54	.495	.461	.433	.409	.389	.372	.358	.345
R_4	.23	.080	.038	.0187	.0104	.0064	.00422	.00273	.00185	.00124	.00091

(IV). DE FOREST'S OTHER FORMULAE.

31. It was stated in par. 27 that De Forest gave three series of formulae satisfying the criterion of smoothness in a greater or less degree. Those just given in the preceding paragraph he considered the best. He adds a warning, however, that the theoretical basis of reduction of mean square error with its assumption that the mean square error of each u_r is e^2 may not always be strictly applicable "because the errors of the given terms will not in general be really equal"— although "it usually happens in statistical and other series that the weights of the terms follow a tolerably regular sequence, so that if for instance the errors e_1, e_2, etc., are greater than e_0, the errors e_{-1} and e_{-2} will be less than e_0, and vice versa, and thus they tend to correct each other" (cf. Lidstone, J. I. A. XLII, 124).

32. The two other series of formulae are given in I, 322-4, and IV, 84-5. The former are especially interesting, as they were founded in 1870 (see I, 275) on exactly the same test of the "flow of the coefficients" which has subsequently come into great prominence, and which, without any knowledge of De Forest's prior exposition, is usually (Sherriff, p. 118; A. W. Watson, J. I. A. XLI, 83; King, J. I. A. XLI, 556, etc.) credited to J. A. Higham (J. I. A. XXIII, 336; XXV, 22), J. Karup (Trans. 2nd Int. Cong. Actuaries, 1899, p. 93), and Lidstone (J. I. A. XLI, 354). Taking as an example of adjustment formulae those of par. 19 in which the coefficients (which De Forest calls "local weights") are in A. P., he says "Further investigation has shown that this arrangement of the weights, although it gives formulas which are very simple and easy of application, is not the best one in theory. To determine what the best arrangement is, we must consider that when one of these formulas is applied at any part of a series, all those terms which are not included by the formula have the weight zero; that as the adjustment progresses, when a term is first included by the formula its weight is negative, it then becomes positive, attains its maximum when the term occupies the middle position, then diminishes till it becomes negative again, and finally resumes the weight zero when the term is no longer included by the formula. To make this transition as unbroken and continuous as

possible, it is evident that if we regard the weights as ordinates to a curve, the form of this curve should be as shown in the annexed figure, for a formula including 7 terms whose positions

1, 2, 3,...7, are laid off equidistantly on the axis of X. The curve is symmetrical with respect to the middle ordinate or axis of Y, and is tangent to the axis of X at the points 0 and 8, which are the positions of the two nearest terms not included by the formula." The curve here given by De Forest is exactly that shown by Mr. Lidstone. Proceeding from this curve, "which has four points of inflexion, so that if it is of algebraic form it must be of a degree not lower than the sixth," and assuming maxima at the points 0 and 8, De Forest finds the 7-term formula, and similarly those up to 15 terms, as in the table below:—

Number of Terms	5	7	9	11	13	15
b_0	.56616	.41476	.32966	.27406	.23466	.20522
b_1	.28923	.29486	.26706	.23737	.21137	.18953
b_2	—.07231	.05494	.12006	.14408	.14934	.14651
b_3		—.05718	—.01198	.03809	.07003	.08755
b_4			—.03997	—.02864	.00195	.02875
b_5				—.02793	—.03005	—.01321
b_6					—.01997	—.02709
b_7						—.01465

These formulae, as would be expected from the manner of their construction, have good values of R_0 and R_4; for example, R_0 for the 7-term formula is .59874 which compares favorably with the minimum value .58 (par. 33), and R_4 for 11 terms is .0205 compared with the minimum .0187.

Another similar method was also suggested by De Forest (I, 324), namely, to give the curve "a contact of the second order, so that its first and second differential coefficients shall both become zero at the points 0 and 8'', thus obtaining for 7 terms the coefficients .45998, .29281, .01752, and —.04032 instead of the above.*

33. The other series of formulae referred to in par. 31 was derived from the following considerations:—The essential distinction between "accuracy" and "smoothness" which has already been emphasized, may be seen very clearly by comparing the "accuracy" (R_0) and "smoothness" (R_4) of the above "R_4 formulae", which are designed to give the smoothest possible results when judged by R_4, with the following values (De Forest, IV, 81) for the "least square" formulae of par. 23 which give the most accurate values when judged by the smallness of R_0:—

Number of Terms	5	7	9	11	13	15	17	19	21	23	25
R_0	.69	.58	.50	.456	.418	.389	.365	.345	.328	.313	.300
R_4	.29	.19	.13	.109	.0907	.0782	.0677	.0600	.0548	.0496	.0456

It is thus clear, as De Forest states (IV, 84), that "it is impossible to get at once the maximum of probable accuracy and the maximum of smoothness. Formulae may, however, be constructed so as to give intermediate and moderately small values to both ratios. Very good results are obtained by assuming that the coefficients b are ordinates to the curve $y=A+Cx^4+Dx^6+Ex^8$, the constant interval between these ordinates being taken as the unit of x. The values of the four constants are found from the four equations

$$(2n+1)A+2[(n^4)C+(n^6)D+(n^8)E]=1$$
$$(n^2)A+(n^6)C+(n^8)D+(n^{10})E=0$$
$$A+(n+1)^4C+(n+1)^6D+(n+1)^8E=0$$
$$2C+3(n+1)^2D+4(n+1)^4E=0$$

"

in which the first two equations are the conditions $c_0=1$ and $c_2=0$ (par. 23) "while the last two require that y and $dy \div dx$ should both become zero when $x=\pm (n+1)$" (see par. 32). The table as given by De Forest is:—

---

* The general principle of both these methods may be compared with that suggested by J. S. Thompson in 1916 (T. A. S. A. XVII, 355).

DEVELOPMENT OF FORMULAE FOR GRADUATION 115

COEFFICIENTS OF THE u's IN DE FOREST'S "INTERMEDIATE FORMULAE"

Number of Terms	5	7	9	11	13	15	17	19	21	23	25
b_0	.44	.32	.24	.206	.176	.154	.136	.122	.112	.102	.094
b_1	.37	.30	.24	.204	.176	.154	.136	.122	.112	.102	.094
b_2	−.09	.14	.18	.177	.162	.146	.133	.121	.110	.102	.094
b_3		−.10	.03	.093	.117	.121	.118	.111	.104	.098	.091
b_4			−.07	.020	.038	.070	.085	.090	.090	.088	.085
b_5				−.057	.037	.005	.037	.056	.067	.071	.073
b_6					−.044	−.040	−.012	.015	.034	.047	.054
b_7						−.033	−.039	−.021	.000	.018	.031
b_8							−.026	−.035	−.025	−.009	.007
b_9								−.020	−.032	−.027	−.015
b_{10}									−.016	−.028	−.026
b_{11}										−.013	−.024
b_{12}											−.011
R_0	.70	.58	.50	.461	.427	.398	.375	.354	.339	.325	.311
R_4	.36	.19	.088	.0516	.0324	.0207	.0146	.0103	.00802	.00589	.00480

The "accuracy" obtained by these formulae is nearly as good as that of the "least square" formulae, while the "smoothness" is much better—although not, of course, as good as that of the "R_4 formulae."

(V). DE FOREST'S FORMULAE FOR "REPEATED ADJUSTMENTS".

34. The suggestion of applying a linear compound graduation formula more than once has often been made—for example, by De Forest in 1873 (II, 337), J. A. Higham in 1882 (J. I. A. XXIII, 336), Altenburger (J. I. A. XLI, 449), and King (J. I. A. XLI, 548); and De Forest in 1878-80 examined the formulae very completely in his "Analyst" papers marked VI, VII, and VIII in the Appendix hereto. Giving first an alegbraical demonstration of the form taken by the coefficients when the general formula (1) or (2) is applied repeatedly, and an indefinite number of times, he shows also the following numerical values of the b's when the group interpolation formula (18), the 5-term "least square" formula, and the 5-term "R_4 formula" (39) are applied 1, 2, 4, 6, and 12 times, together with those of 3, 5, 8, 16, and 32 applications of the "least square" formula:—

COEFFICIENTS OF u FOR REPEATED ADJUSTMENTS

5-term Least Square Formula

Number of Repetitions	1	2	3	4	5	6	8	12	16	32
b_0	.486	.486	.418	.396	.372	.354	.330	.296	.274	.230
b_1	.343	.274	.293	.280	.277	.271	.261	.246	.234	.205
b_2	−.086	.034	.048	.077	.090	.102	.116	.131	.139	.145
b_3		−.059	−.038	−.034	−.023	−.014	.002	.026	.042	.072
b_4		.008	−.019	−.024	−.030	−.032	−.031	−.024	−.015	.014
b_5			.008	.000	−.004	−.009	−.017	−.024	−.026	−.017
b_6			−.001	.004	.004	.003	.000	−.008	−.014	−.022
b_7				−.001	.001	.002	.003	.002	−.002	−.014
b_8					−.001	−.000	.001	.003	.003	−.004
b_9									.002	.002
b_{10}										.003
b_{11}										.001
R_0	.697	.629	.596	.574	.558	.544	.524	.497	.479	.438
R_4	.296	.102		.0513		.0331		.0160		.0058

Number of Repetitions	5-term R_4 Formula (39)					5-term Interpolation Formula (18)				
	1	2	4	6	12	1	2	4	6	12
b_0	.570	.500	.416	.374	.310	.400	.500	.392	.344	.284
b_1	.287	.285	.285	.276	.253	.400	.240	.264	.261	.239
b_2	−.072	.001	.059	.089	.125	−.100	.080	.101	.115	.136
b_3		−.041	−.038	−.023	.015		−.080	−.037	−.008	.035
b_4		.005	−.019	−.029	−.028		.010	−.022	−.030	−.019
b_5			.003	−.004	−.021			−.008	−.016	−.026
b_6			.002	.003	−.004			.008	.004	−.011
b_7				.001	.003			−.002	.002	.000
b_8					.002				.001	.003
b_9									−.001	.001
R_0	.707	.645	.588	.588	.508	.707	.625	.563	.532	.486
R_4	.229	.122	.0622	.0411	.0195	.446	.239	.0829	.0365	.0134

The calculation of these coefficients "is not so difficult or tedious as might be supposed, for it has been found that the coefficients of the extreme terms..... soon become so small as to disappear, that is, they do not affect the first five places of decimals". Modifying the quotations slightly to conform with the nomenclature of this paper, De Forest remarks that "R_0 diminishes slowly as the number of repetitions is increased, while R_4 diminishes much faster"; that "although the coefficients of the original formulae (18) and (39) differ from each other very considerably, yet after 12 applications they approach pretty nearly to each other and also to those of the 12-fold least square formula"; and that "for any large number of repetitions the form of the series of values of b approximates somewhat to that of the 'R_4 formulae'—for instance, the large coefficients of the 12-fold and 32-fold 'least square formulae' approach very near to those of the 11-term and 15-term 'R_4 formulae', and their ratios R_0 and R_4 are also nearly alike'' (VI, 68-69). From this, and the fact that the coefficients from four applications of the 9-term and 13-term least square formulae are also shown to differ ''but little'' from the coefficients of the 17- and 25-term ''R_4 formulae'', De Forest suggests that ''the resultant coefficients approximate to a common form, and they resemble the coefficients of the 'R_4 formulae' a good deal more closely than those of the least square formulae or those of par. 33—which may perhaps be regarded as an argument in favor of the use of the 'R_4 formulae' for purposes of adjustment''.*

(VI). Conclusion.

35. Limitations of space will not permit of anything beyond a mere reference to a number of other matters of interest to actuaries and statisticians which De Forest investigated—such as his decision against the employment of the ''weights'' of the observed u's in graduation by linear compounding (II, 339); his consideration of the construction of mortality tables from censuses only, without any registration of deaths (I, 302); his formula for the force of mortality (II, 343—see also J. I. A. XXII, 231, and his ''Analyst'' letter, vol. VII, p. 54); and his exhaustive inquiry into the tests of a good adjustment (II, 333; III; IV, 107-113; V; VI) which is very closely connected with the subject of this paper. All these matters, and a number of others which may be inferred from the titles of his papers, were treated with great lucidity and a masterly command of exposition which, in combination with unsparing industry and attention to detail, make his writings most attractive. All essential steps and explanations are clearly shown, and nowhere is there any difficulty in following his meaning—with the result

---

* Although outside the scope of this paper, it may be remarked that De Forest subsequently showed (in VII) that the limiting form of the curve of coefficients in the symmetrical case is the normal probability curve when $j=1$, and when $j=2$ or 3 it ''consists of an infinite number of undulations'' of which the principal significant portion ''presents striking analogies to the probability curve''; while in the unsymmetrical case (VIII) he reaches an unsymmetrical probability curve which, as acknowledged in one of Prof. Karl Pearson's papers, is Pearson's ''Type III''.

that, although his investigations cover many pages, they may be read without any hesitation even by a student, and thus possess a merit which is all too frequently sacrificed to brevity by many other writers on such subjects.

36. The recognition obtained by De Forest's writings at the time of publication, however, does not seem to have been very great. This may have been due partly to the rather unusual periodicals in which they appeared, and to the fact that the development of graduation by linear compounding had not then progressed beyond the interpolation methods of section (I). On p. 38 of III De Forest states that "the English mathematical authority to whom my first, second, and third methods of adjustment [covering, apparently, such formulae as (5)-(13), (18)-(19) and a few others (see De Forest, II, 341)] were originally submitted [in 1868] by the Smithsonian Institution expressed certain adverse opinions, but without giving any reasons or specifications so far as I could learn, so that it is impossible to say whether the originality of the work was called in question, or only its probable utility. It is true that my methods were set forth somewhat imperfectly at first, and that they have been much developed and improved since ...". This adverse opinion, which apparently did not cover such improved methods as the least square and R_4 formulae, seems to have caused some delay in publication (III, 39), and to have determined De Forest to publish his later work independently (preface to III). The name of the "English mathematical authority" does not appear. Later, in October, 1878, he had occasion to write to the Editor of the Journal of the Institute of Actuaries with regard to a formula for the force of mortality—the letter being published a year and a half later in April, 1880 (J. I. A. XXII, 231); and the same matter is referred to in his "Analyst" letter of Jan. 8, 1880, mentioned above (vol. VII, p. 54). In this it is stated that "a copy of the Smithsonian Report [of 1873—No. II of the Appendix here—containing the least square and R_4 formulae] was sent to the Editor of the Journal in 1875"; but no recognition seems to have been accorded to it. It is therefore hoped that this paper on his work may result in its receiving that wider recognition which it undoubtedly deserves.

APPENDIX.

The following is a complete list of De Forest's papers (see Royal Society of London, Catalogue of Scientific Papers, vols. VII, IX, and XIV):—

    (a) A method of estimating and correcting the error caused by the unequal length of the calendar months, in reducing observations of temperature: American Journal of Science, XLI (1866), pp. 371-9.

    (b) A second method of correcting monthly means for the unequal length of the months: Ibid., XLII (1866), 154-6 (and p. 289).

    (c) On the reduction of meteorological observations: Ibid., XLIII (1867), 316-322.

I. (d) On some methods of interpolation applicable to the graduation of irregular series, such as tables of mortality, etc.: Smithsonian Report, 1871, 275-339.

II. (e) Ditto: Smithsonian Report, 1873, 319-353.

III. (f) Interpolation and Adjustment of Series (Printed privately by Tuttle, Morehouse, and Taylor Co., New Haven, Conn.; 8vo., pp. 52), 1876.

IV. (g) On adjustment formulas: The Analyst, Des Moines, Iowa ("A monthly Journal of Pure & Applied Mathematics," edited and published by J. E. Hendricks), vol. IV (1877), 79-86 and 107-113.

V. (h) On the grouping of signs of residuals: Ibid., V (1878), 1-6.

VI. (i) On repeated adjustments, and on signs of residuals: Ibid., V (1878), 65-72.

VII. (j) On the limit of repeated adjustments: Ibid., V (1878), 129-140.

    (k) On the develpment of $[p+(1-p)]^\infty$: Ibid., VI (1879), 65-73.

VIII. (l) On unsymmetrical adjustments, and their limits: Ibid., VI (1879), 140-48, 161-170; VII (1880), 1-9 (and postscript p. 22).

    (m) On some properties of polynomials: Ibid., VII (1880), 39-46, 73-82, 105-115.

IX. (n) Letter re "Formula for Force of Mortality": Ibid., VII (1880), 54.

    (o) On a theorem in probability: Ibid., VII (1880), 169-176.

    (p) Law of facility of errors in two dimensions: Ibid., VIII (1881), 3-9, 41-48, 73-82.

    (q) On the elementary theory of errors: Ibid., VIII (1881), 137-148.

    (r) Law of error in the position of a point in space: Ibid., IX (1882), 33-40, 65-74.

    (s) On an unsymmetrical probability curve: Ibid., IX (1882), 135-142, 161-168; X (1883), 1-7, 67-74.

    (t) A method of demonstrating certain properties of polynomials: Ibid., X (1883), 97-105.

X. (u) Letter re "Formula for Force of Mortality": J. I. A., XXII (1881), 231.
    (v) On an unsymmetrical law of error in the position of a point in space: Trans. of the Connecticut Academy of Arts and Sciences, VI (1882-5), 123-138.
    (w) On the law of error in target shooting: Trans. Conn. Academy, VII (1885-8), 1-8.

I am indebted to Miss Todd of the Science Division, New York Public Library, for assistance in checking the above, and to Mr. Keogh, Librarian of Yale University, for the information that Professor Irving Fisher wrote some time ago an unpublished memoir on De Forest. Extracts from this—which Professor Fisher has very kindly allowed me to read—are printed in J. Anderson's "The Town and City of Waterbury, Connecticut", vol. III, which seems to contain a more complete account of De Forest's life than other references which I had previously consulted, namely, "The De Forests of Avesnes (and of New Netherland)" by J. W. De Forest, "A Walloon Family in America—Lockwood De Forest and His Forbears, 1500-1848" by Mrs. Robert W. De Forest, the "Yale Obituary Record, 1880-1890", and "Yale Endowments". From these it appears that De Forest, who nowhere in his writings identifies himself any more completely than as "E. L. De Forest, M.A., of Watertown, Conn.", was born on June 27, 1834, obtained his B.A. at Yale in 1854, and after two years in the engineering department of the Sheffield Scientific School (Yale) received his Ph.B. in 1856, and M.A. in 1867. Going to California in 1857 "he worked for a time in the mines, and then engaged in teaching, and removed from there to Australia, where he became a tutor in the College of Melbourne". Then, after visits to India and Europe, he returned to America, and until his death on June 6, 1888, lived chiefly in New Haven and Watertown, Conn., where he produced the works mentioned in this paper. In 1886 he gave $4,000 to Yale to increase the fund established in 1855 by his father, from which the "De Forest Mathematical Prizes" are still given; and in 1888 he gave $10,000 for the endowment of a mathematical chair.

SOME PRACTICAL FEATURES OF A FIELD TIME DETERMINATION WITH A MERIDIAN TRANSIT

R.S. Woodward

Mr. R. S. WOODWARD read a paper on

SOME PRACTICAL FEATURES OF A FIELD TIME DETERMINATION WITH A MERIDIAN TRANSIT.

[Abstract.]

An important desideratum in all kinds of field work is the adoption of those methods which will secure the accuracy essential in the results sought with the minimum amount of computation. It is in general easier and more conducive to precision to eliminate unnecessary factors involved with the quantity sought than to determine their values and allow for them by computation. Very frequently also a systematic arrangement of observations will secure the maximum precision with the minimum of computation.

In a field time determination for telegraphic longitude the essential quantity is the error of the time-piece used at some determinate epoch, and the unessential factors are the azimuth and collimation of the transit and the rate of the time-piece. It is evident that a minimum of computation will be required if the observations can be so arranged as to eliminate the effect of these factors in the final value of the correction to the time-piece. Although it is usually impossible to eliminate the effect of the azimuth, collimation, and rate completely, it is generally possible to make a close approximation thereto. To show this fact analytically let

Δt = the correction to the time-piece at the epoch t_o;
t = the observed time of a star's transit;
a = the star's right ascension;
a = the azimuth of the transit;
c = the collimation of the transit;
r = the rate of the time-piece;
A = the azimuth factor;
C = the collimation factor;
p = the weight of $(t - a)$;
v = the residual error.

Then the type observation-equation will be

$$\Delta t + Aa + Cc + (t - t_o) r + t - a = v. \qquad (1)$$

The normal equation in Δt, using brackets to indicate summation of like quantities, is

$$[p] \Delta t + [pA] a + [pC] c + [p(t - t_o)] r + [p(t - a)] = 0. \quad (2)$$

This shows that in order to secure the complete elimination of the effect of a, c, and r, we must have

$$[pA] = 0, [pC] = 0, [p(t - t_o)] = 0. \qquad (3)$$

The last of these conditions can always be fulfilled by making

$$t_o = \frac{[pt]}{[p]}. \qquad (4)$$

It may be shown that the value of Δt corresponding to t_o as defined by (4) has a maximum weight. A close approximation to the first two conditions of (3) can be secured by selecting for observation stars of suitable declinations and by reversals of the telescope.

If we put

$$\frac{[pA]}{[p]} = \beta, \frac{[pC]}{[p]} = \gamma, \frac{[p(t - a)]}{[p]} = \Delta t_o,$$

equation (2) gives $\Delta t = - \Delta t_o - \beta a - \gamma c$. \hfill (5)

This shows that in case β and γ are small, as supposed above, an approximate value of Δt is $-\Delta t_0$. After some preliminary observations at a station it is easy to render a and c small, and their approximate values may always be found from the observation equations by a brief inspection; so that with such values of a, c, β, and γ as are nearly always readily attainable Δt may be derived from (5) to the nearest $0^s.01$.

We may thus dispense entirely with the other three normal equations and reach the same result which would follow from their use. The solution may also be checked; for by one or two approximations the values of Δt, a and c which make $[pv] = 0$ can be readily found.

The practical steps in deriving Δt from (5) may be summarized as follows:

1. The mean of the observation-equations for clamp west minus the mean of those for clamp east will give an approximate value of the collimation c.

2. The application of this value of c to each observation equation will give a corrected value of $(t - a)$ for each star.

3. An approximation to the value or values of the azimuth will then result by eliminating Δt from one or more pairs of the corrected observation equations. The azimuth may then be applied to correct the values of $(t - a)$, reached in step 2.

4. The approximate values of a and c will now give an approximate value of Δt from (5), and the application of this value of Δt to the values of $(t - a)$, derived in step 3, will give approximate values of v.

5. Form $[pv]$. If this sum is not zero within $0^s.01$ or $0^s.02$, a brief inspection will show what changes in a and c (and possibly Δt) will make it zero within those limits.

By this process of determining the residuals or their approximate values as soon as possible in the computation any large errors in the values of $(t - a)$ or the azimuth and collimation factors will be easily detected.

In precise longitude determinations it is customary to have for each night's observations two complete time determinations, one immediately preceding and one immediately following the telegraphic comparison of time-pieces. In this case there will be two values of Δt. Calling these $\Delta t'$ and $\Delta t''$ and denoting the corresponding epochs by t_o' and t_o'' the rate of the time-piece will be given with sufficient accuracy for interpolation by the equation

$$r = \frac{\Delta t'' - \Delta t'}{t_o'' - t_o'}.$$

Mr. PAUL thought it an objection to this method that in arranging the groups valuable stars might be lost, so that in a limited time the accuracy of the results would be impaired by the smaller number of observations; moreover, the method did not furnish the computer with a clear idea of the performance of the instrument.

Mr. HALL said that he liked the method, and that he thought it especially good for time work. He had discovered the method once himself, and he knew that it had also been used by Prof. Ormond Stone.

Mr. KUMMELL said that in connection with this subject he had investigated the question of the advisability of using stars towards the pole for time determinations; that is, he had examined the weight co-efficient formulæ to see at what distance north of the zenith a maximum value would be obtained. He found that in general the limit of declination was about 60°.

Mr. PAUL thought that every weight-formula should take account of the increase of atmospheric disturbance with increase of zenith distance.

Mr. KUMMELL then read the following paper entitled

CAN THE ATTRACTION OF A FINITE MASS BE INFINITE?

In Price's Calculus, vol. III, art. 201, discussing the result for the attraction of a thin rectangular plate on a particle external to it and in its own plane it is found that if the attracted particle is at an angle of the rectangle the attraction is infinite. Price's method of determining the attraction of plates consists in integrating between the proper limits the following differentials:

$$d^2X = m\theta\tau \frac{x\,dx\,dy}{(x^2 + y^2)^{\frac{3}{2}}}; \qquad (1_x)$$

$$d^2Y = m\theta\tau \frac{y\,dx\,dy}{(x^2 + y^2)^{\frac{3}{2}}}; \qquad (1_y)$$

where $X = x$ — axial component of attraction;
$Y = y$ — axial component of attraction;
m = mass of attracted particle;
θ = density of attracting mass, supposed homogeneous;
τ = thickness of the plate.

Referring to the attracted particle as origin let (a_0, b_0) be the corner nearest and (a_1, b_1) that farthest from the particle; then

$$X = m\theta\tau \int_{b_0}^{b_1} dy \int_{a_0}^{a_1} \frac{xdx}{(x^2+y^2)^{\frac{3}{2}}}$$

$$= m\theta\tau \int_{b_0}^{b_1} dy \left((a_0^2 + y^2)^{-\frac{1}{2}} - (a_1^2 + y^2)^{-\frac{1}{2}} \right)$$

$$= m\theta\tau l \left\{ \frac{b_1 + (a_0^2 + b_1^2)^{\frac{1}{2}}}{b_0 + (a_0^2 + b_0^2)^{\frac{1}{2}}} \cdot \frac{b_0 + (a_1^2 + b_0^2)^{\frac{1}{2}}}{b_1 + (a_1^2 + b_1^2)^{\frac{1}{2}}} \right\} \quad (2)$$

and a similar expression for Y by exchanging the a's for the b's.

If in this we place $a_0 = b_0 = 0$ then $X = m\theta\tau \infty$.

This is taken by Price to be infinite; yet, since the thickness τ must be taken infinitesimal, this is an entirely unfounded conclusion.

At first, however, I did not suspect this result, and when Mr. Woodward found an infinite attraction of a circular disk on a point at its circumference, which result I checked, it seemed to be possible that the attraction of a finite mass could be infinite. Yet neither Mr. Woodward nor myself was entirely convinced. To settle this question I then resolved to determine the attraction of a right prism and also of a right circular cylinder on a particle at midheight, which, being then moved to the surface and taking the height infinitesimal, would give the attraction of a plate on a particle in a position at which an infinite attraction had been found.

For a right rectangular prism we have, $2h$ being its height,

$$X = m\theta \int_{b_0}^{b_1} dy \int_{a_0}^{a_1} xdx \int_{-h}^{h} \frac{dr}{(x^2+y^2+z^2)^{\frac{3}{2}}}$$

$$= m\theta \int_{b_0}^{b_1} dy \int_{a_0}^{a_1} \frac{2hxdx}{(x^2+y^2)(x^2+y^2+h^2)^{\frac{1}{2}}} \quad (3)$$

Comparing this with (2), putting $2h = \tau$, we readily see that these values are by no means identical, for (2) is of the form $\tau f_2 (a_1, a_0, b_1, b_0)$ while (3) is $\tau f_3 (a_1, a_0, b_1, b_0, \tau)$. Here f_2 is independent of τ while f_3 is not.

Placing $a_0 = b_0 = o$ or supposing the attracted particle at the edge of the prism we have

$$X = 2m\theta h \int_0^{b_1} dy \int_0^{a_1} \frac{xdx}{(x^2+y^2)(x^2+y^2+h^2)^{\frac{1}{2}}}$$

$$= 2m\theta h \int_0^{b_1} dy \int_{\sqrt{y^2+h^2}}^{\sqrt{a_1^2+y^2+h^2}} \frac{dr}{z^2-h^2}$$

$$= 2m\theta h \int_0^{b_1} dy \left[\frac{1}{2h} l \frac{z-h}{z+h} \right]_{\sqrt{y^2+h^2}}^{\sqrt{a_1^2+y^2+h^2}}$$

$$= m\theta \int_0^{b_1} dy \left[l \frac{z-h}{z+h} \right]_{\sqrt{y^2+h^2}}^{\sqrt{a_1^2+y_2+h^2}} \tag{4}$$

This form shows clearly that X vanishes with h; therefore the attraction of a rectangular plate on a particle at one of its corners is not infinite, but, on the contrary, it is infinitesimal.

Similarly the attraction of a circular plate on a particle at its circumference may be found by considering the plate a vanishing cylinder. Using cylindrical co-ordinates to a vertical line through the attracted particle as axis we have, if $\rho =$ horizontal radius vector, $v =$ vectorial angle, for the attraction of a cylinder, radius r, density θ, on a particle at midheight h and distance d from axis of cylinder where

$$\cos v = \frac{\rho^2 + d^2 - r^2}{2\rho d}$$

$$\text{and } \sin v = \frac{1}{2\rho d} \sqrt{[(d+r)^2 - \rho^2][\rho^2 - (d-r)^2]}$$

$$A = m\theta \int_{d-r}^{d+r} \rho^2 d\rho \int_{-v}^{v} \cos v\, dv \int_{-h}^{h} \frac{dr}{(\rho^2+z^2)^{\frac{3}{2}}}$$

$$= m\theta \int_{d-r}^{d+r} \rho^2 d\rho \cdot 2\sin v \cdot \frac{2h}{\rho^2 \sqrt{\rho^2+h^2}}$$

$$= 2m\theta \frac{h}{d} \int_{d-r}^{d+r} \frac{d\rho}{\rho} \sqrt{\frac{[(d+r)^2 - \rho^2][\rho^2 - (d-r)^2]}{\rho^2 + h^2}}$$

$$= 2m\theta \frac{h}{d} \int_{\sqrt{b^2 - h^2}}^{\sqrt{a^2 - h^2}} \frac{d\rho}{\rho} \sqrt{\frac{(a^2 - \rho^2 - h^2)(\rho^2 + h^2 - b^2)}{\rho^2 + h^2}}$$

where $a = \sqrt{(d+r)^2 + h^2}$; and $b = \sqrt{(d-r)^2 + h^2}$ \hfill (5)

Assume $\tan \varphi = \frac{a}{b} \sqrt{\frac{\rho^2 + h^2 - b^2}{a^2 - \rho^2 - h^2}}$ then φ has the limits $\lrcorner$ and 0. We have then $\rho^2 + h^2 = \frac{a^2 b^2}{a^2 \cos^2\varphi + b^2 \sin^2\varphi} = \frac{b^2}{\Delta \varphi^2}$, where $\Delta \varphi$ is an elliptic Δ function to the modulus $\sqrt{1 - \frac{b^2}{a^2}}$. Transforming we have, expressing in Δ functions

$$A = 2m\theta \frac{h}{d} \int_0^{\lrcorner} ab^2 \frac{d\varphi}{\Delta \varphi^3} \cdot \frac{(1 - \Delta\varphi^2)(\Delta\varphi^2 - \frac{b^2}{a^2})}{b^2 - h^2 \Delta\varphi^2}$$

$$= 2m\theta \frac{ab^2}{hd} \left\{ -\frac{h^2}{a^2} \int_0^{\lrcorner} \frac{d\varphi}{\Delta \varphi^3} + \int_0^{\lrcorner} \frac{d\varphi}{\Delta \varphi} - \frac{(d^2 - r^2)^2}{a^2 b^2} \int_0^{\lrcorner} \frac{d\varphi}{(1 - \frac{h^2}{b^2} \Delta\varphi^2) \Delta \varphi} \right\}$$

$$= 2m\theta \frac{ab^2}{hd} \left\{ \int_0^{\lrcorner} \frac{d\varphi}{\Delta \varphi} - \frac{h^2}{b^2} \int_0^{\lrcorner} d\varphi \, \Delta\varphi \right.$$

$$\left. - \frac{(d+r)^2}{a^2} \int_0^{\lrcorner} \frac{d\varphi}{(1 + \frac{4rdh^2}{a^2(d-r)^2} \sin^2\varphi) \Delta\varphi} \right\}$$

$$= 2m\theta \frac{ab^2}{hd} \left\{ F - \frac{h^2}{b^2} E - \left(\frac{d+r}{a}\right)^2 \Pi\left(\frac{4rdh^2}{a^2(d-r)^2}\right) \right\} \quad (6)$$

where F, E, and Π denote quadrantal elliptic integrals of the 1st, 2d, and 3d species, according to Legendre's notation. If our object was to obtain here a formula for convenient computation of the attraction the third term could still be expressed in integrals of the 1st and 2d species, which have been tabulated by Legendre.

Let us now move the particle to the surface of the cylinder, then we have $d = r$; $a = \sqrt{4r^2 + h^2}$; $b = h$, and (6) becomes

$$A' = 2m\theta \frac{ah}{r} (F - E). \qquad (6')$$

Mr. Woodward, as already stated, had found an infinite attraction of a circular disc on a particle on its circumference by using Price's method. Now, since A' is *surely* a finite quantity, we have here the manifest absurdity that the attraction of a cylinder on a particle on its surface would be less than that of a circular disk which is only an infinitesimal part of it.

If we wish to ascertain the attraction of a circular disk of finite small thickness we may, of course, use (6'), and since then F tends to infinity, E may be neglected; therefore

$$A'_{h=0} = \frac{2m\theta}{r} \left[ah \int_0^{\frac{\pi}{2}} \frac{d\varphi}{\sqrt{1 - \frac{4r^2}{a^2} \sin^2 \varphi}} \right]_{h=0}$$

$$= 4m\theta \left[\frac{\int_0^{\frac{\pi}{2}} \frac{d\varphi}{\sqrt{1 - \frac{4r^2}{a^2} \sin^2 \varphi}}}{\frac{1}{h}} \right]_{h=0}$$

$$= 4m\theta \left[\frac{-\int_0^{\frac{\pi}{2}} \frac{4r^2 \sin^2 \varphi \frac{da}{dh}}{a^3 \Delta \varphi^3} d\varphi}{-\frac{1}{h^2}} \right]_{h=0}$$

$$= 4m\theta \left[\frac{\frac{4r^2 h}{a^4} \int_0^{\frac{\pi}{2}} \frac{\sin^2 \varphi \, d\varphi}{\Delta \varphi^3}}{\frac{1}{h^2}} \right]_{h=0}$$

$$= 0$$

Since then the attraction is 0 if $h = 0$ it will be small if h is small, and will continuously grow with h.

The notion of an infinite attraction exerted by a finite mass has thus been dispelled in two special cases in which it seemed to have been proved, and I add some remarks on its general impossibility.

The attraction of a mass element on a particle at distance ρ is

$$m\theta \, \frac{dx \, dy \, dz}{\rho^2}$$

This is then an infinitesimal of the third order, and the summation of its components with respect to a fixed direction requires three integrations, which give surely a finite result if ρ is always finite.

If ρ is infinitesimal there will be *one* element of attraction, which, instead of being of the third order, is only of the *first* order of infinitesimals, and this one element being added to the finite sum of the other elements has no effect. Hence I conclude that a finite mass exerts only a *finite* attraction on a particle.

The paper was discussed by Messrs. HALL, HILL, and WOODWARD. Mr. WOODWARD said that he had arrived at a result similar to Mr. Kummell's by a somewhat different route. The fallacy in Price's Calculus arose from neglecting the thickness of the plate.

NOTE.

The communications and abstracts printed in the proceedings of the Mathematical Section have each been examined by a special committee consisting of the Chairman, the Secretary, and a third member appointed by the Chairman. These third members were as follows:

Title.	Author.	Third Member.
Physical observations of Wolf's comet	W. C. Winlock.	— — —
The Theory of Mercury	G. L. Ravené.	E. B. Elliott.
A group of circles related to Feuerbach's circle	Marcus Baker.	C. H. Kummell.
Some practical features of a time determination	R. S. Woodward.	H. Farquhar.
Can the attraction of a finite mass be infinite?	C. H. Kummell.	R. S. Woodward.

ON THE COMPUTATION
OF
PROBABLE ERROR

ON THE COMPUTATION OF PROBABLE ERROR.

BY T. W. WRIGHT, U. S. LAKE SURVEY, DETROIT, MICHIGAN.

In computing the probable error of the determinations of an observed quantity two forms are in common use, Bessel's and Peters'. If with the ordinary notation we let

$m =$ the number of observations,
$r =$ the probable error of a single observation,
$r_0 =$ the probable error of the final result,
$\rho = 0.4769363$,
$v_1, v_2 \ldots =$ the residual errors of observation, $[vv]$ the sum of their squares and $[v$ their sum without reg'd to sign, then accord'g to Bessel's form,

$$r = \frac{\rho\sqrt{2}}{\sqrt{(m-1)}}\sqrt{[vv]} = \frac{0.6745}{\sqrt{(m-1)}}\sqrt{[vv]},$$

$$r_0 = \frac{\rho\sqrt{2}}{\sqrt{m(m-1)}}\sqrt{[vv]} = \frac{0.6745}{\sqrt{m(m-1)}}\sqrt{[vv]},$$

and according to Peters' form

$$r = \rho\sqrt{\left(\frac{\pi}{m(m-1)}\right)}\cdot[v = \frac{0.8453}{\sqrt{m(m-1)}}\cdot[v,$$

$$r_0 = \rho\sqrt{\left(\frac{\pi}{m^2(m-1)}\right)}\cdot[v = \frac{0.8453}{m\sqrt{(m-1)}}\cdot[v.$$

The latter form can be more rapidly computed and is usually close enough.

The labor of computing the factors $0.6745 \div \sqrt{m(m-1)}, \ldots$ is considerable in any case. Accordingly in practice I have found it convenient to tabulate these quantities for values of m from 2 to 100. The formulæ for the probable error being now written in the simple forms

$$r = \lambda_1\sqrt{[vv]} \qquad r = \lambda'[v$$
$$r_0 = \lambda_2\sqrt{[vv]} \qquad r_0 = \lambda''[v,$$

all that we have to do is to enter the proper table with the argument m, take out the corresponding multiplier λ and perform a single multiplication.

For finding $\sqrt{[vv]}$ a close enough approximation to the square root can be taken out at sight from a table of squares, or if it be preferred the computation may be made logarithmically.

It often happens that we wish to find the probable error of a series of observations roughly and quickly. The following rule will be found convenient in such a case and also as a check on the values found by the preceding method.

A glance at the observed results will show the greatest and least, and their difference will be the *range* of the results. Then for the probable error (r) of a single observation we have, if the number of results is

$$2, \quad r = \tfrac{1}{2} \text{ the range,}$$
$$3, \quad r = \tfrac{1}{3} \text{ "   "}$$
$$\text{between 3 and 8,} \quad r = \tfrac{1}{4} \text{ "   "}$$
$$\text{"   } 8 \text{ "   } 15, \quad r = \tfrac{1}{5} \text{ "   "}$$
$$\text{"   } 15 \text{ "   } 35, \quad r = \tfrac{1}{6} \text{ "   "}$$
$$\text{"   } 35 \text{ "   } 100, \quad r = \tfrac{1}{7} \text{ "   "}$$

The probable error of the final result is found at once by dividing the probable error of a single observation by the square root of the number of observations.

In finding the range caution must be exercised with regard to abnormal results.

Table I.

m	λ_1	λ_2	m	λ_1	λ_2
			40	0.1080	0.0171
			41	.1066	.0167
2	0.6745	0.4769	42	.1053	.0163
3	.4769	.2754	43	.1041	.0159
4	.3894	.1947	44	.1029	.0155
5	0.3372	0.1508	45	0.1017	0.0152
6	.3016	.1231	46	.1005	.0148
7	.2754	.1041	47	.0994	.0145
8	.2549	.0901	48	.0984	.0142
9	.2385	.0795	49	.0974	.0139
10	0.2248	0.0711	50	0.0964	0.0136
11	.2133	.0643	51	.0954	.0134
12	.2029	.0587	52	.0944	.0131
13	.1947	.0540	53	.0935	.0128
14	.1871	.0500	54	.0926	.0126
15	0.1803	0.0465	55	0.0918	0.0124
16	.1742	.0435	56	.0909	.0122
17	.1686	.0409	57	.0901	.0119
18	.1636	.0386	58	.0893	.0117
19	.1590	.0365	59	.0886	.0115
20	0.1547	0.0346	60	0.0878	0.0113
21	.1508	.0329	61	.0871	.0111
22	.1472	.0314	62	.0864	.0110
23	.1438	.0300	63	.0857	.0108
24	.1406	.0287	64	.0850	.0106
25	0.1377	0.0275	65	0.0843	0.0105
26	.1349	.0265	66	.0837	.0103
27	.1323	.0255	67	.0830	.0101
28	.1298	.0245	68	.0824	.0100
29	.1275	.0237	69	.0818	.0098
30	0.1252	0.0229	70	0.0812	0.0097
31	.1231	.0221	71	.0806	.0096
32	.1211	.0214	72	.0800	.0094
33	.1192	.0208	73	.0795	.0093
34	.1174	.0201	74	.0789	.0092
35	0.1157	0.0196	75	0.0784	0.0091
36	.1140	.0190	80	.0759	.0085
37	.1124	.0185	85	.0736	.0080
38	.1109	.0180	90	.0713	.0075
39	.1094	.0175	100	.0678	.0068

Table II.

m	λ'	λ''	m	λ'	λ''
			40	0.0214	0.0034
			41	.0209	.0033
2	0.5978	0.4227	42	.0204	.0031
3	.3451	.1993	43	.0199	.0030
4	.2440	.1220	44	.0194	.0029
5	0.1890	0.0845	45	0.0190	0.0028
6	.1543	.0630	46	.0186	.0027
7	.1304	.0493	47	.0182	.0027
8	.1130	.0399	48	.0178	.0026
9	.0996	.0332	49	.0174	.0025
10	0.0891	0.0282	50	0.0171	0.0024
11	.0806	.0243	51	.0167	.0023
12	.0736	.0212	52	.0164	.0023
13	.0677	.0188	53	.0161	.0022
14	.0627	.0167	54	.0158	.0022
15	0.0583	0.0151	55	0.0155	0.0021
16	.0546	.0136	56	.0152	.0020
17	.0513	.0124	57	.0150	.0020
18	.0483	.0114	58	.0147	.0019
19	.0457	.0105	59	.0145	.0019
20	0.0434	0.0097	60	0.0142	0.0018
21	.0412	.0090	61	.0140	.0018
22	.0393	.0084	62	.0137	.0017
23	.0376	.0078	63	.0135	.0017
24	.0360	.0073	64	.0133	.0017
25	0.0345	0.0069	65	0.0131	0.0016
26	.0332	.0065	66	.0129	.0016
27	.0319	.0061	67	.0127	.0016
28	.0307	.0058	68	.0125	.0015
29	.0297	.0055	69	.0123	.0015
30	0.0287	0.0052	70	0.0122	0.0015
31	.0277	.0050	71	.0120	.0014
32	.0268	.0047	72	.0118	.0014
33	.0260	.9045	73	.0117	.0014
34	.0252	.0043	74	.0115	.0013
35	0.0245	0.0041	75	0.0113	0.0013
36	.0238	.0040	80	.0106	.0012
37	.0232	.0038	85	.0100	.0011
38	.0225	.0037	90	.0095	.0010
39	.0220	.0035	100	.0085	.0008

Example.—In the telegraphic determination of the longitude between St. Paul and Duluth, Minn., June 15, 1871, the following were the corrections found for chronometer No. 176, at 15h. 51m., sidereal time, from the observation of 21 time stars. (See Annual Report of Survey of N. & N. W. Lakes, 1872, p. 40.)

Correction	v	vv
s		
− 8.78	+ 0.04	0.0016
.76	+ .02	4
.85	+ .11	121
.78	+ .04	16
.51	− .23	529
.64	− .10	100
.68	− .06	36
.63	− .11	121
.58	− .16	256
.80	+ .06	36
.75	+ .01	1
.78	+ .04	16
.96	+ .22	484
.64	− .10	100
.65	− .09	81
.83	+ .09	81
.70	− .04	16
.64	+ .10	100
.79	+ .05	25
.90	+ .16	256
− 8.93	+ .19	0.0361
Mean − 8.74	2.02	0.2756

Then
$$m = 21$$
$$[v] = 2.02$$
$$[vv] = 0.2756$$
$$\sqrt{[vv]} = 0.525, \text{ from a table of squares.}$$

Hence from table I,
$$r = 0.525 \times 0.151 = 0.079$$
$$r_0 = 0.525 \times 0.033 = 0.017.$$

From table II,
$$r = 2.02 \times 0.041 = 0.082$$
$$r_0 = 2.02 \times 0.009 = 0.018$$

Approximate Method.—Range = 8.96 − 8.51 = 0.45
$$\therefore r = \frac{0.45}{6} = 0.075$$
$$r_0 = \frac{0.075}{\sqrt{21}} = 0.016.$$

So close an agreement among the determinations of the values of r and r_0 is not always to be expected.

THREE CENTURIES OF SCIENCE IN AMERICA

An Arno Press Collection

Adams, John Quincy. **Report of the Secretary of State upon Weights and Measures.** 1821.

Archibald, Raymond Clare. **A Semicentennial History of the American Mathematical Society: 1888-1938** *and* **Semicentennial Addresses of the American Mathematical Society.** 2 vols. 1938.

Bond, William Cranch. **History and Description of the Astronomical Observatory of Harvard College** *and* **Results of Astronomical Observations Made at the Observatory of Harvard College.** 1856.

Bowditch, Henry Pickering. **The Life and Writings of Henry Pickering Bowditch.** 2 vols. 1980.

Bridgman, Percy Williams. **The Logic of Modern Physics.** 1927.

Bridgman, Percy Williams. **Philosophical Writings of Percy Williams Bridgman.** 1980.

Bridgman, Percy Williams. **Reflections of a Physicist.** 1955.

Bush, Vannevar. **Science the Endless Frontier.** 1955.

Cajori, Florian. **The Chequered Career of Ferdinand Rudolph Hassler.** 1929.

Cohen, I. Bernard, editor. **The Career of William Beaumont and the Reception of His Discovery.** 1980.

Cohen, I. Bernard, editor. **Benjamin Peirce: "Father of Pure Mathematics" in America.** 1980.

Cohen, I. Bernard, editor. **Aspects of Astronomy in America in the Nineteenth Century.** 1980.

Cohen, I. Bernard, editor. **Cotton Mather and American Science and Medicine: With Studies and Documents Concerning the Introduction of Inoculation or Variolation.** 2 vols. 1980.

Cohen, I. Bernard, editor. **The Life and Scientific Work of Othniel Charles Marsh.** 1980.

Cohen, I. Bernard, editor. **The Life and the Scientific and Medical Career of Benjamin Waterhouse: With Some Account of the Introduction of Vaccination in America.** 2 vols. 1980.

Cohen, I. Bernard, editor. **Research and Technology.** 1980.

Cohen, I. Bernard, editor. **Thomas Jefferson and the Sciences.** 1980.

Cooper, Thomas. **Introductory Lecture** *and* **A Discourse on the Connexion Between Chemistry and Medicine.** 2 vols. in one. 1812/1818.

Dalton, John Call. **John Call Dalton on Experimental Method.** 1980.

Darton, Nelson Horatio. **Catalogue and Index of Contributions to North American Geology: 1732-1891.** 1896.

Donnan, F[rederick] G[eorge] and Arthur Haas, editors. **A Commentary on the Scientific Writings of J. Willard Gibbs** *and* Duhem, Pierre. **Josiah-Willard Gibbs: A Propos de la Publication de ses Mémoires Scientifiques.** 3 vols. in two. 1936/1908.

Dupree, A[nderson] Hunter. **Science in the Federal Government: A History of Policies and Activities to 1940.** 1957.

Ellicott, Andrew. **The Journal of Andrew Ellicott.** 1803.

Fulton, John F. **Harvey Cushing: A Biography.** 1946.

Getman, Frederick H. **The Life of Ira Remsen.** 1940.

Goode, George Brown. **The Smithsonian Institution 1846-1896: The History of its First Half Century.** 1897.

Hale, George Ellery. **National Academies and the Progress of Research.** 1915.

Harding, T. Swann. **Two Blades of Grass: A History of Scientific Development in the U.S. Department of Agriculture.** 1947.

Hindle, Brooke. **David Rittenhouse.** 1964.

Hindle, Brooke, editor. **The Scientific Writings of David Rittenhouse.** 1980.

Holden, Edward S[ingleton]. **Memorials of William Cranch Bond, Director of the Harvard College Observatory, 1840-1859, and of his Son, George Phillips Bond, Director of the Harvard College Observatory, 1859-1865.** 1897.

Howard, L[eland] O[sslan]. **Fighting the Insects: The Story of an Entomologist, Telling the Life and Experiences of the Writer.** 1933.

Jaffe, Bernard. **Men of Science in America.** 1958.

Karpinski, Louis C. **Bibliography of Mathematical Works Printed in America through 1850.** Reprinted with Supplement and Second Supplement. 1940/1945.

Loomis, Elias. **The Recent Progress of Astronomy: Especially in the United States.** 1851.

Merrill, Elmer D. **Index Rafinesquianus: The Plant Names Published by C.S. Rafinesque with Reductions, and a Consideration of his Methods, Objectives, and Attainments.** 1949.

Millikan, Robert A[ndrews]. **The Autobiography of Robert A. Millikan.** 1950.

Mitchel, O[rmsby] M[acKnight]. **The Planetary and Stellar Worlds: A Popular Exposition of the Great Discoveries and Theories of Modern Astronomy.** 1848.

Organisation for Economic Co-operation and Development. **Reviews of National Science Policy: United States.** 1968.

Packard, Alpheus S. **Lamarck: The Founder of Evolution; His Life and Work.** 1901.

Pupin, Michael. **From Immigrant to Inventor.** 1930.

Rhees, William J. **An Account of the Smithsonian Institution.** 1859.

Rhees, William J. **The Smithsonian Institution: Documents Relative to its History.** 2 vols. 1901.

Rhees, William J. **William J. Rhees on James Smithson.** 2 vols. in one. 1980.

Scott, William Berryman. **Some Memories of a Palaeontologist.** 1939.

Shryock, Richard H. **American Medical Research Past and Present.** 1947.

Shute, Michael, editor. **The Scientific Work of John Winthrop.** 1980.

Silliman, Benjamin. **A Journal of Travels in England, Holland, and Scotland, and of Two Passages over the Atlantic in the Years 1805 and 1806.** 2 vols. 1812.

Silliman, Benjamin. **A Visit to Europe in 1851.** 2 vols. 1856

Silliman, Benjamin, Jr. **First Principles of Chemistry.** 1864.

Smith, David Eugene and Jekuthiel Ginsburg. **A History of Mathematics in America before 1900.** 1934.

Smith, Edgar Fahs. **James Cutbush: An American Chemist.** 1919.

Smith, Edgar Fahs. **James Woodhouse: A Pioneer in Chemistry, 1770-1809.** 1918.

Smith, Edgar Fahs. **The Life of Robert Hare: An American Chemist (1781-1858).** 1917.

Smith, Edgar Fahs. **Priestley in America: 1794-1804.** 1920.

Sopka, Katherine. **Quantum Physics in America: 1920-1935** (Doctoral Dissertation, Harvard University, 1976). 1980.

Steelman, John R[ay]. **Science and Public Policy: A Report to the President. 1947.**

Stewart, Irvin. **Organizing Scientific Research for War: The Administrative History of the Office of Scientifc Research and Development. 1948.**

Stigler, Stephen M., editor. **American Contributions to Mathematical Statistics in the Nineteenth Century. 2 vols. 1980.**

Trowbridge, John. **What is Electricity? 1899.**

True. Alfred. **Alfred True on Agricultural Experimentation and Research. 1980.**

True, F[rederick] W., editor. **The Semi-Centennial Anniversary of the National Academy of Sciences: 1863-1913** *and* **A History of the First Half-Century of the National Academy of Sciences: 1863-1913. 2 vols. 1913.**

Tyndall, John. **Lectures on Light: Delivered in the United States in 1872-73. 1873.**

U.S. House of Representatives. **Annual Report of the Board of Regents of the Smithsonian Institution...A Memorial of George Brown Goode together with a selection of his Papers on Museums and on the History of Science in America. 1901.**

U.S. National Resources Committee. **Research: A National Resource. 3 vols. in one. 1938-1941.**

U.S. Senate. **Testimony Before the Joint Commission to Consider the Present Organizations of the Signal Service, Geological Survey, Coast and Geodetic Survey, and the Hydrographic Office of the Navy Department. 2 vols. 1866.**